The Handboc
of Environmental Chemistry

Editor-in-Chief: O. Hutzinger

Volume 5 Water Pollution
Part H

Advisory Board:

T. A. Kassim · D. Barceló · P. Fabian · H. Fiedler · H. Frank · J. P. Giesy
R. Hites · M. A. K. Khalil · D. Mackay · A. H. Neilson · J. Paasivirta
H. Parlar · S. H. Safe · P. J. Wangersky

The Handbook of Environmental Chemistry

Recently Published and Forthcoming Volumes

Estuaries

Volume Editor:
Peter J. Wangersky

With contributions by

R. Azzoni · M. Bartoli · S. Bencivelli · N. Berlinsky · A. Binelli
Y. Bogatova · B. Boutier · E. C. V. Butler · G. Castaldelli
J.-F. Chiffoleau · J. M. Z. Comenges · C. Dange · E. A. Fano
I. Ferrari · G. Garkavaya · G. Giordani · C. Gobeil
J.-L. Gonzalez · W. Hamza · T. Jennerjahn · B. Knoppers
J. M. Krest · R. W. Macdonald · P. R. P. Medeiros · M. Naldi
D. Nizzoli · A. Provini · J. M. Smoak · W. F. L. de Souza
P. W. Swarzenski · B. Thouvenin · P. Viaroli · Y. Yu

 Springer

Environmental chemistry is a rather young and interdisciplinary field of science. Its aim is a complete description of the environment and of transformations occurring on a local or global scale. Environmental chemistry also gives an account of the impact of man's activities on the natural environment by describing observed changes.

The Handbook of Environmental Chemistry provides the compilation of today's knowledge. Contributions are written by leading experts with practical experience in their fields. The Handbook will grow with the increase in our scientific understanding and should provide a valuable source not only for scientists, but also for environmental managers and decision-makers.

The Handbook of Environmental Chemistry is published in a series of five volumes:

Volume 1: The Natural Environment and the Biogeochemical Cycles
Volume 2: Reactions and Processes
Volume 3: Anthropogenic Compounds
Volume 4: Air Pollution
Volume 5: Water Pollution

The series Volume 1 The Natural Environment and the Biogeochemical Cycles describes the natural environment and gives an account of the global cycles for elements and classes of natural compounds. The series Volume 2 Reactions and Processes is an account of physical transport, and chemical and biological transformations of chemicals in the environment.

The series Volume 3 Anthropogenic Compounds describes synthetic compounds, and compound classes as well as elements and naturally occurring chemical entities which are mobilized by man's activities.

The series Volume 4 Air Pollution and Volume 5 Water Pollution deal with the description of civilization's effects on the atmosphere and hydrosphere.

Within the individual series articles do not appear in a predetermined sequence. Instead, we invite contributors as our knowledge matures enough to warrant a handbook article.

Suggestions for new topics from the scientific community to members of the Advisory Board or to the Publisher are very welcome.

ISBN 978-3-642-05544-7   e-ISBN 978-3-540-32484-3

ISSN 1433-6863

DOI 10.1007/b89479

Springer is a part of Springer Science+Business Media

springer.com

© Springer-Verlag Berlin Heidelberg 2006
Softcover reprint of the hardcover 1st edition 2006

Cover design: E. Kirchner, Springer-Verlag

Printed on acid-free paper   02/3141 YL – 5 4 3 2 1 0

Prof. Dr. D. Mackay
Department of Chemical Engineering
and Applied Chemistry
University of Toronto
Toronto, Ontario, Canada M5S 1A4

Prof. Dr. A. H. Neilson
Swedish Environmental Research Institute
P.O. Box 21060
10031 Stockholm, Sweden
ahsdair@ivl.se

Prof. Dr. J. Paasivirta
Department of Chemistry
University of Jyväskylä
Survontie 9
P.O. Box 35
40351 Jyväskylä, Finland

Prof. Dr. Dr. H. Parlar
Institut für Lebensmitteltechnologie
und Analytische Chemie
Technische Universität München
85350 Freising-Weihenstephan, Germany

Prof. Dr. S. H. Safe
Department of Veterinary
Physiology and Pharmacology
College of Veterinary Medicine
Texas A & M University
College Station, TX 77843-4466, USA
ssafe@cvm.tamu.edu

Prof. P. J. Wangersky
University of Victoria
Centre for Earth and Ocean Research
P.O. Box 1700
Victoria, BC, V8W 3P6, Canada
wangers@telus. net

The Handbook of Environmental Chemistry
Also Available Electronically

For all customers who have a standing order to The Handbook of Environmental Chemistry, we offer the electronic version via SpringerLink free of charge. Please contact your librarian who can receive a password or free access to the full articles by registering at:

springerlink.com

If you do not have a subscription, you can still view the tables of contents of the volumes and the abstract of each article by going to the SpringerLink Homepage, clicking on "Browse by Online Libraries", then "Chemical Sciences", and finally choose The Handbook of Environmental Chemistry.

You will find information about the

- Editorial Board
- Aims and Scope
- Instructions for Authors
- Sample Contribution

at springer.com using the search function.

Preface

Environmental Chemistry is a relatively young science. Interest in this subject, however, is growing very rapidly and, although no agreement has been reached as yet about the exact content and limits of this interdisciplinary discipline, there appears to be increasing interest in seeing environmental topics which are based on chemistry embodied in this subject. One of the first objectives of Environmental Chemistry must be the study of the environment and of natural chemical processes which occur in the environment. A major purpose of this series on Environmental Chemistry, therefore, is to present a reasonably uniform view of various aspects of the chemistry of the environment and chemical reactions occurring in the environment.

The industrial activities of man have given a new dimension to Environmental Chemistry. We have now synthesized and described over five million chemical compounds and chemical industry produces about hundred and fifty million tons of synthetic chemicals annually. We ship billions of tons of oil per year and through mining operations and other geophysical modifications, large quantities of inorganic and organic materials are released from their natural deposits. Cities and metropolitan areas of up to 15 million inhabitants produce large quantities of waste in relatively small and confined areas. Much of the chemical products and waste products of modern society are released into the environment either during production, storage, transport, use or ultimate disposal. These released materials participate in natural cycles and reactions and frequently lead to interference and disturbance of natural systems.

Environmental Chemistry is concerned with reactions in the environment. It is about distribution and equilibria between environmental compartments. It is about reactions, pathways, thermodynamics and kinetics. An important purpose of this Handbook, is to aid understanding of the basic distribution and chemical reaction processes which occur in the environment.

Laws regulating toxic substances in various countries are designed to assess and control risk of chemicals to man and his environment. Science can contribute in two areas to this assessment; firstly in the area of toxicology and secondly in the area of chemical exposure. The available concentration ("environmental exposure concentration") depends on the fate of chemical compounds in the environment and thus their distribution and reaction behaviour in the environment. One very important contribution of Environmental Chemistry to the above mentioned toxic substances laws is to develop laboratory test methods, or mathematical correlations and models that predict the environmental fate of new chemical compounds. The third purpose of this Handbook is to help in the basic understanding and development of such test methods and models.

The last explicit purpose of the Handbook is to present, in concise form, the most important properties relating to environmental chemistry and hazard assessment for the most important series of chemical compounds.

At the moment three volumes of the Handbook are planned. Volume 1 deals with the natural environment and the biogeochemical cycles therein, including some background information such as energetics and ecology. Volume 2 is concerned with reactions and processes in the environment and deals with physical factors such as transport and adsorption, and chemical, photochemical and biochemical reactions in the environment, as well as some aspects of pharmacokinetics and metabolism within organisms. Volume 3 deals with anthropogenic compounds, their chemical backgrounds, production methods and information about their use, their environmental behaviour, analytical methodology and some important aspects of their toxic effects. The material for volume 1, 2 and 3 was each more than could easily be fitted into a single volume, and for this reason, as well as for the purpose of rapid publication of available manuscripts, all three volumes were divided in the parts A and B. Part A of all three volumes is now being published and the second part of each of these volumes should appear about six months thereafter. Publisher and editor hope to keep materials of the volumes one to three up to date and to extend coverage in the subject areas by publishing further parts in the future. Plans also exist for volumes dealing with different subject matter such as analysis, chemical technology and toxicology, and readers are encouraged to offer suggestions and advice as to future editions of "The Handbook of Environmental Chemistry".

Most chapters in the Handbook are written to a fairly advanced level and should be of interest to the graduate student and practising scientist. I also hope that the subject matter treated will be of interest to people outside chemistry and to scientists in industry as well as government and regulatory bodies. It would be very satisfying for me to see the books used as a basis for developing graduate courses in Environmental Chemistry.

Due to the breadth of the subject matter, it was not easy to edit this Handbook. Specialists had to be found in quite different areas of science who were willing to contribute a chapter within the prescribed schedule. It is with great satisfaction that I thank all 52 authors from 8 countries for their understanding and for devoting their time to this effort. Special thanks are due to Dr. F. Boschke of Springer for his advice and discussions throughout all stages of preparation of the Handbook. Mrs. A. Heinrich of Springer has significantly contributed to the technical development of the book through her conscientious and efficient work. Finally I like to thank my family, students and colleagues for being so patient with me during several critical phases of preparation for the Handbook, and to some colleagues and the secretaries for technical help.

I consider it a privilege to see my chosen subject grow. My interest in Environmental Chemistry dates back to my early college days in Vienna. I received significant impulses during my postdoctoral period at the University of California and my interest slowly developed during my time with the National Research Council of Canada, before I could devote my full time of Environmental Chemistry, here in Amsterdam. I hope this Handbook may help deepen the interest of other scientists in this subject.

Amsterdam, May 1980   *O. Hutzinger*

Twenty-one years have now passed since the appearance of the first volumes of the Handbook. Although the basic concept has remained the same changes and adjustments were necessary.

Some years ago publishers and editors agreed to expand the Handbook by two new open-end volume series: Air Pollution and Water Pollution. These broad topics could not be fitted easily into the headings of the first three volumes. All five volume series are integrated through the choice of topics and by a system of cross referencing.

The outline of the Handbook is thus as follows:

1. The Natural Environment and the Biochemical Cycles,
2. Reaction and Processes,
3. Anthropogenic Compounds,
4. Air Pollution,
5. Water Pollution.

Rapid developments in Environmental Chemistry and the increasing breadth of the subject matter covered made it necessary to establish volume-editors. Each subject is now supervised by specialists in their respective fields.

A recent development is the accessibility of all new volumes of the Handbook from 1990 onwards, available via the Springer Homepage springeronline.com or springerlink.com.

During the last 5 to 10 years there was a growing tendency to include subject matters of societal relevance into a broad view of Environmental Chemistry. Topics include LCA (Life Cycle Analysis), Environmental Management, Sustainable Development and others. Whilst these topics are of great importance for the development and acceptance of Environmental Chemistry Publishers and Editors have decided to keep the Handbook essentially a source of information on "hard sciences".

With books in press and in preparation we have now well over 40 volumes available. Authors, volume-editors and editor-in-chief are rewarded by the broad acceptance of the "Handbook" in the scientific community.

Bayreuth, July 2001   *Otto Hutzinger*

Contents

Foreword

Rivers have always played an important role in human settlement; they have acted as sources of food, of water for people, animals, and crops, as travel routes, and as a means of waste disposal. Estuaries have been particularly important as sources of fish and shellfish. Some larger estuaries, such as the Chesapeake Bay, were so important as food sources that they gave rise to a special way of life, that of the waterman. Man often established settlements at or near estuaries, settlements which later became large port cities.

When the matter was left to Nature, estuaries often changed shape, size, and location, according to the volume of water being carried down by the river, the amount of sediment, and the strength and pathway of the local storms. Once settlements were established along the banks of the lower river, such carefree independence was no longer acceptable. Most of the rivers whose estuaries are discussed in this volume have been subjected to some sort of manipulation, not always with the expected result. In some cases the expected end was achieved, but unexpected side effects have since emerged. It is evident that when one is dealing with natural systems, engineering is not yet an exact science.

What was surprising to me was the degree to which these estuaries differed from one another. To some extent, this was to be expected, since the estuaries were chosen to exhibit particular characteristics. The Amazon, for instance, is a rare example of an estuary where the gradation from fresh water to sea salinities takes place outside the estuary proper. The Mackenzie exhibits ice damming at the mouth, where it enters the Arctic Ocean, at a time of year when its more southerly beginnings are free-flowing and delivering the meltwater from the winter's snows. In the Nile we have the longest record of a river's annual cycle, and an example of how this can change under man's intervention.

Of particular interest to me was the comparison between the Huon and Derwent Rivers in Tasmania. These rivers are not very far apart, as distances go in Australia, and share many of the same biological and geological characteristics. However, only a minor amount of settlement and subsistence farming has taken place along the Huon, while a major city, along with the usual industrial and port facilities, can be found on the banks of the Derwent. The effect of man's intervention is stamped very clearly on that river. Not all of the local industries seem to have left their fingerprints in the waters or sediments; I looked in vain for traces of the chocolate factory.

Possibly no river has been more extensively manipulated than the Danube. We are fortunate in having a record of the changes in river flow and burden of nutrients and particulates brought about by the construction of the various power stations and catchment basins. The effects of these changes on the adjacent coastal areas of the Black Sea are also discussed.

The Po is another river whose heavy nutrient and pollutant load, a result of the highly agricultural Po Valley, affects not only its estuarine life, but also the sea into which it flows. While the North Adriatic can by no means be compared to the Black Sea, over the last twenty years or so the high nutrient levels have given rise to some discomforting blooms of microalgae.

Along the shore, in the extensive region of coastal lagoons, the macroalgae have been the problem. The high nutrient content of the estuarine water has resulted in huge overgrowths of macroalgae, whose subsequent decomposition has led to anoxia and the death of many bottom organisms. Until the nutrient levels were controlled, these blooms were a menace to the thriving clam aquaculture as well as to the tourist industry in the region.

A note which is sounded in several of the chapters concerns the increase in variability in weather over the past ten or twenty years. This variability has brought both drought and flooding to regions of Canada unfamiliar with either phenomenon, and if I am to judge from my morning newspaper, to other regions of the world as well. If the present increase in variability presages a change in global precipitation patterns, perhaps a good place to look for early indicators will be in the geochemistry of estuaries.

Victoria, B.C. (Canada)   *Peter J. Wangersky*

Hdb Env Chem Vol. 5, Part H (2006): 1–49
DOI 10.1007/698_5_022
© Springer-Verlag Berlin Heidelberg 2005
Published online: 8 November 2005

The Tail of Two Rivers in Tasmania: The Derwent and Huon Estuaries

Edward C. V. Butler

CSIRO Marine Research, GPO Box 1538, 7001 Hobart,Tasmania, Australia
Edward.Butler@csiro.au

Abstract The Derwent and Huon Rivers are two relatively small river systems in the southeast of Tasmania (Australia). They terminate in estuaries that are very similar in structure and function. Historically, runoff from their adjoining catchments has been very dilute, but coloured by dissolved organic matter. Their location in cool, temperate latitudes results in a maritime climate that is changeable, but delivers regular rainfall, and therefore river flow, throughout the year. Marked seasonal cycles in nutrient levels and biological activity are seen in neighbouring coastal waters. Discharge from both rivers does not have the same seasonal signature; it is consistently enriched in dissolved organic matter (including nitrogenous forms) and depleted in inorganic nitrogen and phosphorus. Small variations in chemistry of the riverine end-members seem to ensue from intra-catchment differences in geology, soils and vegetation influenced by localised rainfall patterns. Silicon manifestly displays this behaviour in the Huon system.

Both the Derwent and Huon estuaries are drowned river valleys, with a strongly stratified (salt-wedge) water column at their head, tending toward partially mixed at their mouth. They are in a microtidal region. The Derwent estuary conforms to a wave-dominated system; the Huon estuary is intermediate between wave-dominated and tide-dominated. Currents are generally weak for both estuaries ($\leq 0.2\,\mathrm{m\,s^{-1}}$), and weaker in subsurface waters.

With the arrival of Europeans two centuries ago, the Derwent and Huon estuary departed from the same course. The capital city of Hobart established on the western bank of the Derwent estuary, and the catchment was soon modified by agriculture and tree felling, then by damming for irrigation and hydroelectricity generation. During the 20th century, the estuary became more industrialised, and the urban zone around its shores grew to support a population of 190 000, as did the discharges from these activities to the waterway. Development in the Huon catchment was slower and more constrained. Much less of the catchment was taken up with agriculture, and the population has only attained 13 000 in scattered small towns. The only intensive industry currently in the Huon catchment is aquaculture, active in the waters of the lower estuary. However, downstream processing of timber from catchment forests is planned to recommence.

The consequences of these disparate recent histories are that the conditions of the two estuaries are very different. The Derwent estuary has had heavy inputs of organic matter, suspended solids (wood fibre), nutrients, heavy metals and other toxicants. Of all these, the metals (especially zinc, cadmium, lead and mercury) residing in the estuarine sediments pose the greatest threat. They contaminate at levels as severe as seen anywhere worldwide. It might be fortuitous that nutrient inputs do not threaten eutrophication, because the estuary flushes rapidly (~ 15 days), and phytoplankton appear to be light-limited in the middle estuary, where most nutrients enter. Organic toxicants cause localised concerns, but need to be more fully investigated.

In contrast, the Huon estuary has been modified, but its environmental quality remains high with almost all monitoring results below the Australian guidelines. It serves as a useful baseline, though not pristine, against which the contamination of the Derwent estuary can be evaluated. Nutrients from aquaculture and the lower catchment, causing increased phytoplankton biomass and possibly depleted levels of dissolved oxygen in bottom waters, appear to be the only existing challenges for environmental management of the Huon estuary.

Keywords Hydrology · Metals · Nutrients · Organic matter · Phytoplankton

1
Introduction

The Derwent and Huon Rivers are not large on the world scale, nor even compared with rivers in continental Australia. What makes them interesting is that they are examples of rivers that are under-represented in global scientific studies: they are in an island in the Southern Hemisphere, in a cold temperate region, draining catchment soils that are generally poor in nutrients.

The two rivers, and their estuaries especially, are similar in many ways [1]. They can be treated as analogues. However, their modifications since the arrival of Europeans just 200 years ago have followed different paths. The Derwent estuary has become the centrepiece of the capital city of Tasmania, Hobart. Along both its shores is now an urban ribbon, comprising some heavy industry, port facilities and many effluent discharges. The watershed of the Derwent River has been much modified by agriculture and other human activities, and the flow of the river has been modulated and diminished by 16 dams along its course. These changes, and loss of habitat within the estuary itself, have harmed the natural ecosystem (e.g. the endemic spotted handfish is under threat of extinction) and seemingly favoured the introduction of exotic marine pests that have exacerbated the negative trend.

In contrast, the human influence on the Huon River and its estuary has been more moderate. Much of its upper catchment remains as native forest; some tracts are exploited by forestry activities. A patchwork of agriculture – horticulture and livestock grazing – pervades the lower catchment. Human settlement remains sparse with series of small townships along the watercourse. A recent development is salmonid, and to a lesser extent shellfish, aquaculture in the marine end of the estuary.

As well as critically examining the available physico-chemical data for each estuary and its context in regard to catchment and regional environment, I discuss briefly the possibility of the Huon estuary providing an operational baseline for management of the more polluted Derwent estuary. The ecology of the estuaries, apart from phytoplankton, is not examined here; the reader is referred to Edgar et al. [1], Jordan et al. [2], and the NSR [3] reports.

2
Broad Environment

Tasmania has a cool temperate maritime climate dominated by a zonal westerly wind regime [4, 5]. The westerly winds are present throughout the year, but fluctuate from weekly intervals (with the passage of anticyclones), seasonally (strongest in late winter/early spring) and inter-annually. The marine influence results in mild winters and cool summers with a typical daily temperature range of 7 °C, but with weather that is very changeable. The island is sufficiently large to have a slight continental effect on climate in its central highlands, much of which form the upper catchment of the Derwent River.

Rainfall is broadly highest near the west coast and declines steadily to the east. The rugged topography of many parts of the island creates microclimates; higher altitude features produce localised rainfall patterns superimposed on the general trend. In the west of the Huon catchment, annual average rainfall peaks near 3000 cm; it is 500 cm and less in the southeast of the Derwent catchment [6]. Precipitation is generally uniform throughout the year, although toward the west coast it peaks in late winter. In the east, it is slightly higher in autumn and spring, affected by intermittent cyclonic depressions off the east coast [5]. Atmospheric deposition, wet or dry, throughout Tasmania is typically uncontaminated, because to the west, over the Southern Ocean, is one of the longest fetches anywhere.

The Derwent and Huon catchments (Fig. 1) are located in the same geological province characterised by post-Carboniferous cover (Permo-Triassic mudstone, sandstone and shales into which large sheets of Jurassic dolerite have intruded) over pre-Carboniferous basement rocks [7, 8]. Only in their headwaters do they differ, with the Derwent draining highlands of similar geology to its lower catchment, but with scattered outcropping of Tertiary basalt, and in the far northwest Quaternary glacial and periglacial deposits overlying the dolerite. A few western tributaries of the Derwent River have their sources in an Ordovician limestone sequence. The upper Huon catchment sits in an adjacent geological province of mostly Precambrian rocks (unmetamorphosed sandstone, siltstone, conglomerate, etc., with occasional metamorphic quartzite). The hills between Cygnet (Fig. 2a) and the Huon estuary are a local anomaly, with Upper Carboniferous glacio-marine sediments intruded by alkaline, Cretaceous syenite. The Central Plateau and the higher altitudes of the western Huon catchment were repeatedly glaciated during the Quaternary period, and classical glacial and periglacial landforms are common.

Geological faulting during the early Tertiary period has established the dominant NW–SE alignment of valleys seen in the course of the Derwent and Huon Rivers. The soils reflect the partitioning of the region's geology and some influence from vegetation. The lower catchments of both rivers have typically acidic, organic-rich podzols [10]. Skeletal soils and moor podzol peats overlie the ancient rocks that give rise to the Huon River. The upper Derwent catchment is different with alpine humus soils and moor peats. The soils overlying the Permo-Triassic sedimentary rocks are particularly impoverished in both nitrogen and phosphorus [11]. Those with Jurassic dolerite as parent material are marginally better (typically low N and medium P), but it is only the soils derived from Tertiary basalt in the upper Derwent catchment that have adequate nutrients (typically medium N and high P) to be described as fertile.

Vegetation of the catchments comprises three main assemblages: alpine (austral-montane), temperate rain forest and sclerophyll forests dominated by *Eucalyptus* spp. [12]. Within these assemblages are many different vegetation types – Kirkpatrick and Dickinson [13] identify about 30 over the Derwent and Huon catchments. This diverse mosaic is a result of the complexity of topography, regional geology and soils, overlain with microclimatic variability. The Derwent River marks an approximate transition between rain forest and wet sclerophyll forests to the west, and dry sclerophyll forest, grassy woodland and grassland to the east. The latter two vegetation types were probably maintained by burning practices of the indigenous population in the past; little now remains after clearing for agriculture and grazing. More than

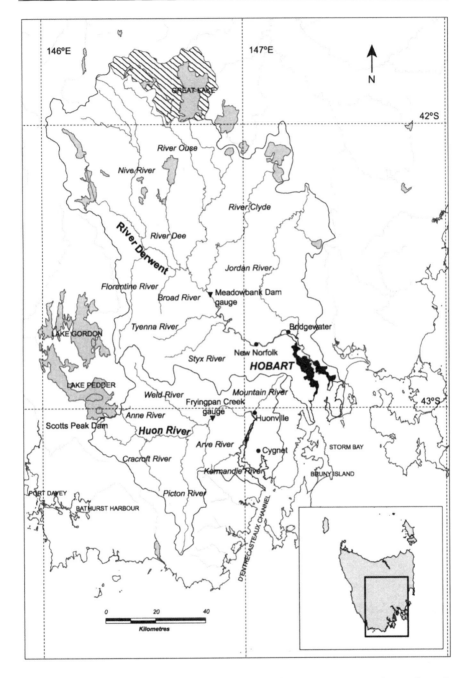

Fig. 1 Map of Derwent and Huon catchments in southeast Tasmania. Catchment boundaries, major towns and tributary rivers are indicated. The *cross-hatched* area of the Great Lake/Ouse River subcatchment is that part diverted to the South Esk River system. The urban zone of Hobart is also indicated along the shore of the Derwent estuary

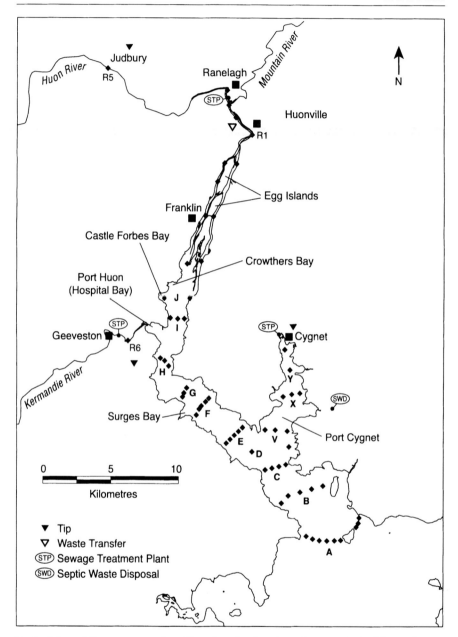

Fig. 2 a Map of Huon estuary. Sampling sites for the Huon Estuary Study [9] are shown, as are the locations of the major land-based discharges to the water body. Individual sites on cross-estuary transects are numbered from western to eastern shores (e.g. *A*1 -- *A*9); axial sites at the head of the estuary are numbered upstream (*R*1 -- *R*5). *R*6 is on the tributary Kermandie River

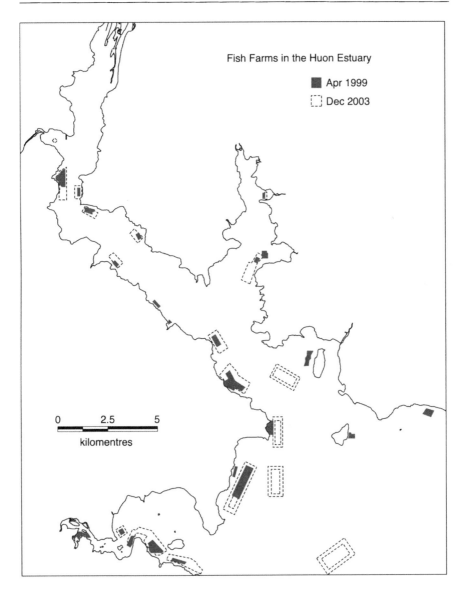

Fig. 2 b Map of the lower Huon estuary and adjacent D'Entrecasteaux Channel depicting the location and area of marine farm leases at two junctures – April 1999 (toward the end of the Huon Estuary Study) and December 2003

a quarter of the Derwent River catchment has been cleared (27.6% including urban areas and water storage impoundments), while a much smaller fraction of the Huon River catchment (5.6%) is so affected [1].

Coloured dissolved organic matter (CDOM) is leached readily from some vegetation types (e.g. buttongrass moors *Gymnoschoenus sphaerocephalus*), plant species (e.g. tea-tree *Leptospermum* spp.) and also soils (e.g. peats) in the Huon catchment, and to a lesser extent the Derwent catchment. Consequently, the waters of many tributaries of both rivers are strongly coloured. Generally, the concentration of the nutrients nitrogen and phosphorus in runoff to Australian rivers is relatively low, especially if catchments retain their original forest cover [14]. This is further accentuated in rivers of southeast Tasmania by low nutrient concentrations in precipitation and naturally impoverished catchment soils.

2.1
Human Influence

Aborigines are believed to have arrived in Tasmania more than 35 000 years ago. The two tribes of the southeast coast (Oyster Bay and South East) were hunter-gatherers, who used the estuaries of the region extensively as a source of food, particularly shellfish [15]. Inland in the Derwent catchment were the Big River Tribe. They hunted terrestrial animals as a staple of their diet, but also ventured to the coast regularly for shellfish. The imprint of these first Tasmanians upon the land and its drainage systems was minimal (base population estimate of 3000–4000 for the entire island), except perhaps where they modified catchment vegetation by use of fire.

From 1804 at Sullivan's Cove (Fig. 3) – still a hub of present-day Hobart – the first European settlement in Tasmania rapidly expanded. The population was $\sim$ 5000 by the 1820s, $\sim$ 35 000 by 1901, and is presently just under 200 000 in Greater Hobart (not all within the Derwent catchment). During development, the western shoreline of the Derwent estuary was modified by land reclamation, as were a number of the side bays later [17]. Further upstream, convict labour built the Bridgewater causeway in the 1830s. Settlement of much of the farmed land in the Derwent River catchment had happened by the 1850s. The first-class deepwater port became one of the most important in Australia from about that time, and maintained that status for more than a century. Industrialisation of the Derwent estuary in the 20th century was promoted by the advent of hydro-electricity. First came the electrolytic zinc refinery at Risdon just north of the city in 1917; the newsprint mill at Boyer followed in 1941. Smaller industries such as abattoirs, tanneries, foundries and horticulture (e.g. hops and market gardens) have operated from the mid-to-late 1800s. These were joined during last century by textile manufacturing, fertiliser production, timber milling and food processing.

The Huon region has always been sparsely settled from the arrival of the first Europeans about 160 years ago. The present population is $\sim$ 13 000; most are dispersed in small towns. It was not until the second half of the 1800s that land in the lower catchment was seriously taken up for forestry and agriculture, mainly horticulture (apple and pear orchards), grazing and dairying [18]. By the late 1800s these activities extended into sawmilling,

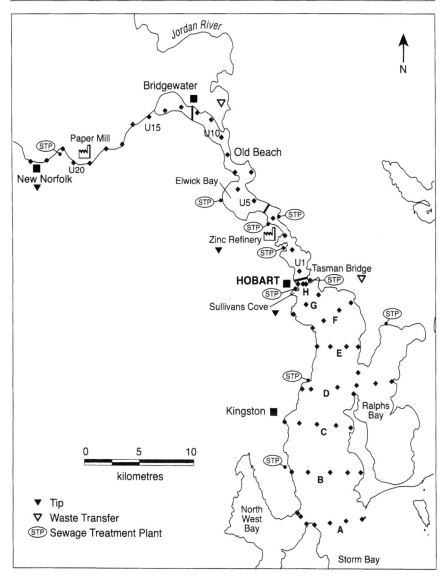

Fig. 3 Map of Derwent estuary. Sampling sites for the Derwent Estuary Study (CSIRO unpublished data, 1993–1994) and also for several more recent surveys (see [16]) are shown, as are the locations of the major land-based discharges to the water body. Individual sites on cross-estuary transects are numbered from western to eastern shores (e.g. A1 -- A7); single sites in the river channel above transect H are numbered upstream (U1 -- U23)

shipbuilding and fruit processing along the banks of the Huon estuary. Transport for the region from settlement until well into the last century was mostly provided by the waterway. For part of that time, it had two important ports (Port Huon and Port Cygnet, Fig. 2a) serving national and international markets. Modification of the estuary included the draining and reclamation of wetlands around Franklin, and the dredging of the western channel around Egg Islands to allow water traffic to Franklin and Huonville. The Huon estuary has not experienced major industry, apart from three stints on the shore of Hospital Bay (Port Huon). The "largest sawmill in Australasia" was cutting local timbers there from the early 1900s until its closure in 1929 [19]. On the same site, a neutral sulfite semi-chemical pulping mill (processing eucalypt hardwoods) operated from 1962–1982 and 1986–1991. Marine farming began in the lower estuary (Fig. 2b) in the mid-1980s with salmonid finfish (now exclusively Atlantic salmon *Salmo salar*) and to lesser degree shellfish (mostly blue mussels *Mytilus edulis*). A major wood processing plant upstream at the junction of the Huon and Arve Rivers (Fig. 1) is planned to begin operation in late 2004.

3
River and Estuary Hydrology

3.1
Rivers and Catchments

Characteristic data for both rivers and their catchments are presented in Table 1. During the last four decades, headwaters from an area of $\sim 10\%$ of each catchment have been diverted for hydroelectricity generation. This amounts to 12% of the total water yield from the Derwent catchment. Discharge at the river mouth has fallen $\sim 30\%$ from the 1920s as a result of diversions, extractions for irrigation, fish hatcheries and other changed land practices, as well as drier climate in the intervening years [20]. Median flows at Meadowbank Dam have decreased from 100 to $75 \, \mathrm{m^3 \, s^{-1}}$. Moreover, the 16 impoundments along the Derwent River, and their use in hydroelectricity generation, have smoothed out the river flows, both seasonally and annually. In comparison with pre-dam records, current high flow rates (5th percentile) have halved, but low flows (90th percentile) have doubled [21].

In the Huon catchment, diversion of its headwaters above Scotts Peak Dam (1972) has reduced the annual discharge of the Huon River system from 3000 to $2600 \, \mathrm{E6 \, m^3}$ (Gallagher [23], and references therein). Livingston [24] estimated that after the dam was built, the median flow decreased by 15% and low flows by about 8%. The Huon River at the flow gauging station at Frying Pan Creek (annual median $41 \, \mathrm{m^3 \, s^{-1}}$) presents a monthly flow record similar to the Derwent River, with the higher flows in late winter/early spring in line with the rainfall pattern (Fig. 4). No power stations are sited within the modified Huon catchment. The uniform summer flow is possibly maintained by either reserves of groundwater, or retention in the button-grass plains of the upper catchment, or both.

Table 1 Characteristics of the catchments and estuaries of the Derwent and Huon Rivers

Characteristic	Derwent	Huon
Catchment		
Total estuarine catchment area (km^2)	9076	3042
River length (km, source to sea)	215	135[a]
Topographic relief (m)	0–1449	0–1425
Catchment land use (%)[b]		
– Natural/semi-natural vegetation	71 (28)[c]	94 (28.5)[c]
– Cleared for agriculture	25	5.5
– Lakes/dams	3.5	< 0.5
– Urban/roads	0.5	< 0.5
Human population ($\times 10^3$)	~ 190	~ 13
Hydrology[d]		
Mean annual rainfall (mm)	1014	1676
Minimum annual rainfall (mm)	445	734
Maximum annual rainfall (mm)	2801	2993
Mean run-off coefficient	0.41	0.60
Mean annual run-off (mm)	417	999
Mean annual run-off volume (m$^3 \times 10^6$)	3784[e]	3038
Estuary		
Length (km)	52	39
Area (km^2)	191	83
Mean depth (m)	14.7	16.6
Volume (m$^3 \times 10^6$)	2815	1380
Mean tidal amplitude (m)	0.8	0.9

[a] Historically 169 km before dam at Scotts Peak
[b] Edgar et al. [1]
[c] Value in parenthesis is percentage of catchment as exploited crown land (mostly wood production)
[d] Rainfall data derived from Bureau of Meteorology [6]; flow data derived from DPIWE stream gauging, except where indicated
[e] Coughanowr [22]. Edgar et al. [1] suggest a mean annual run-off of 510 mm (volume of 4720×10^6 m^3), which gives a run-off coefficient of 0.51

In each case, it is the main tributary that overwhelmingly dominates the discharge to the estuary. Streams in the estuarine discharge area of the Huon system normally contribute less than 3% of the Huon River at Judbury [9]. Whereas typical flows from the Jordan River and other small streams entering the Derwent estuary directly amount to less than 1% of the discharge of the Derwent River [25]. This situation can be ascribed to the low rainfall zones around the estuaries.

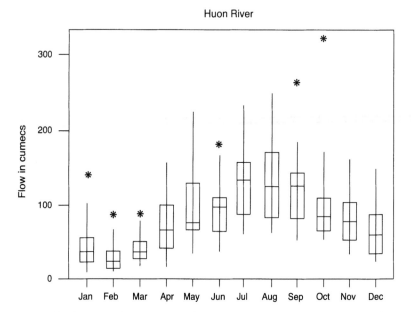

Fig. 4 Monthly flow record (1948–1996) for the Huon River at Frying Pan Creek gauge (Fig. 1) showing regular flow throughout the year, peaking in winter and early spring. *Boxes* encompass upper and lower quartiles with median, *whiskers* indicate full range. Extreme flows are indicated by *asterisks*

3.2
Estuaries

Southeast Tasmania has a wave-dominated coast, on account of the refraction of the Southern Ocean swells around the island [26]. The estuaries of both the Derwent and Huon Rivers are drowned river valleys. They are microtidal systems, strongly stratified at the head of the estuary and becoming partially mixed toward the mouth. Although neither estuary has a classical barrier, the South Arm complex of the Derwent estuary is a depositional shoreline feature (tombolos, curved beaches, etc.). It can be classified as a wave-dominated estuary [27]. The Huon estuary is more sheltered by Bruny Island. The elongate islands near the mouth of the Huon estuary could be viewed as a residual bay-head delta. However, its general morphology and sedimentary environment are closer to a tide-dominated, drowned river valley (see Fig. 2 in Roy et al. [28]). It is perhaps most fittingly classified as an estuary intermediate between wave-dominated and tide-dominated.

Using the Hansen and Rattray [29] classification, both estuaries are salt-wedge (Type 4) in character in the upper reaches. In the lower Derwent estuary, it is a partially mixed water column (Type 2b [30]), while the lower Huon verges on fjordic (Type 3a/3b, Hunter JR in CSIRO [9], Chap. 3), possibly because it is deeper and narrower in its lower reaches. Very close similarities for the two estuaries, among Tasmania estuaries generally, are noted in cluster analysis of their "physical variables" (catchment area, total annual runoff, estuarine perimeter length, etc., see Edgar et al. [1] and Table 1). They are also characterised by three hydrographic regions, the first two usual for a drowned river valley: (i) a shallow brackish zone in the upper half of the estuary, (ii) a deeper marine lower estuary, and (iii) a prominent side-arm off the lower estuary (Ralphs Bay in the Derwent estuary; Port Cygnet in the Huon estuary, Figs. 2 and 3). The mixing zone between the brackish and marine sections in the Huon estuary is an abrupt transition at the "elbow" of the estuary [9]. However, the mixing zone in the Derwent estuary is more elongated as a meandering section flanked on either side by a series of small bays. It could arguably be considered as a distinct middle estuary section. Indeed, Thomson and Godfrey [31] in their study of the circulation dynamics of the Derwent estuary propose that mixing is pronounced and dominantly wind-driven from Old Beach to the Tasman Bridge (the seaward limit of their observations, see Fig. 3). Tidal mixing was observable, but small, from the head of the estuary down to the Tasman Bridge; it was dwarfed by wind mixing in the middle estuary.

Freshwater tends to flow out on the eastern side of both lower estuaries – attributable to a combination of Earth's rotation (Coriolis effect) and prevailing winds from the west. The mid-stream islands (Egg Islands) in the upper Huon estuary can complicate this generalisation with changeable freshwater flow down the channels on either side. The overturning buoyancy-driven circulation that defines these estuaries has a compensating landward, subsurface flow of salty water along the western shore. Relatively greater penetration of this salty water into the upper Derwent estuary is observed owing to its greater cross-sectional area (see CSIRO [9], Sect. 3.3.3). Nevertheless, outcropping of high salinity water ($S > 30$) in the small bays and inlets of the west and southwest shores of the Huon estuary (from Crowthers Bay to Surges Bay) was commonly observed. The Huon estuary is seemingly a more mature system than the Derwent, with greater deposition in the fluvial delta exemplified by the Egg Islands (see [28]).

The microtidal regime and the lack of a large basin in the upper part of both estuaries result in small horizontal tidal excursions. The maximum calculated for the Huon estuary is ~ 600 m upstream and downstream of a location just below the Egg Islands; more typical is an oscillation of ± 150 m (CSIRO [9], Chap. 3). Davies and Kalish [32] estimated the tidal excursion in the Derwent estuary to be 1–2 km.

Observed and modelled currents in surface waters are generally weak (≤ 0.2 m s^{-1}) from data in the Huon estuary [9] and the Derwent estuary [25, 30, 33]. The upstream flow of saline bottom waters is even weaker, estimated in the Derwent estuary as 0.02–0.05 m s^{-1} [3]. The general circulation pattern for southeast Tasmania is presented in Fig. 5.

The average flushing period for the Huon estuary is ~ 7 days [9], while that of the Derwent estuary is ~ 15 days (Coughanowr [25] and Hunter JR, per-

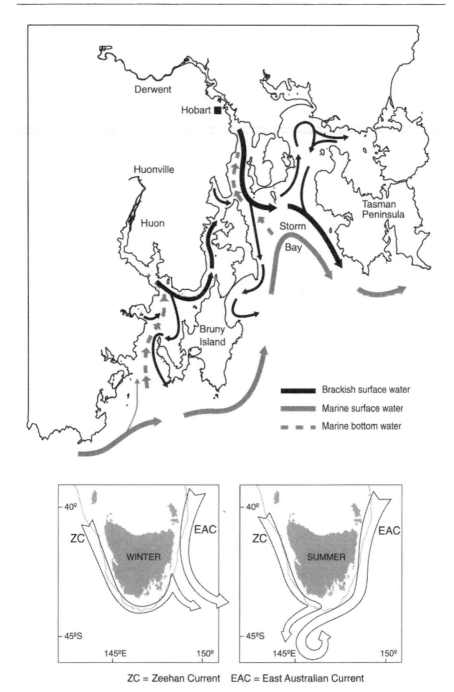

Fig. 5 General circulation pattern for coastal waters in southeast Tasmania, modified from Cooper et al. [34] and updated with information from Herzfeld M (personal communication). *Inset* shows major currents affecting Tasmania: extension of the East Australian Current (in summer) and the Zeehan Current (in winter) [35]

sonal communication). Greater variation applies when different water types are considered: the surface layer in both estuaries is likely to be flushed in a day or two, but isolated deeper waters near the head of either estuary could require much longer to be flushed (e.g. 20–35 days for the upper Derwent estuary [32]). In contrast to the regular, slower mechanisms of tidal action and entrainment in surface brackish waters, salty waters beneath the halocline in the upper Derwent estuary are known to also be precipitately displaced by higher river flows. For discharges $> 75\,\mathrm{m^3\,s^{-1}}$ the toe of the salt wedge begins to move downstream from New Norfolk. At the critical flow of about $150\,\mathrm{m^3\,s^{-1}}$ [32], the saline bottom layer is expelled downstream as far as Bridgewater. Such "critical flushing" behaviour has been observed in at least one other Australian estuary (Hacking in New South Wales [31]); it is also conceivable for the Huon estuary, but has yet to be recorded.

4
Estuarine Chemistry

4.1
Source Waters

Before focussing on the two estuaries, it is worthwhile remarking on the characteristics of the source waters, those from the land via the rivers and those from the sea encircling Tasmania.

4.1.1
Freshwaters

Runoff from the Tasmanian landmass is expected to be intrinsically pure, because precipitation is derived from the Southern Ocean, and historically in the island's south, its path to streams has been filtered by well-vegetated catchments on generally impoverished soils. Surface waters are, thus, low in conductivity and suspended solids. Buckney and Tyler [36] observed that the ionic fraction of surface waters, in the provinces taking in the Derwent and Huon catchments, are determined by atmospheric precipitation and regional geology. In the upland source waters of the Huon River, as seen in small lakes and tarns, precipitation dominates. The ionic proportions are that of very dilute seawater; moorland vegetation and soils cause them to be coloured and acidic (pH 4–6). Small quantities of the divalent cations calcium and magnesium, along with bicarbonate, appear to be leached from streambed sediments throughout the Derwent catchment and in the lower Huon system. Consequently, river waters are near neutral pH and modified from the ionic proportions of seawater, but still very dilute. Only in the forested tributaries draining Ordovician limestone substrate (Florentine, Styx, Weld, and possibly Tyenna) is the conductivity raised, with higher concentrations of calcium, magnesium and bicarbonate ions. Coughanowr [22] also reports higher levels of nitrate in these streams.

Analyses of Huon River water for nutrients and trace elements (CSIRO [9], Chaps. 4 and 9) confirm that the water quality is very similar to the pristine rivers entering Bathurst Harbour and Port Davey ([37], Butler and Plaschke, unpublished results) in the wilderness area of southwest Tasmania. Appreciable levels of total nitrogen (10–20 μM) have been recorded in the Huon [9] and Derwent Rivers [25], most of it as dissolved organic nitrogen. This riverine export of organic nitrogen from unpolluted, forested catchments could be a more universal attribute in temperate zones (see [38]). Correspondingly high levels of dissolved organic carbon (DOC, 10–15 mg L^{-1}) were observed in the Huon River entering its estuary (HES, unpublished data); DOC in the Derwent River at the same junction was lower (2–8 mg L^{-1}, [16]). Of the trace elements, iron and aluminium are sometimes naturally higher, associated with CDOM, in the runoff from sub-alpine moorlands and sedgelands [37, 39].

Modification of the historic condition of Tasmanian waterways has resulted from recent human activity – agriculture, forestry, mining, urbanisation and heavy industry. Agriculture (grazing and limited zones of more intensive horticulture and cropping) and forestry (especially the practice of clearfelling) have directly degraded the quality and yield of water in some tributaries of the Derwent and Huon systems. Moreover, both rivers have felt an indirect influence from the last three activities through generation of hydroelectricity. Bobbi [40] observed that several streams in the drainage area adjacent to the Huon estuary had higher conductivity, turbidity and concentrations of nutrients (N and P). They stood out as those with catchments and riparian zones damaged by agricultural practices, and contaminated by human wastes.

The concentrations of total suspended solids – as a coarse measure of contamination – are very low in the Derwent (average 3 mg L^{-1} [41]) and the Huon (0.8–3.0 mg L^{-1} [9]) Rivers entering their estuaries compared with others worldwide (see elsewhere in this volume). The influence of a large region of agriculture in the eastern half of the middle and lower catchment of the Derwent River appears to be masked by a series of four "run-of-the-river" dams, the last of which is Meadowbank Dam (Fig. 1) trapping suspended material along the river's course to the estuary. The eastern tributaries (Dee, Ouse, Clyde and Jordan) have higher total suspended solids, turbidity and total nutrients than the other Derwent tributaries [42]. The Clyde River also gains dissolved organic forms of nitrogen and phosphorus from the outflows of Lakes Sorell and Crescent. These two lakes are enriched in organic matter because they were not scoured out during the last glaciation. The Jordan River suffers further in its lower reaches with increased salinity owing to its diminished flow and the geology of the region.

4.1.2
Seawaters

The coastal seawaters of southeast Tasmania are the marine source for the Derwent and Huon estuaries. The subtropical convergence zone is the principal oceanic feature at this latitude; it ensures that the region is highly variable throughout the year, as well as interannually [43]. The seasonal cycle in southeast Tasmania alternates between an incursion of subtropical waters in summer, as an extension of the East Australian Current (EAC, Fig. 5) down the island's east coast, and their displacement by sub-Antarctic waters from the south in winter. This wintertime condition is complicated somewhat by waters from the region of the Great Australian Bight moving down the west coast, and around the south of Tasmania in the Zeehan Current [35]. Runoff from the high rainfall regions in the southwest can dilute waters of the Zeehan Current in transit.

With the oscillation of water masses between summer and winter, their physico-chemical signature changes. The subtropical waters are saltier, warmer, lower in dissolved oxygen and depleted in nutrients, when compared with sub-Antarctic waters [44]. The nutrient nitrate has a pronounced seasonal cycle (Fig. 6). It peaks during the period from the onset of autumn and winter cooling from March through to the spring warming in September. Nitrate concentrations at this time are usually > 3 µM. The spring microalgal blooms deplete the nutrient so that during late spring and summer its concentration is < 0.5 µM, often much less. An influx of offshore waters (varying

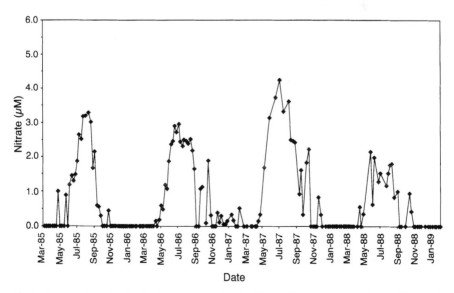

Fig. 6 Seasonal cycle of nitrate concentration in Storm Bay waters, southeast Tasmania (1985–1989 [44] and unpublished data)

interannually) can augment winter concentrations to $\sim 6\,\mu M$. The same process during the warmer months, or possibly the influence of mesoscale eddies, causes spikes in concentration of $1-2\,\mu M$ [43]. Levels of filterable reactive phosphorus (hereafter referred to as phosphate) respond to seasonality, but to a lesser extent, mostly ranging from about 0.2 to 0.7 μM. Chlorophyll a concentrations vary over the course of a year from 0.5 to 4 $\mu g\,L^{-1}$ [44].

4.2
Huon Estuary

4.2.1
Inputs

Historically, the Huon estuary has received small loadings of materials from human activities, more often than not diffuse inputs in runoff after modification of catchment land cover and riparian zones. Hospital Bay (Fig. 2a) was more seriously degraded: first from sawdust washed into the bay during the time of the sawmill, and later from direct discharge of wastes from the pulp mill. Deposits of sawdust have changed the western foreshore of the bay markedly, filling in what was originally watercourse. While the pulp mill was operational, its discharge caused the surface sediments of Hospital Bay to be soft, sticky mud that reeked of hydrogen sulfide. This "anaerobic blanket" has declined from a recorded 72 ha of bay sediments in 1982 to about 1 ha off the pulp-mill wharf in 1996[1]. Other inputs to the estuary have included sewage, often untreated from some townships until the last 10–20 years. Currently, there are three sewage treatment plants (STPs), all with some degree of secondary treatment. However, outside the larger towns, septic tanks are the rule. The use of chemicals in agriculture in the catchment has received some study (e.g. accumulation of residues in orchard soils [45]), but transfer to streams has scarcely been looked at, apart from very limited pesticide monitoring [23] and an older study of a few organochlorines [46]. Wotherspoon et al.[2] provide detail on annual usage rates of pesticides in the catchment. Marine farming of finfish, although demanding a clean aquatic environment, also brings problems with wastes and nutrient enrichment. Woodward et al. [47] made an early assessment of the effects of fish-farm waste in the Huon estuary with a view to improving management practices. Later work by McGhie et al. [48] looked at the degradation of fish-farm wastes on the floor of the Huon estuary.

---

[1] Hyde RG (1996) Environmental report to APM: Hospital Bay survey 22-3-96. Unpublished technical report PH96/38

[2] Wotherspoon K, Phillips G, Morgan S, Moore S, Hallen M (1994) Water quality in the Huon River and potential sources of pollution. Unpublished report, Centre for Environmental Studies, University of Tasmania, Hobart

Gallagher [23] has made a useful and extensive review of the regional characteristics and water quality in the Huon catchment and estuary up until 1996. In addition, he has considered the human pressures on the system, and provided an annotated bibliography of past studies of water quality throughout the catchment.

Much of the information in the next sections comes from the CSIRO Huon Estuary Study [9]. Observations were made at a range of temporal and spatial scales [49], including seasonal, estuary-wide surveys of the full water column, weekly monitoring at a small sub-set of key sites, and an intensive sampling of the estuary's surface sediments.

4.2.2
Dissolved Oxygen, Organic Matter and Suspended Solids

On account of the naturally high levels of dissolved organic matter present in riverine outflow, stratified Tasmanian estuaries in the west and south are susceptible to depletion of dissolved oxygen (DO) in their upper sections. This condition has been recorded in the pristine Bathurst Harbour–Port Davey system [50]. In the Huon estuary, we routinely observed that DO was undersaturated in bottom waters at the top of Egg Islands, but in our observations it did not fall below 40% saturation in the upper estuary. Otherwise, estuary waters are generally well oxygenated (80–100% DO saturation). Only sporadically in localised bottom waters has the influence of human activities caused serious decline in DO. Gallagher [23] notes occasional reports of very low DO concentrations in Hospital Bay and waters outside at the time when the pulp mill was operating. Likewise, DO levels down to 3% saturation were seen in the vicinity of fish farms during summer [9]. It is surmised that high respiration in the organic-rich bottom sediments in the lower estuary make this area vulnerable to DO depletion with any additional loading of organic matter.

Dissolved organic matter (DOM, determined by solid-phase extraction) in the estuarine waters had an inverse relation with salinity (Fig. 7), indicating that the source was riverine [9]. The depression below the linear mixing relation indicates that there was an apparent loss of DOM at intermediate salinities, possibly as flocculated colloidal material. ^{13}C : ^{12}C isotope ratios in the DOM suggested that the material arriving in the freshwater had a terrestrial C3 plant source, while that at the seawater end was consistent with marine microalgae. The light absorption coefficient by coloured dissolved organic matter, $a_{CDOM}(440)$, in the Huon River as it entered the estuary ranged from about 7–14 m^{-1}, very high for an Australian waterway. Unlike bulk DOM, CDOM had a conservative relation with salinity, or even possibly some input at the fresh/brackish water interface. Further analysis of the riverine DOM (McGhie TK, unpublished data) pointed to a significant fraction of polyphenolic tannins ($\sim$ 20–40%). Particulate organic matter (POM) was also characterised concurrently using the lipid biomarker tech-

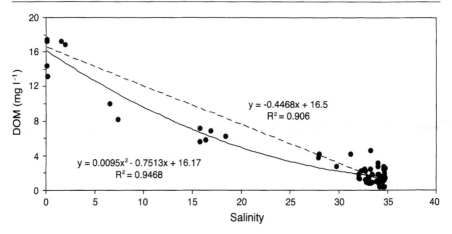

Fig. 7 Relation of dissolved organic matter (DOM – determined by solid-phase extraction) to salinity in the Huon estuary, July 1996. Regression data are shown for the curve of best fit and also for linear mixing

nique, an effective method for tracing diverse sources of organic matter in estuarine sediments [51]. In contrast to DOM, POM appeared to be derived from aquatic sources (in an October 1996 survey), mainly diatoms and other microalgae outweighing the bacteria. Rather than a situation where the DOM is derived from the breakdown of POM (as proposed for marine systems), terrestrially sourced DOM may influence microalgal blooms in the Huon estuary. Such a hypothesis is consistent with one of the criteria for blooms of the toxic dinoflagellate *Gymnodinium catenatum* in the estuary [52, 53].

Results from a 1997 survey [9] indicated that organic matter in the sediments follows the same general pattern observed in the water column. In the upper estuary, and certainly as far as Hospital Bay and the elbow of the estuary, the dominant source was terrestrial (delineated by $^{13}C : ^{12}C$ isotope ratios). Whereas in the lower estuary, the majority of organic matter originated from microalgal production in the water column, which eventually deposits on the sediments. The organic carbon and organic nitrogen content of the sediments is predictably associated with its grain size. C_{org} varied from < 1% in the sandy upper reaches and mouth to a maximum of 8% at the elbow of the estuary with a secondary peak in Port Cygnet. N_{org} paralleled this distribution, ranging from < 0.2 to 0.75%. The C_{org} in surface sediments is sufficiently enriched overall to suggest the Huon estuary was mesotrophic. Anomalously high C_{org} in and just outside Hospital Bay coincided with peak values of sitosterol (a terrestrial plant biomarker) and also overlapped previously reported distribution of wood fibre waste [54] from past sawmill and pulp mill operations. Other specific lipid biomarkers were useful in tracing natural, as well as human-influenced (see below), processes in the Huon estuary. The microalgal biomarkers diatomsterol (for diatoms) and dinosterol (for dinoflagellates) were more prominent in the lower estuary. In the 1997 sampling [9], diatomsterol was abundant and dinosterol low, in accord with recent bloom history at that time where the regular diatom blooms (*Pseudonitzschia* spp. and *Chaetocerus* spp.) had not been displaced by dinoflagellate blooms (especially *G. catenatum*) in the preceding three years.

Apart from a very few instances in 1996–1998, concentrations of suspended solids in the estuary [9] were less than $6 \, mg \, kg^{-1}$. Both the freshwater and marine endmembers had very low concentrations: the former $\leq 3 \, mg \, kg^{-1}$, and the latter $< 2 \, mg \, kg^{-1}$. The tributary Kermandie River was a consistent source of suspended solids. The concentration was regularly measured at $6–12 \, mg \, kg^{-1}$ just before it entered Hospital Bay, but on one occasion it reached $39 \, mg \, kg^{-1}$. Slightly higher levels were also observed in the main river around the Egg Islands, arising either from local inputs, or possibly from flocculation of river-borne material at the brackish salinities (5–10) common in that section. The latter postulate would be coherent with the apparent removal of DOM mentioned above, and the coincident observation of a peak in the detrital absorption coefficient.

4.2.3
Nutrients, Chlorophyll and Microalgae

Interest in these aspects of water quality in coastal and estuarine regions of southeast Tasmania was raised in the 1980s and early 1990s with blooms of harmful microalgae, particularly *G. catenatum* [52, 55]. Along with physical conditions, nutrients and other chemical compounds were seen as necessary pre-requisites for bloom formation. From investigation of the estuary in 1996–1998, the nutrients nitrogen, phosphorus and silicon in their various forms were generally found to have natural distributions [9], apart from localised effects of human activities or dense microalgal blooms.

Dissolved organic nitrogen was the main form of nitrogen. It verged on conservative behaviour with salinity; higher concentrations ($9–18 \, \mu M$) at the river end derived from terrestrial runoff. DON in marine waters was $3–10 \, \mu M$. Particulate nitrogen ($< 1.5–5 \, \mu M$) was a small and fairly uniform fraction of nitrogen across the estuarine gradient. Therefore, total nitrogen had the characteristics of DON (inverse with salinity), but the mixing line was displaced up or down depending upon the seasonal influence of dissolved inorganic nitrogen. Sporadic high concentrations of particulate nitrogen coincided with high-density microalgal blooms.

Dissolved inorganic nitrogen (DIN, the sum of nitrate, nitrite and ammonia concentrations) was usually dominated by nitrate, especially in winter. During this non-productive season, nitrate increased linearly with salinity up to about $5 \, \mu M$. Riverine concentrations were always low at $< 0.05–0.7 \, \mu M$. Over spring, summer and early autumn (particularly in 1997/1998) nitrate was strongly removed from estuarine waters by biological activity, replicating the cycle seen in Storm Bay [44]. A residual store of nitrate was present throughout the year in bottom waters at the marine end of the Huon estuary. Deeper marine waters entering via southern D'Entrecasteaux Channel were the likely source, but small concentrations might arise from internal recycling of nitrogen during periods of high biological production. Nitrite behaved similarly to nitrate, but was at 10% or less of nitrate's concentration, apart from late autumn 1998, when it was over a third. Remineralisation of detrital organic nitrogen from the dense microalgal blooms of earlier that year is a credible explanation for this outlying observation. Ammonia was also at low, but patchy, concentrations ($< 1.0~\mu M$) in the estuary. Higher levels were observed at times just above the sediments (notably in spring/summer 1997/1998, as an early step in the sequence leading to the raised nitrite levels), near to finfish farms and downstream of STPs (Fig. 8).

In contrast, total phosphorus was mostly in the inorganic form of phosphate, and it increased from very low levels in freshwaters ($< 0.08-0.5~\mu M$) to higher concentrations in marine waters ($0.2-0.8~\mu M$). Dissolved organic phosphorus and particulate phosphorus were consistently low, except for rare peak concentrations of the latter during microalgal blooms. Higher surface concentrations of total phosphorus near towns were possibly the only lingering signs of sewage or stormwater input, because nitrogen species were rapidly stripped out of estuarine waters by microalgal uptake. Phosphate was influenced by seasonal uptake by microalgae, but it was not removed as rapidly or as extensively as nitrate (Fig. 9). Almost without exception, phosphate concentrations in at least some bottom waters of the middle or lower estuary exceeded those in seawaters along the marine boundary (A transect in

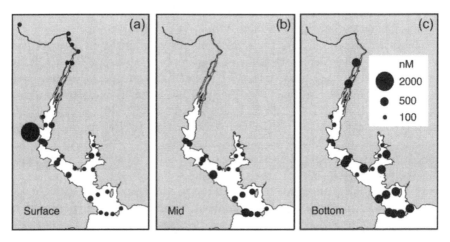

Fig. 8 Distribution of ammonia in the Huon estuary in waters at **a** surface, **b** mid-depth chlorophyll *a* maximum, and **c** about 1 m off the bottom. Concentration is indicated by the size of the *dot* at each site. The site on Kermandie River downstream of an STP outfall was regularly high

Fig. 2a). Similar observations in other estuaries have been put down to efflux of phosphate from reducing sediments [56]. On occasions, phosphate appears to be able to escape a thin capping layer of oxidised sediments, where it would normally bind to iron oxyhydroxides.

Filtered reactive silicon (hereafter referred to as silicate) in the estuary was dominated by terrigenous sources, and so it had the expected inverse relation with salinity (Fig. 10). Much higher concentrations were found in the Kermandie River (100–308 μM; median: 239 μM) than in the Huon River (50–101 μM; median: 72 μM). When the former's discharge was sufficient, it added silicate to the middle estuary, and disrupted the more usual conservative relation with salinity. During summer in the lower Huon estuary, silicate concentrations of ≤ 3 μM, or their relative proportion to nitrogen and phosphorus, were possibly limiting to diatoms [57, 58].

Seasonal variation in N : P ratios suggested nitrogen limitation ($< 5 : 1$) in the summer months, even though winter-time ratios ($\sim 10 : 1$) did not attain the Redfield ratio of 16 : 1. It was even more pronounced ($< 3 : 1$) during the dense microalgal blooms recorded in the 1997/1998 summer. Definitive proof of nitrogen limitation can only come from culturing experiments in the laboratory, or from bioassay techniques (e.g. [59]).

Microalgal dynamics in the Huon estuary are characterised unusually by blooms of various diatoms interspersed with dinoflagellate blooms, and they often co-exist [60]. Irregular blooms of the PST-producing *G. catenatum* have been recorded from 1986 in the estuary [52]. They have been known to close shellfish leases for up to nine months. Over the 1996–1998 period of the Huon Estuary Study [9], we observed four key microalgal taxa: *Chaetoceros* spp. and *Pseudonitzschia* spp. for the diatoms; and *Ceratium* spp. and the toxic species *Gymnodinium catenatum* for the dinoflagellates. These were the

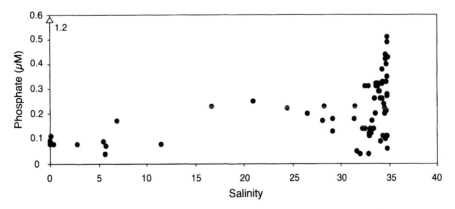

Fig. 9 Relation of phosphate (filterable reactive phosphorus) to salinity in the Huon estuary, December 1997. The Kermandie River (site *R6*) value is off-scale, as indicated. The greatest depletion of phosphate (for $S = 30$–35) occurs in Port Cygnet

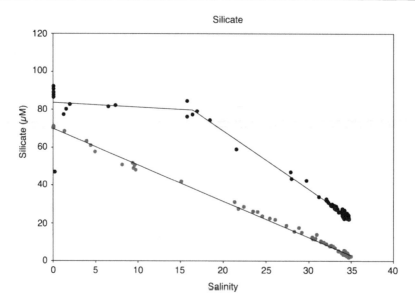

Fig. 10 Relation of silicate (filterable reactive silicon) to salinity for representative winter surveys, July 1996 (*upper*, offset by $+20\,\mu mol\,L^{-1}$) and June 1997 (*lower*). Silicate concentrations at freshwater site *R6* on the lower Kermandie River are off-scale, typically $200–300\,\mu mol\,L^{-1}$. The input of this tributary at higher flows (e.g. July 1996) modifies the standard inverse linear plots for surveys such as represented by that of June 1997

bloom species against a fairly constant background of small flagellates (classified by HPLC marker pigments as a mixture of haptophytes and prasinophytes, with some cyanobacteria – not discussed further here).

Chlorophyll *a*, used as a measure of microalgal biomass, was uniformly low ($\sim 0.3\,\mu g\,L^{-1}$) in winter. Since nutrients are replete at this time, it is adjudged that light and estuarine flushing limit primary production. Blooms can occur from early spring through summer to late autumn. We observed that chlorophyll *a* concentrations exceeded $4\,\mu g\,L^{-1}$ about 10% of the time. Annual mean and median levels both were $\sim 1\,\mu g\,L^{-1}$. In concert with the earlier assessment using organic matter in sediments as the index, the chlorophyll and nutrient data suggest that the estuary is mesotrophic. There were strong interannual variations in the pattern of blooms, reflected in the chlorophyll *a* seasonal mean concentrations (Fig. 11). *G. catenatum* was not present in the phytoplankton in 1996 and early 1997 (nor in the two years preceding), but developed extensive blooms in the summer and again in the autumn of 1997/1998, with chlorophyll *a* in spot measurements as high as $50\,\mu g\,L^{-1}$. The reason for a bloom of the toxic dinoflagellate after an absence for several years is not readily explained by physical and chemical conditions over the two study years 1996/1997 and 1997/1998. Certainly, *G. catenatum* is not limited by levels of (macro-)nutrients, because of its strong capacity for diur-

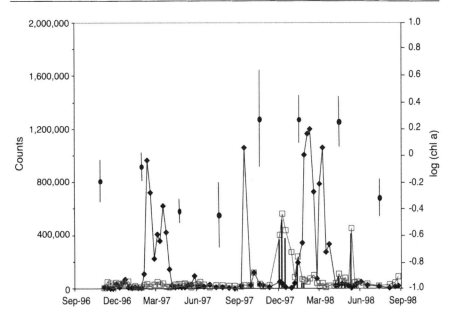

Fig. 11 Seasonal mean concentrations (●) and 95% confidence intervals of chlorophyll *a* in the Huon estuary. Data are derived from depth-integrated samples from five representative stations in the middle and lower estuary. Superimposed are time-lines of cell counts for key phytoplankton classes (♦ diatoms, ☐ dinoflagellates) obtained by net sampling (20-μm mesh). *Vertical bars* indicate the proportion of *Gymnodinium catenatum* in dinoflagellate counts

nal vertical migration (demonstrated definitively in situ by vertical profiling instruments [9])[3] to access reserves of nutrients in bottom waters immediately overlying the sediments.

In the terminology of Sharp, Church and colleagues [61], the Huon estuary is dominated by the "biochemical filter". The small tides and the low concentrations of suspended solids mean that the distribution, behaviour and speciation of chemical solutes are determined by biological activity rather than being influenced by interactions with particulate matter (the "geochemical filter"). Consistent with this model, the peak of phytoplankton biomass, as portrayed by chlorophyll *a* concentrations, was routinely located in the middle to lower estuary. A corollary to this distribution was that the greatest depletion of nutrients (and other assimilated minor elements) was also observed in this region of the waterway (see Fig. 9 for phosphate, Dec 1997). Port Cygnet, too, was a biologically active zone. Although strong depletion of nutrients was observed here, it was only in late autumn and winter that the highest concentrations of chlorophyll *a* were observed in this side arm. One

[3] Clementson LA, Sherlock MJ, McKenzie DC, Parslow JS, Butler ECV (2004) An autonomous profiler for monitoring water quality in sheltered coastal waters. Unpublished report from CSIRO Marine Research, Hobart

possibility for this apparent disparity is that the high numbers of shellfish, both farmed and wild, in this side arm were harvesting the microalgae during the productive times so that they did not reach their potential peak biomass (see Appendix in [62]).

In estuaries with both the geochemical and biochemical filters operating, the turbidity in the upper estuary strongly limits the light for microalgal photosynthesis. In the Huon estuary, the light is diminished instead by the high concentrations of CDOM. An additional influence is possible; nutrients released by the processing of CDOM in the microbial loop have been observed to alter microalgal community structure by favouring some taxa over others [53, 63].

Occasionally, chlorophyll *a* peaks were observed in the western bays (e.g. *Mesodinium rubrum* bloom, Castle Forbes Bay in 1996) of the upper to middle estuary. These isolated occurrences are attributed to the fact that these bays often have outcropping of saline, and hence relatively transparent, water at the surface. In the upper estuary, where waterborne primary producers are constrained by light, other species use alternative niches to fill the gap. Microphytobenthos (MPB) are active on the mudflats. Cook [64] studied two mudflats adjacent to saltmarsh in the Huon estuary, one in Castle Forbes Bay and the other at the head of Port Cygnet. He estimated primary production by MPB as a percentage of total production by phytoplankton (throughout the estuary) to be 2.5% at the more marine Port Cygnet and 12% at the brackish, upper-estuary site at Castle Forbes Bay. Primary production on the Port Cygnet mudflat was suggested to be limited by the greater wave energy there, and it was evaluated as a net heterotrophic system. In contrast, the Castle Forbes Bay mudflat was net autotrophic. Primary production on its higher, more exposed region was greater than on the lower, more submerged portion. Here, the CDOM in the water apparently limited photosynthesis of immersed MPB. Small patches of seagrass are also known in the upper estuary, as well as in parts of Port Cygnet. The extent of their contribution to primary production has not been evaluated.

In 1996–1998 surveys of the Huon estuary [9], we found chlorophyll *a* levels not to be significantly higher in the proximity of finfish farms, at either surface or mid-depth. It was concluded generally that that there was no evidence of a consistent effect of farm proximity on chlorophyll *a*. The result was not unexpected given the short flushing times for the estuary, comparable with time scales for phytoplankton growth. Nutrients from farms should be dispersed broadly within the estuary before there is any biomass response from phytoplankton. Nevertheless, nutrient budgeting and modelling suggest that nutrients from finfish farms were broadly influencing primary production in the estuary. At 1997–1998 stocking levels, nutrients from the farms were fuelling 25% of the estuarine biomass of microalgae. The effects of increasing the stocking of farmed fish on microalgal biomass has been assessed ([9] and Parslow JS, unpublished results). At four times the 1997–1998 input of dissolved inorganic nitrogen to the system, the Huon estuary could be on the path to eutrophication, with the model indicating the likelihood of prolonged microalgal blooms.

4.2.4
Trace Metals and Other Elements

Earlier monitoring of trace metals in the water column of the Huon estuary and its tributaries has been reviewed [23], but sampling was haphazard and the results are too few to draw any conclusions. Cooper et al. [34] determined metals in mussels and oysters at two locations in the lower Huon estuary (Police Point and Garden Island Bay) as reference sites for a study of shellfish in the contaminated Derwent estuary. Zinc (20–30 $\mu g\,g^{-1}$, wet weight), cadmium ($< 0.2\,\mu g\,g^{-1}$) and lead ($< 0.2\,\mu g\,g^{-1}$) were considered to be at baseline values, typical of uncontaminated sites. Chesterman[4] obtained metal concentrations (among 70 elements) in sediments associated with Hospital Bay and the decommissioning of the pulp mill (Table 2). The samples were collected about the APM wharf (the main discharge point for mill effluents), although results were also reported for a "background" site at Castle Forbes Bay, a small bay 3 km upstream of Port Huon on the western shore. Higher concentrations were found in the surface sample at site 13 alongside the wharf. Moderately high levels were recorded for some metals in the northern part of D'Entrecasteaux Channel [65]. For example, metal concentrations in a sediment sample from Oyster Cove were 36 mg Pb kg^{-1}, 8.3 mg Cu kg^{-1} and 172 mg Zn kg^{-1}.

We made a contaminants survey at a subset of estuary-wide sites shown in Fig. 2a following a regular hydrological survey in September 1998 [9, 68]. Surface sediments were collected at a smaller group of six representative sites. In an independent survey the previous year (Feb 97), Jones et al. [66] took grab samples of sediments at 18 sites throughout the Huon estuary, as background data for a more intensive geochemical survey of Derwent estuary sediments. Although our analyses were of acid-extractable metals determined by ICP-SMS, and those of Jones et al. [66] were total metals by XRF and NAA methods, the concentration ranges found were in good agreement (Table 2). Slightly raised levels of zinc, and possibly copper and cadmium, were observed in Hospital Bay and Port Cygnet. They indicate persistence of localised contamination from historic port and industrial activities, and possibly continuing discharges from adjacent human settlements.

Trace element results in the water column from our contaminants survey (Sept 1998) are presented in Table 3. All of the metals from the thirteen sites

---

[4] Chesterman R (1995) AMCOR Paper Group Port Huon mill decommissioning plan. Unpublished report to AMCOR, March 1995. Environmental Scientific Services, Hobart

Table 2 Trace element concentration ranges (mg kg^{-1}, dry matter basis) in surface sediments of the Huon estuary

Trace element	Chesterman[a]	Jones et al. [66][b]	CSIRO [9][c]	Guidelines ANZECC/ ARMCANZ [67]	
	Hospital Bay[d] 3 sites (1995)	Entire estuary 18 sites (1997)	Entire estuary[e] 6 sites (1998)	ISQG Low	ISQG High
Arsenic	18–22	4–25	1.9–28	20	70
Cadmium	0.2–0.4	< 10	0.02–0.26	1.5	10
Chromium	34–53	50–80	–	80	370
Cobalt	9–20	15–35	3.7–12.5	–	–
Copper	17–252	7–32	3.6–47.0	65	270
Iron (%)	2.3–4.1	1.73–4.05	0.6–4.3	–	–
Lead	10–88	< 2–48	3.6–29	50	220
Manganese	98–116	–	27–87	–	–
Mercury	–	< 5		0.15	1
Nickel	24–36	< 2–28	5.6–21.6	21	52
Vanadium	59–112	–	–		
Zinc	51–1002	< 2–66	14–92	200	410

[a] Acid-extractable metals (conc. HNO$_3$/HCl); also reported Ti, Mo, Th and U (not shown)
[b] Total metals (XRF and NAA); also reported Sb and Au (not shown)
[c] Acid-extractable metals (conc. HNO$_3$, ICP-SMS)
[d] Includes a site in Castle Forbes Bay
[e] Includes a site in Hospital Bay

in the Huon River and estuary were below the latest national [67] trigger levels. The Huon estuary, when compared with the pristine Bathurst Harbour estuary [37], had similar concentrations of some metals (Cd and Cu), slightly raised concentrations of others (Mn, Co and Ni), and lower concentrations of the remainder (Fe and Zn). Waters entering in the Huon River, when compared with the Old River (main tributary to Bathurst Harbour), had marginally higher concentrations of cobalt, copper, and nickel, a similar level of cadmium, and lower concentrations of iron, manganese and zinc [37]. The results are perhaps not surprising, because these rivers are draining back-to-back catchments. The Huon River catchment above Judbury is affected by just logging of a sequence of coops in a central corridor; the catchment of the Old River is in the South West National Park, and so is devoid of human activity.

In contrast, the Kermandie River in the lower catchment had cobalt (5.6 nM), copper (26 nM), nickel (19 nM) and zinc (44 nM) above their respective guideline trigger concentrations (Table 3). This river carries a substantial load of suspended solids (see above). It is likely that much of its trace

Table 3 Trace element concentrations (nM) in estuarine waters

Trace element	Bathurst [a] Mackey et al. [37] 10 sites	Huon [b] CSIRO [9] 13 sites	Derwent [c] CSIRO 1993–1994, unpublished 10–12 sites	*Guidelines* ANZECC/ ARMCANZ [d] [67] Fresh	Marine
Arsenic	–	0.5–23	–	32 (As^V)	32 (As^V)
Cadmium	0.02–0.05	0.02–0.05	0.01–14.7	0.12 [f]	15.1
Cobalt	< 0.7–0.9	< 0.2–2.4	0.3–19.1	4.1	15.3
Copper	1.7–6.7	1.7–3.8	1.8–232	5.2 [f]	5.0
Iron	336–7900	161–5370	75–67 300	–	–
Lead	–	[0.08–0.40]	–	5.8 [f]	3.9
Manganese	21–124	5.3–293	8–1910	855	855
Mercury	–	< 0.002	< 0.002–1.72 [e]	0.065	0.150
Nickel	1.7–3.7	[3.7–7.2]	2.7–27.8 [e]	11.6 [f]	555
Vanadium	–	[9.2–35]	–	118	196
Zinc	10–58	3.8–34	6.2–5374	36.7 [f]	41.3

[a] Old River – Bathurst Harbour estuary (Jan–Feb 1990 survey)
[b] Huon River, at Judbury, and estuary (Sep 1998 survey)
[c] Derwent River, at New Norfolk, and estuary (8 surveys, 1992–1994)
[d] Guidelines for Aquatic Ecosystems – all concentrations, formally µg L^{-1}, have been converted to nM
[e] Mercury determined for six of eight surveys; nickel determined only for four of eight surveys.
[f] Hardness-dependent
Values reported in square brackets are provisional, having been determined by an insufficiently validated method (ICPMS)

metal load is associated with this material. Indeed, cobalt and zinc exceeded the trigger level when unfiltered concentrations were considered, but fell below it when instead filtered concentrations were used. Copper was still two to three times the trigger level in even the filtered Kermandie River sample. Likely sources were individually or in unison: the effluent from the Geeveston STP, leaching of soils on land used in the past for horticulture, or desorption from particulate matter in DOM-rich stream waters.

A couple of other trace elements have been studied separately in waters of the Huon system. Featherstone et al. [62] have looked at arsenic after its earlier, intensive use as a pesticide on apple orchards in the lower catchment from the 1920s to the early 1980s. No evidence of contamination was found in waters or sediments. Inorganic arsenic concentrations in the Huon River were very low (0.023–0.057 µg L^{-1}), even compared to other pristine freshwaters worldwide. In the estuary, arsenic exhibited mostly linear mixing with salinity up to concentrations typical of unpolluted coastal seawater (1.4–1.7 µg L^{-1}).

In summer, strong removal of inorganic arsenic was observed in a pattern very similar to phosphate (Fig. 9). The extent of depletion was strongly related to cell numbers of the diatom *Pseudonitzschia* spp.

The behaviour of the iodine was investigated in the Huon estuary to discern the effect of humic-rich freshwaters upon the speciation of this element with multiple oxidation states [69]. Total dissolved iodine was conservative with salinity increasing from < 50 nM in river water to ~ 400 nM in coastal seawater. Iodate, the oxidised form of iodine, was undetectable ($\leq$ 30 nM) in freshwater up to a threshold brackish salinity (8–12). From there up to the seawater end-member, it had a linear relation with salinity, reaching a concentration of 250–300 nM. In the brackish waters, iodate originating from coastal waters was both reduced and converted to organic forms. The transformation was most likely mediated by a combination of chemical, biochemical and photochemical reactions, with humic material as a pivotal substrate.

4.2.5
Organic Contaminants

The deposition of organic waste (feed and faeces) from finfish cages is very localised. Using a variety of lipid biomarkers (fatty acids and sterols), significant contamination was restricted to sediments within a radius of ~ 10 m around each cage [9, 48]. The cholesterol:phytol ratio proved quite sensitive to fish-farm waste. In one instance a high ratio measured in the lower estuary (near site F3, see Fig. 2a,b) appeared to indicate an occurrence of a fish cage being mistakenly located outside a lease area.

Earlier faecal coliform (*Escherischia coli*) monitoring of 49 tributaries of the Huon estuary identified that 11 of these tributaries (mostly small creeks/rivulets) were contaminated in summer, and that three remained that way in winter [40]. Examination of sediments with biomarkers can indicate if sewage contamination regularly extended into the estuary. The selective biomarker for human faecal contamination is coprostanol (5β(H)-cholestan-3β-ol), which constitutes ~ 60% of the total sterols found in human faeces [70]. To avoid environmental artefacts seen in anoxic sediments, the ratio of coprostanol to 5α-cholestanol (5α(H)-cholestan-3β-ol), in place of simply coprostanol, was used to identify zones contaminated with human faecal matter. A ratio greater than 0.5 characterised a contaminated sediment. Using this criterion, it was only the northern half of Port Cygnet that showed signs of sewage contamination [9]. The source would be the discharge from the Cygnet township STP at the head of the side bay. The herbivore faecal biomarker 24-ethylcoprostanol was detected at trace levels in many estuarine sediments, but in insufficient quantities to suggest categorically a source from livestock.

Concerns about synthetic organic chemicals in the Huon region have centred on pesticides associated with horticulture, since their gradual displacement of traditional mineral-based treatments (e.g. lead arsenate, Bordeaux mixture and copper oxychloride) from about 50 years ago. DDT and its breakdown product DDE, as well as DDD/Υ-BHC, were detected in a muddy sediment and biota (zooplankton, turritellid gastropods, portunid crabs, southern cardinal fish and a marine conger eel) of the Huon estuary in 1969 [46]. This was six to seven years after the last wide-scale application of DDT and DDD in the Huon Valley. The crabs just outside Hospital Bay had the highest concentrations (19.8–24.5 ppm dry weight for combined DDT/DDE). It was also found that the organochlorine pesticide residues were routinely higher in Mountain River roach, trout and eels, relative to the same species further inland in the Russell River (e.g. in eels, DDE up to 11 ppm versus 3.65 ppm, and DDT 10.1 ppm versus 1.3 ppm). The Mountain River had a large expanse of orchards along its course, while the Russell River valley has never had orchards.

The only other organic chemical for which there is some historical monitoring information is atrazine (triazine herbicide). It has been used in forestry operations to control competing or pest species. The last reported application in the Huon Valley was on newly established plantations in the catchments of the Arve and Russell Rivers in 1993. Data from around that time suggest that atrazine measured in stream waters immediately after application were typically in the range 1–10 μg L^{-1} (guideline trigger level: 0.5 μg L^{-1} [67]), but occasional concentrations in the tens of μg L^{-1} were noted, and one as high as 380 μg L^{-1} [23]. It has been noted that atrazine can continue to leach from treated soils in Tasmania, with levels of 2 μg L^{-1} reported more than a year later in stream waters.

A standard suite of pesticides – organochlorine, organophosphorus and other (including representatives from triazine and pyridine classes) – was measured in water and sediment samples collected in September 1998 [9]. All of the pesticides were below their detection limit (organochlorine and organophosphorus < 0.1 μg L^{-1} and other < 0.05 μg L^{-1}) in waters at six sites estuary-wide. The same was true of measurements in sediments (organochlorine and organophosphorus < 0.01 mg kg^{-1}, triazine < 0.05 mg kg^{-1} and pyridine group < 0.5 mg kg^{-1}), apart from DDT and DDD, which were present in the Hospital Bay sediment sample at 180 μg kg^{-1} and 140 μg kg^{-1}, respectively. These last results support the earlier findings of Steen [46], and highlight the persistence of this group of organochlorine compounds. The recent ANZECC/ARMCANZ [67] guidelines would now require the pesticides to be measured down to detection limits an order of magnitude lower in several instances. Nevertheless, the Huon estuary does not appear to have serious pesticide contamination in water or sediments. The localised exception is DDT and its degradation products, which are above the ISQG-High level (Total DDT 46 μg kg^{-1}, DDD 20 μg kg^{-1}) for the new guidelines, in the vicinity of Hospital Bay.

4.3
Derwent Estuary

4.3.1
Inputs

With the city of Hobart on its shores, the Derwent estuary receives many of the city's waste discharges (Fig. 3). These span two major industrial sites (an electrolytic zinc refinery in the middle estuary and a pulp/paper mill in the upper estuary) through light industrial operations to STPs and urban runoff. In 1996, most of the biochemical oxygen demand (total $\sim 8000\,t\,yr^{-1}$) and the bulk of metals (e.g. zinc $\approx 220\,t\,yr^{-1}$) originated from industry. Whereas, most of the nutrient load of $\sim 550\,t\,yr^{-1}$ dissolved inorganic nitrogen (DIN) and $\sim 150\,t\,yr^{-1}$ total phosphorus came from STPs [25].

Government and industry have done much in the last decade to improve the environmental quality of the estuary. Sewage discharge into the estuary has been reduced in quantity (thirteen STPs in 1996 down to ten currently) and improved in quality with upgrading of treatment procedures (secondary treatment with disinfection is now the minimum standard). Heavy industry has tackled contaminant release specifically with effluent polishing stages and groundwater treatment; other smaller manufacturing operations have re-routed waste streams away from the estuary to STPs.

Management of urban runoff has also been improved for the thirteen rivulets and > 270 stormwater outlets. As a result, faecal bacterial loads have fallen by more than 95% and heavy metal loads have been halved over five years [16]. The input of total suspended solids has decreased by almost 20% over the same interval, although 71% of this input is now associated with stormwater drainage. Better water quality in the estuary has ensued from these management improvements, but at the same time, BOD and nutrient loads have increased marginally and estuarine sediments remain a major source of contaminants to the water column. A recent, thorough audit of pollution sources and loads to the Derwent estuary is provided in Green and Coughanowr [16].

4.3.2
Dissolved Oxygen, Organic Matter and Suspended Solids

In the same manner as the Huon estuary, bottom waters of the upper Derwent estuary are prone to DO depletion, especially during the warmer months of the year. This condition is exacerbated by artificial introduction of organic matter. A pulp and paper mill has operated at the head of the Derwent estuary, near New Norfolk (Fig. 3), since 1941. Its effluent (including wood fibre, $\sim 100\,t$ solids, and total oxygen demand (BOD) $35\text{--}40\,t\,day^{-1}$) was discharged directly to estuarine surface waters until 1988 [32], when primary treatment was instituted. Before the upgrade in late spring/summer of 1978/1979, Ritz and Buttemore [71] observed a serious oxygen sag during two 24-hour stations in bottom waters immediately downstream of the mill. It was also accompanied by significant levels of sulphide (up to 11 mg L^{-1}). Davies and Kalish [32] confirmed these results. Their surveys during 1988–1989 found that sub-halocline waters ($S > 20$) were below 10% oxygen saturation for around seven months of the year from late spring to autumn, depending upon river flows. Concomitantly, they observed instances of sulphide concentrations > 5 mg L^{-1} in bottom waters of the upper estuary. Even following improved treatment procedures at the mill, observations in 2000 and 2001 revealed DO concentrations < 2 mg L^{-1} in the same location and upstream to New Norfolk from November or December through to April or May [3, 16]. Nevertheless, it was contended that this represented a significant improvement from the year-round DO depletion of a decade earlier [3].

CSIRO Coastal Zone Program (unpublished data, 1993-1994) found suspended solids were generally low (< 6 mg L^{-1}), away from effluent discharges, and rivulet and stormwater outflows. They were universally lower (≤ 4 mg L^{-1}) in the upper estuary around New Norfolk and in the lower estuary (below Tasman Bridge). Bottom waters in the middle estuary (Stations U2–U10) had greater concentrations of suspended solids, with occasional instances of high levels (> 10 mg L^{-1}). Coughanowr [41] reported the same general distribution of suspended solids in the Derwent estuary, but a slightly higher mean concentration of 8 mg L^{-1}. Results from more recent monitoring of SPM [3, 16] remain essentially unchanged.

4.3.3
Nutrients, Chlorophyll and Microalgae

The impetus to establish the nutrient status of the Derwent estuary was provided by it being the site for the first observed bloom of the toxic G. catenatum in Tasmania [52]. This observation was linked with the incidence of high ammonia levels around the aggregation of STPs in the middle estuary. As expected many of the forms of nitrogen and phosphorus – in particular ammonia, dissolved reactive phosphorus, total phosphorus – peak in the middle estuary coinciding with the cluster of inputs in the vicinity [16]. Nitrate has been found at close to natural levels (< 0.3–$6 \mu M$) throughout the estuary with seasonal cycling imprinted in the data (see Fig. 6); the seasonality decayed up the estuary. Nitrate remained high (~ 1–$3.5 \mu M$) at the head of the estuary, above and below New Norfolk, throughout the year (CSIRO unpublished data, 1993–1994). There is some discharge of nitrate from the Boyer Mill (combined effluent stream median concentration: $2.9 \mu M$, 95%ile: $26 \mu M$ [3]), but it is also naturally high in the lower Derwent River, entering from some western tributaries [22]. The nitrate levels at New Norfolk did not correlate well with Derwent River discharge [41], but this is not unexpected given the reported link between runoff and season for nutrient concentrations in streams [72]. Evidence of some minor nitrate input in the middle estuary (either directly from STP discharges or from nitrification) was observed as a slight peak in the mixing curve with salinity at non-productive times of the year (CSIRO unpublished data, 1993–1994). Nitrite alone has only been measured in a single estuarine survey (Jul 1994); it was just a few percent of the nitrate concentration at that time.

Information on the levels and distribution of ammonia in earlier reports on the Derwent estuary (and even some more recent findings [3]) are to be treated with caution owing to a combination of likely contamination of samples and unsuitability of analytical methods [41]. Ammonia concentrations over an annual cycle (Mar 1993–Mar 1994) have been estimated at ~ 1–$1.5\,\mu M$ in the lower estuary, throughout water column (these levels are elevated when compared with marine end-member concentrations reported for the Huon estuary – see above). Surface waters were low in the upper estuary ($\sim 1\,\mu M$). They reached peak values in the middle estuary (1.5–$3\,\mu M$); even higher values in the side bays pointed to the source being effluents from STPs and stormwater. Ammonia concentrations in bottom waters increased with penetration up the estuary, augmented by an efflux from sediments (arising from remineralisation of organic substrate). They attained nearly $4\,\mu M$ at depth in the upper estuary.

Coughanowr [41] reported that mean concentration of total nitrogen (TN) in surface waters increased up the estuary from about $8\,\mu M$ at the mouth to $17\,\mu M$ in the river at New Norfolk. This inverse relation with salinity is consistent with observations in the Huon estuary. Again, the bulk of the nitrogen was in the dissolved organic form. The pulp/paper mill at Boyer adds TN, so that the concentration in its vicinity reaches $22\,\mu M$. More recent results from surveys in 2000 and 2001 [16] complicate the picture. Approximately the same range of concentrations was observed (7–$25\,\mu M$), but this was the spread of measurements for *all* stations down the main axis of the estuary and all bays, apart from Prince of Wales Bay (14–$40\,\mu M$). The variability of the data make it difficult to make further interpretations, but marginally higher concentrations in the middle estuary suggest that urban inputs are a factor in the distribution of all species of nitrogen.

Phosphate concentrations were low ($< 0.1\,\mu M$) in the river and upper estuary stations (above Bridgewater) throughout the year. This section of the waterway was deemed to be phosphorus-limited [3], consistent with the higher concentrations of nitrate there. Seawater concentrations varied between about 0.3 and $0.5\,\mu M$, but they could be depleted sometimes in summer to $\sim 0.1\,\mu M$ [41], probably associated with ingress of East Australian Current (EAC) water. At all times, there has been evidence of addition of phosphate to surface waters of the middle estuary, consistent with a source in STP outflow. Bottom waters in the middle and upper estuary also have higher phosphate concentrations (up to $\sim 1.1\,\mu M$); the origin may be a flux of phosphate from sediments to overlying waters as suggested for the Huon estuary. From limited measurements, total phosphorus parallels the distribution of phosphate: low levels at the head of the estuary (typically 0.1–0.5 μM), reaching a maximum in the middle estuary (~ 1.5 μM around site U2, Fig. 3), before subsiding slightly to seawater concentrations (typically 0.6–1.2 μM).

Silicon as the third essential microalgal macronutrient – for the diatoms at least – has the expected inverse relation with salinity for the bioavailable form silicate. We observed the concentration in the Derwent River at New Norfolk, during eight surveys in 1992–1994, to vary from 68–112 μM (CSIRO unpublished data, 1993–1994). From the few observations, the freshwater end-member concentration did not appear to be a function of flow rate, nor particularly of season (when both Derwent and Huon data taken together). The influence of the Kermandie subcatchment in the Huon system suggests that some tributaries more than others might deliver silicate to the main river. Therefore, rainfall onto Si-rich subcatchments or flow of these tributaries, rather than overall rainfall or flow rate, is likely to correlate with silicate levels at the head of the estuary. The mixing line for silicate with salinity over the estuarine gradient was mostly close to linear. At the marine end of the estuary, the silicate concentrations were universally very low, mostly 1–3 μM, in agreement with equivalent data from the Huon estuary.

Recent reports of chlorophyll *a* in the Derwent estuary indicate a range of 0.2–7.8 μg L^{-1} [16]. These results were in line with our eight estuarine surveys in 1992–1994 (0.1–8.5 μg L^{-1}, CSIRO unpublished data, 1993–1994); the highest values observed were during a *G. catenatum* bloom (Fig. 12). As in the Huon

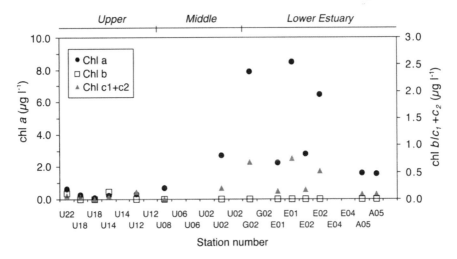

Fig. 12 Chlorophyll pigment concentrations in a survey of the Derwent estuary in April 1993 (CSIRO unpublished data, 1993–1994), during a *Gymnodinium catenatum* bloom (Jameson I, personal communication)

estuary, wintertime concentrations were generally low, but in some locations in the middle and lower estuary could still exceed 1 μg L^{-1}. Irregular microalgal blooms were observed from spring through summer into autumn. Low chlorophyll *a* concentrations were typical of the upper estuary, it being light-limited and rapidly flushed. The highest concentrations were seen in the middle estuary, several of the adjacent bays and sometimes into the lower estuary.

The phytoplankton species composition of the estuary seems a little more diverse than the neighbouring Huon estuary and Storm Bay. A compilation of information in the 1990s listed around 83 diatom species, 73 dinoflagellates and 30–40 small flagellates [59]. Several of the key species in the Huon estuary – *Pseudonitzschia* spp. (diatoms), *Ceratium* spp. (dinoflagellates) and *G. catenatum* (toxic dinoflagellate) – were prominent again here. A difference was the episodic "green water blooms" (*Tetraselmis* sp. and *Pyramimonas* spp.) in the middle estuary, most likely triggered by sewage nutrients. Nevertheless, seasonal monitoring of oceanic invader species (subantarctic dinoflagellate *Dinophysis truncata* and the EAC diatom *Bacteriastrum furcatum*) at nine sites in southeastern Tasmanian waters, including Derwent and Huon estuaries, shows that they were pervasive throughout the region [60]. Individual species were not necessarily recorded coincidentally at all sites.

Apart from the introduced *G. catenatum*, the phytoplankton community has not changed significantly since surveys in the late 1940s. Although the ample nutrient levels could suggest a trend toward eutrophication in the middle Derwent estuary, this is not borne out by the microalgal biomass or dynamics [59]. Primary production appears to be limited there by light and perhaps some other less obvious factors (e.g. trace metal pollution or residence time). The light penetration is attenuated through absorption by CDOM, which may also act indirectly by modifying the chemical reactivity of micronutrients through complexation, and redox reactions [73]. In the upper estuary with its mudflats and marshes, primary production by microphytobenthos has been reported to exceed that of phytoplankton by 10–30 times, and aquatic macrophytes were greater again by a factor of three[5].

4.3.4
Trace Metals

With an electrolytic zinc refinery operating on its shores for almost 90 years, the question of the build up of metals in the Derwent estuary from this and other sources became more pressing over the decades. It turned into a priority in the early 1970s with consumers of local, farmed Pacific oysters prone to vomiting after a meal of the shellfish. Prompt scientific investigation revealed very high levels of zinc, cadmium and copper (up to 21 000, 63 and 450 mg kg^{-1} wet

---

[5] Roberts S, Beattie G, Beardall J, Quigg A (2001) Primary production, phytoplankton nutrient status and bacterial production in the Derwent River, Tasmania. Unpublished report to Norske Skog Paper. Department of Biological Sciences, Monash University, Melbourne

weight, respectively) in oysters from leases in Ralphs Bay [74, 75]. The amount of zinc and cadmium in a half-dozen oysters were sufficient to function as an emetic. Studies of a range of finfish caught in the estuary showed that levels of zinc, cadmium and copper were acceptable in these species [75]. Nevertheless, subsequent measurements of mercury in 16 finfish species identified that levels of this element were above the guideline value of 0.5 mg kg^{-1} [76].

At this stage, Bloom and co-workers [65, 77, 78] instituted an intensive and wide-ranging survey of heavy metals in the Derwent estuary and environs. 51 surface water samples were collected throughout the estuary in a single day. Other samples included bottom sediments from 121 sites in and adjacent to the estuary, 21 shellfish species from 42 sites, 16 species of finfish from 13 sites (representing the complete annual catch of an amateur fisherman), sewage sludge and liquid effluent from four treatment plants, and dust fallout at 18 sites on both shores of the estuary. Salient results were: 1500 μg L^{-1} Zn, 15 μg L^{-1} Cd and 14–16 μg L^{-1} Hg as extremes in filtered waters, and peak values of > 10 g kg^{-1} Zn, 0.86 g kg^{-1} Cd and 1.1 g kg^{-1} Hg in sediments. In addition, dried oyster flesh was found to have up to 100 000 mg kg^{-1} Zn and > 200 mg kg^{-1} Cd, corroborating the earlier work of Thrower and Eustace [74, 75]. After comparisons with other heavily polluted systems globally [78], Bloom branded the Derwent as "the worst polluted river in the world" [79]. This statement attracted strong condemnation locally, but Förstner and Wittmann [80] echoed it in their perspective on global metal pollution. The reliability of the analytical measurements produced in the survey were also confirmed by later work [81].

Studies of metals in the Derwent estuary following the initial alarm have focussed upon trends in levels and distributions. They have looked at waters [16, 25, 70, 82], sediments (Table 4) [16, 25, 33, 66, 83–85] and biota [16, 25, 34, 86, 87]. Monitoring of all three sample types have shown initial sharp decreases from the extreme levels of the early 1970s (e.g. [34]), but latterly concentrations of many of the metals have plateaued ([25], e.g. plots of Zn in mussels or Hg in Flathead).

We looked at a suite of metals (Fe, Mn, Cu, Cd, Co, Ni, Hg and Zn) in the water column of the estuary during 1992–1994 ([70, 82], CSIRO unpublished data, 1993–1994). Results for all metals, with the possible exception of Ni, showed evidence of input in the mid-salinity range. Representative examples are shown in Fig. 13. Very high concentrations of all metals were associated with particulates in the bottom waters from the vicinity of Elwick Bay up to Bridgewater (upper to middle estuary transition). The estuary is shallower ($\leq$ 7 m) for much of this reach, and it is plausible that wind-induced mixing or high flows both can induce resuspension of fine metal-loaded sediments. As noted above, higher concentrations of suspended solids are observed in bottom waters of this part of the estuary. Jones et al. [66] observe that flood scouring intermittently displaces metal-laden sediments from the upper estuary, but below Bridgewater freshwaters are always separated from the channel floor by the salt-water wedge and so scouring has not been observed downstream of this point.

Table 4 Trace element concentration ranges (mg kg^{-1}, dry matter basis) in surface sediments of the Derwent estuary (modified from Green and Coughanowr [16])

Trace element	Bloom [65]	Pirzl [85]	Jones et al. [66]	Green et al. [16][a]	Guidelines ANZECC/ ARMCANZ [67]	
					ISQG Low	ISQG High
	102 sites (1975)	40 sites (1996)	69 sites (1996/97)	123 sites (2000)		
Arsenic	–	0.001–20.9	1–657	1–1400	20	70
Cadmium	0.3–1400	0–134	< 10–180	1–477	1.5	10
Chromium	1.1–258	–	< 5–183	–	80	370
Cobalt	0.4–137	–	8–95	3–85	–	–
Copper	1.5–10 050	7–530	< 2–1182	1–1490	65	270
Lead	0.7–41 700	10.5–2078	4–3866	8–8120	50	220
Manganese	0.8–8900	6.5–781	–	2–7740	–	–
Mercury	0.01–111	0.023–55.7	< 5–36	0.02–130	0.15	1
Nickel	0.05–36	–	< 2–35	–	21	52
Zinc	22–104 000	26.8–19 201	< 2–22 593	24–59 000	200	410

[a] Compilation of data from surveys by Tasmanian Government agencies, University of Tasmania and NSR Environmental Consultants (the latter for Norske-Skog Paper)

The metals exhibited a number of groupings. Zn, Cd and Hg had low concentrations at both estuarine end-members, and increased almost linearly from either end to a mid-salinity peak in surface waters. As expected on the basis of likely sources and chemical behaviour, Zn and Cd were correlated closely, either as total dissolved or filterable metal. Hg was associated with the other two metals, but it was also strongly linked to the levels of suspended solids. Another group of metals (Fe, Mn and Co) was at higher concentrations in river water than in marine waters, quite probably as a result of natural processes. All showed a hump in the mixing curve with salinity in the mid-range. Co was highly correlated with Fe in the particulate phase, presumably because it adsorbs to hydrous iron oxide particles. Mn is not so closely correlated; although often also in a hydrous oxide phase, it has different redox behaviour than Fe. For example, it was found to be readily reduced from Mn^{IV} to soluble Mn^{II} ions under hypoxic conditions in the upper estuary (Fig. 14). Cu appeared to be intermediate in behaviour between the Zn and Fe groups. In contrast, Ni was at a near constant concentration (~ 5 nmol kg^{-1}) over the full salinity range, and not markedly different from the levels in the estuaries of the Huon and Bathurst Harbour.

Process improvements and pollution abatement programs at industrial sites have reduced extreme localised contamination. This is borne out, when comparing the levels of metals in the water column reported by Bloom and

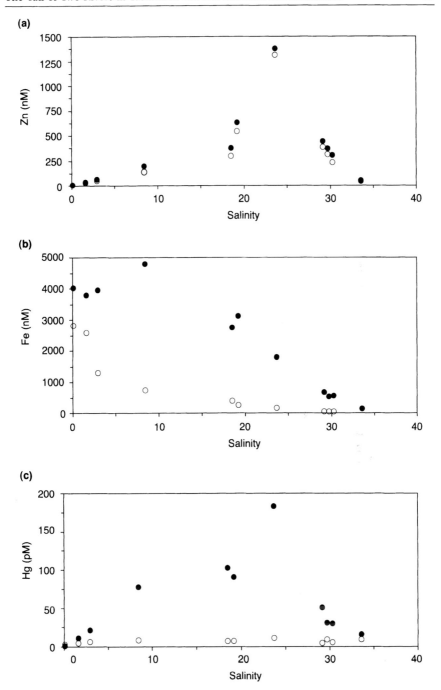

Fig. 13 Representative plots of surface trace metal concentrations versus salinity in the Derwent estuary, January 1994: **a** zinc, **b** iron, and **c** mercury. • Unfiltered, acid-labile. ◦ Zn and Fe, filtered; Hg, unfiltered, $SnCl_2$-reactive. See text for discussion of behaviour of individual trace metals

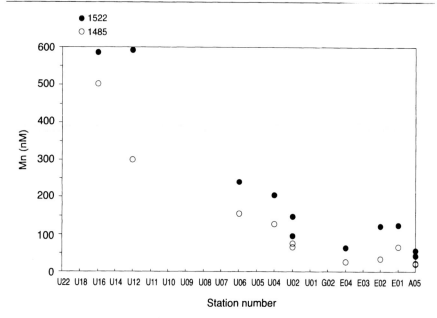

Fig. 14 Relation of manganese concentration in subsurface waters with station location in the Derwent estuary, January 1994. • Unfiltered, acid-labile – including hydrous oxides of Mn^{IV}. ○ Filtered – mostly dissolved Mn^{II}). Stations to the *left* are further upstream (see Fig. 3). Station *U*18 regularly has bottom waters strongly depleted in oxygen at this time of the year

colleagues [65, 78] to our results almost 20 years later [70]. Similar evidence can be found in sediment results [85], and particularly metal distributions down cores [33, 83, 84]. However, recent surveys suggest that metals have also redistributed upstream and downstream of their original hotspots around the zinc refinery [16, 66]. Moreover, for the metals most closely linked to the refined ores (Zn, Pb, Cd and Hg), the Derwent estuary remains one of the most contaminated around the world. The maximum coefficient of enrichment for Derwent estuary sediments (referenced to the Huon estuary) was determined to be 565 for Zn and 110 for Pb [66]. A maximum coefficient of 3000 is estimated for Cd using the Derwent estuary sediment data of Jones et al. [66], compared to our mean value of Cd ($0.06\,mg\,kg^{-1}$) in the lower Huon estuary [68].

4.3.5
Organic Contaminants

Much less monitoring has been done for organic compounds in the Derwent estuary. The impetus for studies has arisen in three areas: (i) contaminants in the paper mill waste stream, (ii) organic matter from treated sewage discharge, and (iii) contaminants in urban runoff (rivulets and stormwater). Sediments have received most attention, although resin acids and chlorinated organic compounds have also been determined in estuary waters.

The pulp/paper mill discharge has also had a remote physical effect in the upper and middle reaches of the Derwent estuary with significant deposition of particulate organic matter forming rafts of sludge on the surface of the estuarine sediments. Of 8.5×10^5 m^3 of sludge coating the sediment between New Norfolk and Green Island (near site U16, Fig. 3), almost 95% was swept out by a flood (760 m^3 s^{-1}) in October 1988. However, 4.0×10^6 m^3 is estimated to remain in the lower stretch between Bridgewater and the Bowen Bridge. Sludge was generally observed to be strongly reducing, reeking of hydrogen sulphide, gelatinous and fibre-rich with high organic content ($> 30\%$ LOI). It also had high levels of metals (up to 11 μg g^{-1} Hg, 3.8 μg g^{-1} Cd, and 1590 μg g^{-1} Zn). From 1990, with primary treatment of the discharge, the volume of sludge on the estuary bottom has decreased through the combined action of scouring by the river and natural decomposition [3, 16].

Leeming and Nichols [88] in a September 1993 survey of the Derwent estuary used sterol biomarkers to resolve the sources and distribution of pollution originating from pulp fibre and sewage. Plant sterols were used to demarcate the extent of contamination from the Boyer paper mill; it was worst in the upper estuary and extended into the middle estuary. The faecal biomarker was detected throughout the estuary in suspended particulate matter from Bridgewater to near the mouth of the estuary in excess of 60 ng L^{-1} (equivalent to Australian primary contact limit for faecal coliforms). The extent of contamination was most severe in the middle estuary, where for example the level of 900 ng L^{-1} at Kangaroo Bay was estimated to represent just a 1 : 255 dilution of the whole effluent. Levels of coprostanol in sediments confirmed the extent of the contamination (only falling to baseline levels outside the estuary), and its severity in relation to other examples nationally and internationally.

Aliphatic hydrocarbons decreased from high levels (400–4600 μg g^{-1}) in wood-fibre sludges and sediments of the upper and middle estuary [89, 90] to background levels (0.8–6.4 μg g^{-1}) at the mouth of the estuary and beyond [91]. Very high levels (up to 10 100 μg g^{-1}), on a worldwide rating, were recorded in Prince of Wales Bay [90]. Petrogenic (lubricating and fuel oils from urban runoff) and sewage-derived hydrocarbons dominated the upper and middle sections of the estuary; lower estuary hydrocarbons were primarily biogenic. Polycyclic hydrocarbon (PAH) concentrations were also measured in the survey of sediments in Prince of Wales Bay [90]. They were very irregular, varying from 21–27 040 μg g^{-1}, as a result of localised pollution from stormwater outflows and marinas. PAHs could not be detected (< 0.5 μg L^{-1}) in surface or bottom waters of the upper estuary [3].

Recent monitoring in the upper estuary of "adsorbable organic halogens" (AOX, a bulk measure of chlorinated organic compounds) showed nearly uniform concentrations of about 40 μg L^{-1} from New Norfolk to Bridgewater [3].

AOX in the paper mill waste stream was found to be $20-70\,\mu\text{g}\,\text{L}^{-1}$. Earlier measurements, before decommissioning of the mill's chlor-alkali bleaching plant in 1993, yielded a similar AOX concentration of $40-50\,\mu\text{g}\,\text{L}^{-1}$ at Bridgewater, but up to $260\,\mu\text{g}\,\text{L}^{-1}$ in surface waters about 0.5 km downstream of the paper mill outfall (Volkman JK, Mackey DJ, Butler ECV, unpublished results).

The pulp/paper mill at Boyer processes both hardwoods and softwoods, in a ratio of about $1:2$. The latter contain resin acids, a toxicant at low levels ($> 0.2-0.5\,\text{mg}\,\text{L}^{-1}$) for both native and exotic fish [92]. Two groups of resin acids have been identified in the waste stream: labile – abietane resin acids (palustric, levopimaric, abietic and neoabietic acids) and refractory – pimarane resin acids (pimaric, sandaracopimaric and isopimaric acid) and the aromatic resin acid, dehydroabietic acid. Only dissolved forms of the refractory group persist in the water column more than 3 km downstream of the mill [3]. In the early 1990s, surveys of resin acids in the upper Derwent estuary gave variable results (e.g. $< 0.01-0.78\,\text{mg}\,\text{L}^{-1}$ total resin acids at 0.5 km below mill) depending upon the behaviour of the effluent plume and the state of the tide [93]. Again, more recent surveys indicate improved treatment procedures with the maximum concentration of resin acids attaining $0.08\,\text{mg}\,\text{L}^{-1}$ at the effluent discharge point; at Bridgewater the concentration was $0.01\,\text{mg}\,\text{L}^{-1}$ [3]. Nevertheless, a persistent record of the past discharge of resin acids is the very high concentrations (often $> 200\,\text{mg}\,\text{kg}^{-1}$ and up to $3250\,\text{mg}\,\text{kg}^{-1}$) seen in the sediments 0.5 km upstream and downstream of the mill's effluent outfall. The peak concentrations are doubtless associated with zones of wood-fibre deposition.

Synthetic organic compounds, such as pesticides (organochlorine, organophosphorus or others), polychlorinated biphenyls (PCBs), plasticisers and surfactants (e.g. phthalates and alkylphenols), and antifoulants (e.g. tributyltin), have not been examined in the Derwent estuary, apart from a single survey of seafood. A range of seafood types were taken from the estuary and analysed for organochlorine pesticides and PCBs; none were found above the analytical detection limits [16].

5
Conclusions

The two estuaries and their river systems in a broad categorisation have more in common with Northern Hemisphere temperate estuaries with their pronounced seasonal behaviour, general climatic conditions and ecosystem types, than they do with many other continental Australian estuaries. Yet, this is an overly simplistic analysis because the catchments and their influence upon the waterways, when undisturbed, are very Australian – particularly with runoff poor in nutrients and many minerals. They are different again for the large amounts of organic matter that are extracted into subalpine tributary streams from peats, moorland and other vegetation, while at the same time having naturally very low loads of suspended solids.

In their recent history, the Derwent and Huon estuaries would have functioned alike in the manner of physical, chemical and biological processes. However, with the arrival of European settlers, they were embarked on different courses.

The Derwent estuary, with Hobart as the capital city and an important national port, has been a focus for development. Consequently, it has had a chequered environmental history, which was shown up in sharp relief in the latter decades of last century. First, severe contamination in waters, sediments and biota from metals associated with the refining of zinc ores was discovered. Subsequently, the levels and distribution of nutrients and organic matter originating from sewage, urban runoff and paper production were a further concern. Finally, the question of pollution by individual or groups of organic toxicants has only been partly answered. Resin acids have been examined in the paper-mill waste stream, and organochlorine compounds and PAHs have been monitored occasionally at a limited number of estuary sites. Information is lacking on phenolics, non-chlorinated pesticides, and several other organic toxicants listed by the National Pollutant Inventory (www.npi.gov.au.index.html) in emission data from industrial facilities in the Derwent catchment.

In general, the Derwent's environmental problems are well studied and documented. Improvements in waste control for industrial operations, and latterly also for urban drainage (rivulets and stormwater), have brought about improvements in water quality. Nevertheless, a daunting legacy in the form of grossly contaminated sediments are possibly now constraining the scale of further improvements that might be gained by better waste management. Although high sedimentation rates might be a localised problem in the sub-catchments of side bays, upstream damming means that the main channel of the estuary receives too little suspended material for natural capping to play a part in isolating the contaminated sediments.

Science is being actively used to assist with environmental management of the River Derwent from headwaters to the sea, and the mitigation of contaminated sites. Some interesting challenges for this approach are the ecological effects ensuring from the degradation of the waterway. The first is the conundrum concerning possible eutrophication of the estuary. Sufficient nutrients (and possibly rising levels) enter the middle estuary to fuel more frequent and denser blooms of phytoplankton, including some harmful species. Steady concentrations of chlorophyll a in the estuary during the last two decades match those of the mesotrophic Huon estuary. This observation suggests that the phytoplankton are held in check by an intangible balance of low light penetration (i.e. a shallow euphotic zone), a surfeit of organic matter and high levels of metals and possibly other toxicants. Changes in any one of these properties, even seeming environmental improvements, could unsettle this poise. Such uncertainties are increasingly being investigated by applications of environmental models (e.g. [30], and application of LOICZ nutrient model to Derwent estuary by JS Parslow, http://data.ecology.su.se/MNODE/Australia/Derwent/derwent.htm) and ecological risk analysis (e.g. [3]). A comprehensive appraisal requires a mix of these scenario-testing tools with carefully targeted field studies for key process identification and validation.

Another challenge is associated with the degraded environment perhaps favouring invasion by exotic species (see [94] for elaboration) on one hand, and harming endemic populations on the other. I have already mentioned the arrival of the toxic dinoflagellate *G. catenatum*. Other invaders have included the Northern Pacific seastar (*Asterias amurensis*), the Japanese seaweed wakame (*Undaria pinnatifida*), and the New Zealand turritellid gastropod (*Maoricolpus roseus*) among a total recorded list of 43 species, and potentially up to 70 [16]. The Northern Pacific seastar and possibly the condition of the estuary are contributing to the decline of the endangered species, the spotted handfish (*Brachionichthys hirsutus*), which is endemic to just the Derwent estuary.

The Huon estuary is a modified water body, but it is a much less degraded system than its neighbour. Particularly valuable is the pristine upper catchment within in the South West Wilderness World Heritage Area, which is the source of much of the volume of freshwater, and its high quality. Not only did the development of the region occur later than that along the Derwent River, but the population pressure has also always been light. Moreover, historical industrial activity – notably in the Hospital Bay precinct – has left but a small and fading "footprint". Some signs of neglect are evident in the lower catchment, and they sound a cautionary note. I am referring to the degradation of a number of tributary streams. Our standout example was the Kermandie River, which through destruction of the riparian zone of feeder streams, unsuitable land management practices and poor waste control has deteriorated water quality. The Kermandie River in itself has a discharge too small to cause anything but a localised blemish in the estuarine system. However, too many similar occurrences will add up to a more serious problem. Further clearfelling of native forests in the middle catchment (e.g. along the Picton River, Fig. 1) is planned to supply the new wood-processing facility at the junction of the Huon and Arve Rivers. Both the forest clearing and the industrial site need to be carefully assessed and monitored as to their affect upon water quality and quantity in the Huon River.

The obvious challenge lying ahead for the Huon estuary itself is sustainable development of aquaculture within the estuary and outside in the waters of D'Entrecasteaux Channel. Finfish farming contributes waste in the form of localised deposition of organic matter, and more widely in the form of raised nutrient concentrations in estuarine waters. This needs to be balanced against other sources of these substances. In addition, all aspects of farming should be constantly under review for improved environmental outcomes.

Finally, the two estuaries are useful analogues for each other – really two sides of a coin. The Derwent providing an indication for the Huon of the result if development does not pay sufficient attention to the estuarine ecosystem, or worse uses it just as a waste repository. The obverse is that the Huon provides an indication of what sort of waterway can be rediscovered if environmental managers of the Derwent can institute effective effluent controls and successful remediation of contaminated sites. In combination with historical and other remnant indicators in the Derwent estuary (e.g. the sedimentary record [95] and the persistence of some habitats/zones of the estuary in relatively unscathed condition – for instance the upper Derwent marshes [2]), the Huon estuary provides a useful pointer to background conditions existing before urbanisation and industrialisation of the Derwent. However, it should be noted that the Huon estuary does not provide a true historic baseline. Fulfilling this role is a possibility for two river systems further to the south – the Catamaran and the Lune – which are both effectively undeveloped.

Acknowledgements I am grateful to the following for the insights that they provided on the Derwent and Huon estuaries: Denis Mackey, Mike Pook, Pat Quilty, Peter Harris, Mike Herzfeld, Ian Jameson and David Leaman. Thanks also to Ros Watson and Lea Crosswell in the preparation of the figures, and I appreciate the input from Graham Green in reviewing an earlier version of this chapter. Without the unstinting contributions of the Derwent and Huon Estuary Study teams from CSIRO Marine Research, this work would not have been possible. Finally, I thank Pete Wangersky for his guidance and encouragement to write about these two Tasmanian estuaries.

References

1. Edgar GJ, Barrett NS, Graddon DJ (1999) A classification of Tasmanian estuaries and assessment of their conservation significance using ecological and physical attributes, population and land use. Technical report series no 2. Tasmanian Aquaculture and Fisheries Institute, Hobart www.utas.edu.au/tafi/PDF_files/Tech_Report_2_Edgar_Barrett.pdf
2. Jordan A, Lawler M, Halley V (2001) Estuarine habitat mapping in the Derwent – integrating science and management. NHT final report. Tasmanian Aquaculture and Fisheries Institute, Hobart www.utas.edu.au/tafi/seamap/pdf/Derwent_habitat%20mapping.pdf
3. NSR (2001) Boyer Mill ecological risk assessment. Final report to Norske Skog Paper. NSR Environmental Consultants, Melbourne
4. Langford J (1965) Weather and climate. In: Davies JL (ed) Atlas of Tasmania. Lands and Survey Department, Hobart, p 2
5. Bureau of Meteorology (1993) Climate of Tasmania. Australian Government Publishing Service, Canberra
6. Bureau of Meteorology (2000) Climatic atlas of Australia: rainfall. Bureau of Meteorology, Melbourne
7. Davies JL (1965) Landforms. In: Davies JL (ed) Atlas of Tasmania. Lands and Survey Department, Hobart, p 19

8. Department of Mines (1976) Geological Map of Tasmania (1:500000). Hobart
9. CSIRO (2000) Huon Estuary Study – environmental research for integrated catchment management and aquaculture. Final report to Fisheries Research and Development Corporation. Project no 96/284, June 2000. CSIRO Division of Marine Research. Marine Laboratories, Hobart www.marine.csiro.au/ResProj/CoasEnvMarPol/huonest/hesreport_pdf/index.html
10. Nicholls KD, Dimmock GM (1965) Soils. In: Davies JL (ed) Atlas of Tasmania. Lands and Survey Department, Hobart, p 27
11. Grant JC, Laffan MD, Hill RB, Neilsen WA (1995) Forest soils of Tasmania. A handbook for identification and management. Forestry Tasmania, Launceston
12. Jackson WD (1965) Vegetation. In: Davies JL (ed) Atlas of Tasmania. Lands and Survey Department, Hobart, p 30
13. Kirkpatrick JB, Dickinson JM (1984) Tasmania, vegetation map, 1:500000. Forestry Commission of Tasmania, Hobart
14. Harris GP (2001) Mar Freshw Res 52:139
15. Ryan L (1996) The aboriginal Tasmanians. Allen and Unwin, St Leonards, NSW
16. Green G, Coughanowr C (2003) State of the Derwent estuary 2003: a review of pollution sources, loads and environmental quality data from 1997–2003. Derwent Estuary Program, DPIWE, Hobart www.dpiwe.tas.gov.au/inter.nsf/WebPages/SSKA-5TA84R?open
17. Hudspeth A, Scripps L (2000) Capital port: a history of the Marine Board of Hobart 1858–1997. Hobart Ports Corporation, Hobart
18. Hammond D (1995) The Huon Valley: yesterday and today. Southern Holdings, Lucaston
19. Row M (1980) The Huon timber company and the crown: a tale of resource development. In: The history of the Huon, Channel and Bruny Island region, vol 20. Tasmanian Historical Research Association, Hobart, p 87
20. Derwent Estuary Program (2003) Environmental flows: linking the catchment and the estuary. DPIWE, Hobart www.derwentriver.tas.gov.au/content/derwent_factfile/environmental_flows.html
21. Hydro Tasmania (2003) Derwent water management review. Environmental review report. Hydro Tasmania, Hobart www.hydro.com.au/environment/waterreviews/pdf/Derwent-Env-review.pdf
22. Coughanowr C (2001) Nutrients in the Derwent estuary catchment. Final report to the Natural Heritage Trust. May 2001. Department of Primary Industries, Water and Environment, Hobart
23. Gallagher S (1996) Huon Catchment Healthy Rivers Project: water quality assessment report. Prepared for National Landcare Program, Huon Valley Council and Tasmanian Department of Primary Industry and Fisheries, Tasmania. DPIF, Hobart
24. Livingston A (2001) Hydrological and engineering issues associated with draining and restoring Lake Pedder. In: Shaples C (ed) Lake Pedder: values and restoration. Occasional paper no 27. Centre for Environmental Studies, University of Tasmania, Hobart, p 131
25. Coughanowr C (1997) State of the Derwent estuary: a review of environmental quality data to 1997. Supervising Scientist Report no 129. Supervising Scientist, Canberra
26. Jennings JN, Bird ECF (1967) Regional geomorphological characteristics of some Australian estuaries. In: Lauff GH (ed) Estuaries. Am Ass Adv Sci, Washington DC, p 121
27. Harris PT, Heap AD, Bryce SM, Porter-Smith R, Ryan DA, Heggie DT (2002) J Sed Res 72:858

28. Roy PS, Williams RJ, Jones AR, Yassini I, Gibbs PJ, Coates B, West RJ, Scanes PR, Hudson JP, Nichol S (2001) Estuar Coast Shelf Sci 53:351
29. Hansen DV, Rattray M (1966) Limnol Oceanogr 11:319
30. Hunter JR, Andrewartha JR (1992) A modelling study of an effluent discharge at Selfs Point, Tasmania. Report OMR-48/54. CSIRO Division of Oceanography, Hobart
31. Thomson JD, Godfrey JS (1985) Aust J Mar Freshw Res 36:765
32. Davies PE, Kalish SR (1994) Aust J Mar Freshw Res 45:109
33. Jordan A, Samson C, Lawler M, Halley V (2003) Assessment of seabed habitats and heavy metals in Cornelian Bay. Technical report series no 18. Tasmanian Aquaculture and Fisheries Institute, Hobart www.utas.edu.au/tafi/seamap/pdf/Tech_Report_18_CornelianBay.pdf
34. Cooper RJ, Langlois D, Olley J (1982) J Appl Toxicol 2:99
35. Cresswell G (2000) Pap Proc R Soc Tasmania 133:21
36. Buckney RT, Tyler PA (1973) Int Revue Ges Hydrobiol 58:61
37. Mackey DJ, Butler ECV, Carpenter PD, Higgins HW, O'Sullivan JE (1996) Sci Total Environ 191:137
38. Perakis SS, Hedin LO (2002) Nature 415:416
39. Koehnken L (2001) Gordon River water quality assessment. In: Basslink integrated impact assessment statement: potential effects of changes to hydro power generation, Appendix 3. Basslink and Hydro Tasmania, Hobart
40. Bobbi C (1998) Water quality of rivers in the Huon catchment. Report series WRA 98/01. Tasmania Department of Primary Industry and Fisheries, Hobart
41. Coughanowr C (1995) Derwent estuary nutrient program, technical report. Department of Environment and Land Management. Tasmanian Printing Authority, Hobart
42. Berry K (2001) Water quality in the Jordan catchment, 3rd progress report. Department of Primary Industry, Water and Environment, Hobart
43. Harris G, Nilsson C, Clementson L, Thomas D (1987) Aust J Mar Freshw Res 38:569
44. Clementson LA, Harris GP, Griffiths FB and Rimmer DW (1989) Aust J Mar Freshw Res 40:25
45. Merry RH, Tiller KG, Alston AM (1983) Aust J Soils Res 21:549
46. Steen JW (1969) An investigation of chlorinated hydrocarbon residues in the aquatic fauna of the Huon Valley. BSc (Hons) Thesis, University of Tasmania
47. Woodward IO, Gallagher JB, Rushton MJ, Machin PJ, Mihalenko S (1992) Salmon farming and the environment of the Huon estuary, Tasmania. Technical report no 45. Sea Fisheries Division Department of Primary Industry, Fisheries and Energy, Hobart
48. McGhie TK, Crawford CM, Mitchell IM, O'Brien D (2000) Aquaculture 187:351
49. Butler ECV, Blackburn SI, Clementson LA, Morgan PP, Parslow JS, Volkman JK (2001) ICES J Mar Sci 58:460
50. Edgar GJ, Cresswell GR (1991) Pap Proc R Soc Tasmania 125:61
51. Laureillard J, Saliot A (1993) Mar Chem 43:247
52. Hallegraeff GM, McCausland MA, Brown RK (1995) J Plankton Res 17:1163
53. Doblin MA, Blackburn SI, Hallegraeff GM (1999) J Exp Mar Biol Ecol 236:33
54. Jones G, Lawson B (1980) Environmental report to the APM on the Huon River estuary, January 1980. School of Zoology, University of New South Wales
55. Harmful Algal Bloom Task Force (1994) Report to the Minister for Primary Industry and Fisheries. Department of Primary Industry and Fisheries Tasmania, Hobart
56. Maher WA, DeVries M (1994) Chem Geol 112:91
57. Martin-Jézéquel V, Hildebrand M, Brzezinski MA (2000) J Phycol 36:821
58. Leblanc K, Quéguiner B, Garcia N, Rimmelin P, Raimbault P (2003) Oceanol Acta 26:339

59. Hallegraeff GM, Westwood KJ (1994) Identification of key environmental factors leading to nuisance algal blooms in the Derwent River. Report for Department of Environment and Land Management. University of Tasmania Department of Plant Science, Hobart

60. Jameson I, Hallegraeff GM (1994) Algal bloom monitoring for the Tasmanian salmonid farming industry: 1992–1994 report. Reports from the SALTAS 1993–1994 research and development programme, Hobart

61. Sharp JH, Pennock JR, Church TM, Tramontano JM, Cifuentes LA (1984) The estuarine interaction of nutrients, organics, and metals: a case study of the Delaware estuary. In: Kennedy VS (ed) The estuary as a filter. Academic, New York, p 241

62. Featherstone AM, Butler ECV, O'Grady BV (2004) Estuaries 27:18

63. Carlsson P, Granéli E, Tester P, Boni L (1995) Mar Ecol Prog Ser 127:213

64. Cook PLM (2003) Carbon and nitrogen cycling on intertidal mudflats in a temperate Australian estuary. PhD Thesis, University of Tasmania

65. Bloom H (1975) Heavy metals in the Derwent estuary. Chemistry Department, University of Tasmania, Hobart

66. Jones BG, Chenhall BE, Debretsion F, Hutton AC (2003) Aust J Earth Sci 50:653

67. ANZECC/ARMCANZ (2000) Australian and New Zealand guidelines for fresh and marine water quality, vol 1. The Guidelines. National water quality management strategy, paper no 4. Canberra www.deh.gov.au/water/quality/nwqms/volume1.html

68. Townsend AT, O'Sullivan JE, Featherstone AM, Butler ECV, Mackey DJ (2001) J Environ Monitor 3:113

69. Cook PLM, Carpenter PD, Butler ECV (2000) Mar Chem 69:179

70. Butler ECV, Green GJ, Higgins HW, Holdsworth DG, Leeming R, Mackey DJ, Morgan PP, Nichols PD, O'Sullivan JE, Plaschke RB, Revill AT, Volkman JK, Watson RJ (1996) Contaminants in the Derwent estuary – 20 years on. Proceedings 23rd hydrology and water resources symposium, Hobart, p 679

71. Ritz DA, Buttemore RE (1984) Pap Proc R Soc Tasmania 118:109

72. Webb BW, Walling DE (1985) Water Res 19:1005

73. Rashid MA (1985) Geochemistry of marine humic compounds. Springer, Berlin Heidelberg New York

74. Thrower SJ, Eustace IJ (1973) Aust Fish 32:7

75. Eustace IJ (1974) Aust J Mar Freshw Res 25:209

76. Ratkowsky DA, Dix TG, Wilson KC (1975) Aust J Mar Freshw Res 26:223

77. Ayling GM, Bloom H (1976) Atmos Environ 10:61

78. Bloom H, Ayling GM (1977) Environ Geol 2:3

79. Bennett B (1999) Ecos 100:10

80. Förstner U, Wittmann GTW (1981) Metal pollution in the aquatic environment, 2nd edn. Springer, Berlin Heidelberg New York

81. Noller BN, Bloom H, Dineen RD, Johnson MG, Hammond RP (1993) Environ Monit Assess 28:169

82. Plaschke RB, Dal Pont G, Butler ECV (1997) Mar Pollut Bull 34:177

83. Hanslow S (1984) Metals in Derwent sediments. Graduate Diploma of Environmental Studies Thesis, University of Tasmania

84. Wood JM, Horowitz P, Cox H (1992) Sci Total Environ 125:253

85. Pirzl HR (1996) Distributions and changes of heavy metal concentrations in sediments of the Derwent estuary. BSc (Hons) Thesis, University of Tasmania

86. Langlois D, Cooper RJ, Clark NH, Ratkowsky DA (1987) Mar Pollut Bull 18:67

87. Dineen RD, Noller BN (1995) Toxic elements in fish and shellfish from the Derwent estuary. Department of Environment and Land Management, Hobart

88. Leeming R, Nichols P (1998) Mar Freshwat Res 49:7

89. Volkman J, Deprez P, Holdsworth D, Rayner M, Nichols P (1989) Extractable organic compounds in sludge samples from the River Derwent, Tasmania. Proceedings of Derwent River habitat quality seminar. University of Tasmania, Hobart

90. Green G (1997) Hydrocarbons and faecal material in urban stormwater and estuarine sediments: source characterisation and quantification. PhD Thesis, University of Tasmania

91. Volkman J, Rogers G, Blackman A, Neill G (1988) Biogenic and petroleum hydrocarbons in sediments from the D'Entrecasteaux Channel near Hobart, Tasmania. AMSA Silver Jubilee Commemorative Volume. Wavelength, Chippendale, NSW

92. Davies PE, Fulton W, Kalish S (1988) The environmental effects of effluent from the ANM newsprint mills at Boyer. Tasmania Inland Fisheries Commission, Hobart

93. Volkman JK, Holdsworth DG, Richardson DE (1993) J Chromatogr 643:209

94. Ruiz GM, Hewitt CL (2003) Toward understanding patterns of coastal marine invasions: a prospectus. In: Leppäkoski E, Gollasch S, Olenin S (eds) Invasive aquatic species of Europe: distribution, impacts and management. Kluwer, Dordrecht, p 529

95. Samson C (2002) Using sediment cores to document changes in benthic habitats and species in the Derwent estuary: demonstrating the benefits of historical perspectives for management. School of Zoology, University of Tasmania, Hobart

Hdb Env Chem Vol. 5, Part H (2006): 51–70
DOI 10.1007/698_5_026
© Springer-Verlag Berlin Heidelberg 2005
Published online: 2 December 2005

The São Francisco Estuary, Brazil

Bastiaan Knoppers[1] (✉) · Paulo R. P. Medeiros[2] · Weber F. L. de Souza[1] ·
Tim Jennerjahn[3]

[1]Departamento de Geoquímica, Universidade Federal Fluminense,
Morro do Valonguinho s/n, 24210-455 Niterói, Brazil
geoknop@geoq.uff.br, geowfls@vm.uff.br

[2]Departamento de Geografia e Meio Ambiente/LABMAR,
Universidade Federal de Alagoas, 570021-090 Maceió, Brazil
prpm@fapeal.br

[3]Zentrum für Marine Tropenökologie, Fahrenheitstrasse 1, 28359 Bremen, Germany
tjenner@zmt.uni-bremen.de

Abstract This is a first account of the physical and biogeochemical characteristics of
the tropical São Francisco (SF) estuarine system, East Brazil, western South Atlantic.
The estuary (Lat. $10°36'$S Long. $36°\ 23'$W) is fed by the humid to semiarid SF river
basin (AB = 634×10^3 km^2, L = 2700 km), the second largest of Brazil's territory. Since
the 1950s, SF has evolved into a unique system almost solely impacted by a cascade
of dams, which now control 98% of the basin and reduced discharge to the estuary by
35%. The recent Xingó dam, operating since 1994 at 180 km from the coast, regulated
the formerly unimodal seasonal discharge (range 800 to 8000 m^3 s^{-1}) to a constant flow
of around 2000 m^3 s^{-1}. The formerly turbid river waters have become transparent and
oligotrophic due to drastic material retention by the dams. The young Xingó reservoir ex-
erted significant changes in the relative composition of inorganic and organic dissolved

and particulate constituents being delivered to the estuarine mixing zone and thus also to the composition and sustenance of phytoplankton biomass and production. The more or less constant river flow eliminated the former seasonal migration pattern of the estuarine mixing zone and its lower saline portion (S>5 to <15) is now largely positioned over its pro-delta shoals. Relict muddy deposits beyond the river mouth, eroded and resuspended by intense wave mixing, have become the main source of suspended matter to the mixing zone and maintain the coastal plume more turbid than the river. These processes seem to control the behavior of several dissolved inorganic constituents, except for dissolved silicate, which behaves conservatively proportional to the mixing of fresh and marine waters. The extremely low chlorophyll *a* concentrations indicate that nutrient uptake by primary production along the mixing zone is minor. The coastal plume generally disperses southwestwards at an oblique angle to the coast. Its oligotrophic conditions are maintained by both the low material yields of the basin and efficient flushing by the oceanic South Equatorial Current (SEC), which impinges directly upon the coast.

Keywords Dam impacts · Nutrients · Organic matter · Suspended matter · Tropical estuary

1
Introduction

Estuaries are ecosystems characterized by the exchange of water and materials derived from the land and the sea, and generally sustain moderate to high primary production rates and fisheries yields [1]. Material exchange is controlled by the estuarine flushing time of water and a variety of physical-chemical and biological processes, which modify the concentration, character, and fate of materials [2–7]. River-borne materials react due to the sudden changes in salinity, pH, turbidity, respiration, and photosynthesis. A fraction of the particulates and associated chemical constituents are retained by accumulation in the sediments and others are bypassed to the sea without reaction [8, 9]. Estuaries thus play an important role in controlling the transfer of matter at the land–sea interface and also ocean margin processes [10–12].

Much of the detailed information on the fundamental processes operating in estuaries has been gathered in small-to-medium sized temperate estuaries and the tropical dispersal systems of the world's large rivers [13–15]. Less is known about the tropical humid estuaries fed by medium sized rivers, which due to their large number are also thought to contribute significantly to the global water and material input to the ocean [16]. However, it is becoming increasingly difficult to systematize knowledge on these estuaries as the majority are being altered by human interventions, particularly since the accelerated development of tropical countries from the 1950s onwards. Urbanization, industrialization, deforestation, agriculture, mining, and engineering works (i.e. dams) have changed the hydrological balance, material yields, and the water quality of systems [13, 17–20], including those of the tropical Brazilian coast [21].

The tropical coast of Brazil (Lat. 2°S and 22°S), excluding the Amazon in the north, harbors about 50 small and seven medium sized river estuaries subject to either humid or semiarid climates (Fig. 1, [22]), including the São Francisco river estuary with the second largest drainage basin of Brazil's territory. It is a unique example of an estuary almost solely impacted by hydrological alterations and extreme reduction in sediment and nutrient discharge from dams, with hardly any compensation of materials by downstream replenishment from natural or cultural sources. Although most estuaries impacted by dams upstream also generally retain nutrients and suspended matter, discharge of nutrients from domestic effluents downstream more than make up for what is trapped in the reservoirs [19, 23, 24]. This article presents a first account of the physical and biogeochemical characteristics of the estuarine and coastal waters of the São Francisco river, particularly after construction of the recent Xingó dam in 1994, set 180 km from the coast.

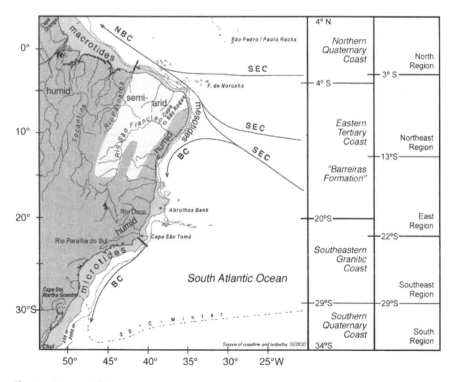

Fig. 1 Climate, tidal regimes, western boundary currents, geology, and geographical regions of the Brazilian coast and shelf. Adapted from [21, 22, 27, 28, 39]

2
Geography of the São Francisco River, Estuary, and Shelf

2.1
Physiography and Climate

The São Francisco estuary is located at (Lat. 10°25'S and Long. 36°23'W) the boundary between the Northeast and East geographical regions of the tropical coast of Brazil (Fig. 1, [22]), western South Atlantic. It is fed by the large São Francisco river basin which covers an area of 0.634×10^6 km^2 between Lat. 7° and 21°S and Long. 35° and 47°40'W, corresponding to 7.5% of Brazil's territory. The 2700 km long São Francisco river is drained by ten sub-basins and the watershed comprises four main geomorphological sectors: the upper, middle, middle-lower and lower, including the delta plain (Fig. 2). The river is born in the south at about 1800 m altitude in the Canastra mountain range, traverses from south to north an extensive depression of the Atlantic Altiplain and the Central Brazilian Plateau, and in the Northeast it diverts towards the Southeast and delta plain [25, 26].

Due to its large extension, the river is subject to several climatic conditions (Fig. 1) [27]. The upper sector has a climate of Köppen type Awa (warm, hu-

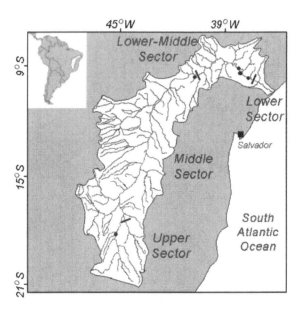

Fig. 2 Regional watershed classification with the upper, middle, middle-lower, and lower sectors (*bars*) and the cascade of dams (*dots*) in the São Francisco basin. Note that the majority of dams are set in the middle-lower sector between 200 and 700 km from the estuary [26]

mid, summer rainfall) and the delta plain of type As (warm, humid, winter rainfall). The intermediate portion, partially set within the "drought polygon" of the Northeast, is characterized by a semiarid climate of type Bhw (tropical dry, semiarid, winter rainfall). The negative precipitation to evaporation balance of this stretch is enhanced by the presence of seven large dam reservoirs (Fig. 2). The basin's average annual precipitation is 916 mm year^{-1} and evapo-transpiration 774 mm year^{-1}, but the upper humid sector and the delta plain receive more rainfall, by a factor of two, than the middle and middle-lower sectors. The overall climatic setting of the estuary is governed by the Equatorial Atlantic Air Mass and the Intertropical Convergence Zone (ITCZ), which expands and contracts around the Equator, causing the seasonal variability of rainfall and the E and NE trade winds, which impinge upon the coast at an angle.

2.2
Geomorphology and Evolution

The entire tropical coastal zone of Brazil, excluding the Amazon Quaternary plain, is fringed by Tertiary tablelands with incised flat-bottom valleys of unconsolidated Late Tertiary alluvial fans, denominated Tertiary Barreira Formations [28]. Fossil bluffs of the tablelands outcrop all along the shoreline intercalated by narrow stretches of Quaternary plains. The São Francisco plain has an area of 800 km^2 and sedimentary deposits are Pleistocene alluvial fans and beach-ridge terraces (120 ka) at the internal portion of the plain, Holocene paleo-lagoons (5730 ± 200 years BP) in the low-lying areas, Pleistocene and Holocene beach-ridge terraces, fresh water marshes and mangrove swamps, and inactive and active Holocene dunes (Fig. 3) [29]. The deposits are linked to sea-level history marked by two main transgressive episodes. The older one, denominated as the Penultimate Transgression, reached its maximum at 120 ka when sea-level reached a highstand of 8 ± 2 m above the present. The younger being the Last Transgression with a maximum at 5.1 ka when sea level rose 5 m above that of today. Sea-level lowered thereafter under high frequency oscillations up to a recent standstill and today's slight rise [29–32].

2.3
Bathymetry and Sedimentology

The facies of the active migrating dunes north of the river mouth represent the updrift side, where sands, fed by the preferential NE-SW longshore drift maintained by the predominant easterly waves, are trapped due to the hydrodynamic barrier imposed by river discharge (i.e., "jetty" or "groyne effect", [33]). The mangrove swamps are set at the downdrift side, traditionally being nourished by river-borne sediments [29]. At present, however, the

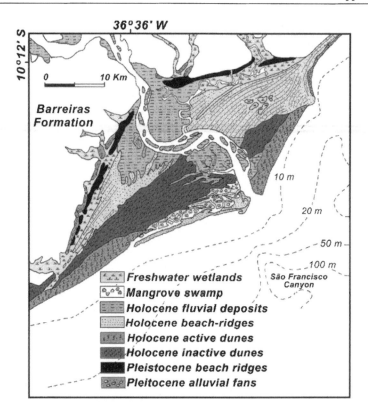

Fig. 3 Geological and sedimentological chart for the São Francisco river mouth. Adapted from [29]

downdrift side is suffering from acute coastal erosion, due to the lack of re-plenishment of river-borne sediments induced by retention in dam reservoirs upstream [25, 34].

The bathymetry of the river mouth reflects the asymmetry between the updrift and downdrift sides and, with its upstream islands and pro-delta shoals, fits into the category of an estuarine-delta (Fig. 4) [29]. The estuary is composed of an internal channel up to 12 m deep, one subaqueous bar perpendicular to the coast at the updrift side, and another arch near to parallel to the coast along the downdrift side. Waves break over the bars and induce marked resuspension of sediments. The predominant direction of waves is from the east and northeast, and southeast in winter, all of which reach the river mouth at an angle, and have periods of 5–7 s with heights of 1.0–1.5 m [29, 35].

The shelf sediments at the updrift side are mainly authigenic carbonates (from Halimeda and Coralline algae), with a narrow band of siliclastic sediments close to shore. The downdrift side, beyond the river mouth shoals and the inner shelf, is characterized by siliclastic (partially muddy) and the

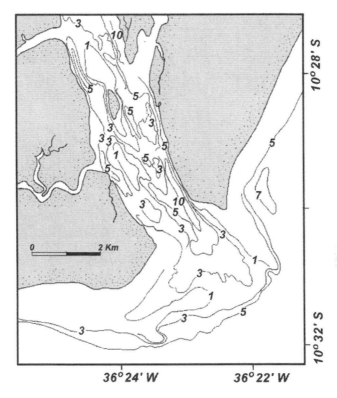

Fig. 4 Bathymetry of the São Francisco estuary and coast. Adapted from [29]

outer shelf by carbonatic sediments. The siliclastic sediments originate from the river, coastal erosion of the fossil bluffs of the Tertiary Tablelands, and from earlier deposits during lower sea-level stands. The width of the shelf varies between 20 and 40 km with gradients of 1 : 700 to 1 : 100. The shelf margin closest to the river mouth is located at depths of 50 to 70 m and is characterized by the São Francisco canyon (Fig. 3), where muddy deposits prevail [29, 36].

2.4
Human Interventions

The SF basin has evolved into a unique system almost solely impacted by a cascade of seven major dams and reservoirs, which have reduced runoff, eliminated the seasonal pattern of discharge, and control 98% of the basin (Fig. 2). The first, Três Marias, was set in the humid upper basin and entered operation in 1952. The remainder, in downstream order, are Sobradinho (1980), the Paulo Afonso I–IV complex (1955–1980), Itaparica (1988), Moxotó (1978), and Xingó (1994). These were constructed along a stretch of about

700 km in the semiarid middle-lower portion of the basin to cover the energy demand of the Northeast, spur agriculture via irrigation, and for flood control. Sobradinho harbors the largest reservoir ($V = 34$ km^3, $L \cong 350$ km), which significantly alters the precipitation–evaporation balance of the river and retains the largest fraction of suspended solids from upstream. It is in part responsible for the almost oligotrophic conditions of the river and other reservoirs downstream to the coast [26, 34, 37].

The introduction of contaminants from urban, agricultural, and industrial sources along the lower sector downstream of the Xingó dam to the estuary is minor, considering the volume of the river, demographic density, settlement size, per capita effluent input, and the lack of significant industries. The more populated and industrialized region is encountered in the upper basin, which due to its long distance from the coast and the potential retention of materials by the dams downstream, does not imminently affect the estuary [25].

3
Hydrography of the River and Sea

3.1
River Runoff

The first dam (Três Marias of 1952) set in the upper basin hardly affected the flow of the distant lower basin, as shown by the long-term hydrographic record gauged 120 km from the estuary (Fig. 5) [26]. Since the 1970s, however, the dams of the middle-lower basin reduced average annual river discharge to the estuary by 35% and consecutively distorted the natural interannual and unimodal seasonal pattern of flow. The average annual discharge for the predam period of 1938–1973 was 3010 ± 50 m^3 s^{-1} and exhibited a seasonal range from 800 to 8000 m^3 s^{-1}, with frequent peaks up to 15 000 m^3 s^{-1} or more. The natural longer term interannual oscillations of flow are related to changes in the precipitation regime of the Northeast, characterized by 2.5, 6.4, and 9.8 year cycles [38]. From 1985 onwards, distortions became more obvious with substantially truncated peak flows. During the period 1995–2001 discharge was finally leveled out by the Xingó dam to 1760 ± 235 m^3 s^{-1}. Variability includes the daily, weekly, and monthly changes of flow regulation by the dam, in accordance to the demand for electrical energy generation. Also, an extreme drought event in 2001 reduced flow down to about 1200 m^3 s^{-1} for a few months. The replenishment of water and materials along the 180 km stretch between the Xingó dam and the estuary is on average of minor significance due to the semiarid conditions and the low land relief of the watershed, as well as the intermittent nature of its tributaries. Estimates of water input during winter rainfall by the tributaries closer to the more humid coast lie around 5% of total flow [34].

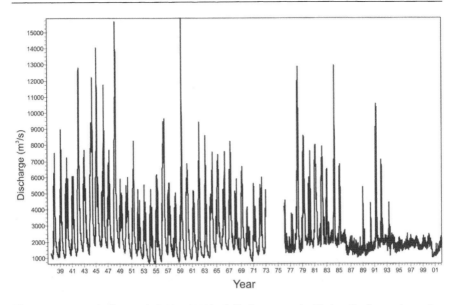

Fig. 5 Long term hydrograph (1938–2002) of discharges at the Traipu fluviometric station located 120 km from the coast [26]. Note the distortion of river flow by the dams, particularly since the 1970s and the regulation since the late Xingó dam (180 km from the coast) from 1994 onwards. In 2001, flow was reduced drastically due to an extreme drought

3.2
Estuarine, Shelf, and Oceanic Waters

The São Francisco estuarine-delta lies within the direct pathway of the tropical oligotrophic South Equatorial Current (SEC, $T > 20\,°C$, $S > 36.9$), which impinges with three branches upon the shelf and coast between 7° and 17°S (Fig. 1). The region is of particular oceanographic interest, as it is here where Brazil's western boundary currents are born. Between 7° and 10°S, a total of 12 Sv ($1\,Sv = 1 \times 10^6\,m^3\,s^{-1}$) or more from SEC flows northwestward forming the North Brazil Current (NBC). At around 15°S about 4 Sv forms the weak southward meandering Brazil Current (BC) (Fig. 1) [39]. The Tropical Surface Waters (TSW, $S > 35.9$) of SEC, which represent the marine end-member, efficiently dilute the inner shelf waters. The estuarine mixing zone is now generally set externally from and beyond the river mouth along the coast. This is corroborated by the temperature versus salinity diagram in Fig. 6 and the distribution of salinity against the distance from the river mouth (Fig. 7), obtained from monthly runs performed over an annual cycle in 2000/2001 [34].

An example of the overall setting of the estuarine-coastal plume is demonstrated by the LANDSAT 7 image of Fig. 8 and the corresponding in situ calibrated concentration of total suspended solids in Fig. 9. The oligotrophic river waters are driven out 1–2 km perpendicular to the river mouth, forming a jetty effect [33], spread out as a bulge over the arch of subaqueous bars of

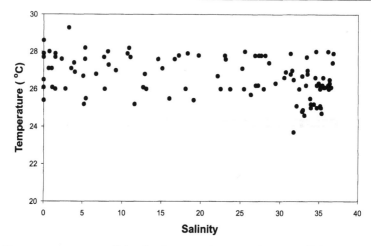

Fig. 6 Temperature versus salinity for the estuarine-coastal waters off the São Francisco river mouth, obtained from nine runs between Nov 2000 and Sept 2001 [34]

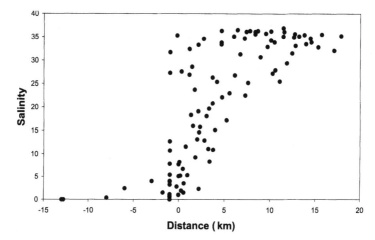

Fig. 7 Salinity distribution against distance from the river mouth (0 km) between the fresh and marine end-members, obtained from nine runs during the period Nov 2000 to Sept 2001 [34]

the pro-delta and proliferate to the SW along the coast, which also represents the most common dispersal pattern. TSW of SEC are encountered directly at the updrift side and about 10 km perpendicular and 30 km downdrift from the river mouth, and form the bottom waters of the entire estuarine-coastal plume.

The estuarine mixing zone lost its natural migration pattern and the low salinity portion (0 < S < 5) oscillated inwards and outwards along the river mouth in accordance with the dampened variability of river flow, tidal pumping, and the wind regime. Its salinity portion between 5 and 15 generally covers the stretch from the river mouth to over the pro-delta shoals, where intense

Fig. 8 LANDSAT 7 TM image, orbit point 214/67 of the São Francisco coast. Date 5 September 2001. Time 12:18:09 GMT. Gray scale transformation of color composition RGB (TM 3,2,1) (Lorenzetti, personal communication)

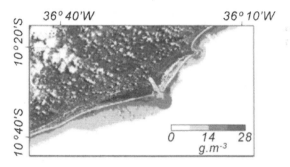

Fig. 9 Concentrations of total suspended solids ($g\,m^{-3}$) calibrated from reflectance with the logarithmic algorithm of the LANDSAT 7 TM data for 5 September 2001 (Lorenzetti, personal communication)

wind-wave induced mixing homogenizes the shallow water column. Under certain conditions, however, a salt wedge may intrude up to a maximum of 10 km into the river mouth and maintain oligohaline conditions of surface waters within, as observed on two occasions during an extreme drought event in 2001. On these occasions, the dam reduced river flow from about 1750 $m^3\,s^{-1}$ down to around 1200 $m^3\,s^{-1}$, the semidiurnal mesotides attained their maximum range of 2.7 m and stronger easterly winds and waves were predominant [34].

4
Hydrochemistry of the River and Sea

4.1
The River End-Member

Information on the composition, concentration, loads, and yields of matter for the river end-member of the estuary prior to the Xingó dam are extremely scant and restricted to sporadic measurements of some water quality parameters [26, 40, 41]. For the postdam period, a more in-depth water quality study was performed from 2000 to 2001 with a monthly sampling frequency at a station set 80 km from the estuary [34]. The results are summarized in Table 1 and compared to prior studies.

The scarcity of dissolved inorganic nitrogen (DIN) and phosphorous (DIP), total suspended solids (TSS), particulate organic carbon (POC) and nitrogen (PON), total phosphorous (TP), and chlorophyll a (Table 1) clearly indicates the transparent and oligotrophic nature of the fresh water endmember of the estuary today. All water quality parameters, except for dissolved silicate (DSi), also represent extremely low concentrations when compared to other rivers of the east coast of Brazil [42] and other medium sized rivers of the humid tropics [13, 43].

The consecutive retention of materials along the dam cascade since the 1970s is responsible for the near to total impoverishment of the river water reaching the estuary. The average TSS concentrations and loads in 1970 were 70 mg L^{-1} and 7×10^6 t year^{-1} [40], respectively. In 1984/1985 they were threefold lower with 27 mg L^{-1} and 2.6×10^6 t year^{-1} [41]. In 2000/2001 values reached a low of 5 mg L^{-1} and 0.3×10^6 t year^{-1} [34].

The corresponding TSS yield changed from 4.2 to 0.2 t km^{-2} year^{-1}, being extremely small today in comparison to other rivers of the east coast of Brazil [42]. A similar trend was also found for DIN loads with 70×10^3 t year^{-1} in 1984/1985 and 4×10^3 t year^{-1} in 2000/2001. The trend was also found for DIP. The young

Table 1 Average concentrations and standard deviations of physical-chemical parameters of the riverine and marine end-members of the São Francisco estuary and coastal waters, obtained from nine runs during the period Nov 2000 to Sept 2001 [34]

Parameter	Freshwater end-member	Marine end-member
pH	8.0 ± 0.1	8.3 ± 0.2
TSS (mg L^{-1})	5.2 ± 2.3	6.6 ± 7.7
DIP (μM)	0.1 ± 0.1	0.1 ± 0.1
NO$_3^-$ – N (μM)	3.7 ± 4.3	0.5 ± 0.8
DSi (μM)	313 ± 77	27 ± 30
DIN:DIP$_{at}$	49 ± 64	11 ± 30
DSi:DIN$_{at}$	203 ± 189	76 ± 62
POC (μg L^{-1})	373 ± 32	186 ± 56
PON (μg L^{-1})	39 ± 4	21 ± 2
TP (μg L^{-1})	22 ± 13	20 ± 10
Chl a (μg L^{-1})	1.6 ± 1.1	0.4 ± 0.3
POC:PON$_{wt}$	10 ± 2	7 ± 2
POC:Chl a_{wt}	305 ± 50	612 ± 565

age and oligotrophic nature of the Xingó reservoir is responsible for the low delivery of TSS, DIN, DIP, POC and PON to the estuary. The lack of significant human effluent sources downstream does not compensate for the reduction of the nutrient mix, as observed in many other river-estuarine systems subject to cultural eutrophication downstream of dams [17, 19, 23, 24].

DSi concentrations, however, have not been strongly affected by removal due to, for example, uptake by siliceous phytoplankton and sedimentation of frustules in the impoundment and downstream to the estuary. The DSi loads diminished more or less in proportion to river discharge from 650×10^3 t year^{-1} (1984/1985) to 450×10^3 t year^{-1} (2000/2001) and the phytoplankton biomass indicator chlorophyll a maintained low concentrations. The crucial changes that probably affected the behavior of nutrients along the estuarine mixing zone and the already low potential primary productivity of the system are the overall lowering of nitrogen and phosphorous loads and the drastic increase of the DSi : DIP ratio (e.g., from 5 : 1 to over 200 : 1). Alterations in the nutrient mix could lead to changes in the composition of phytoplankton in the coastal waters, as found in the Danube system of the Black Sea [23]. However, the Iron Gates dam of the Danube river decreased silicate inputs and the Xingó dam of the São Francisco decreased the nitrogen and phosphorous loads to the coast.

4.2
The Marine End-Member

The average concentrations of the water quality parameters obtained from the runs in 2000/2001 for the marine end-member (TSW of SEC) are given in Table 1 and corroborate the presence of extreme oligotrophic conditions, characteristic for the entire northeast shelf of Brazil [21, 34]. The general oligotrophic nature of both sources in terms of their nutrient mix (except DSi) and chlorophyll a is, however, one of the remarkable features of the entire São Francisco dispersal system, making it also rather difficult to quantify the behavior of these constituents along the estuarine mixing zone.

5
Estuarine Mixing Zone and Dispersal System

The simplest and traditional approach to elucidate material dynamics of the estuarine mixing zone involves the comparison of concentrations between non-reactive (i.e., salt) and reactive inorganic and organic, dissolved and particulate materials. Composite plots of salinity, (giving the proportion of mixing of water between the fresh and marine end-members) against concentrations of the reactive constituents at the encountered salinity are used to exemplify material behavior. The plots can quantify the degree to which the estuary serves as a source or a sink of materials (e.g., non-conservative behavior) or whether materials are passed without reaction (e.g., conservative behavior), being merely diluted in proportion to mixing [5]. Based on this approach, manifold whole system mass balance analyses have been made [8, 44]. The available information for the São Francisco, however, only permits a semiquantitative evaluation of the behavior of the materials along the estuarine mixing zone as it is now set from and off the river mouth. It therefore corresponds to a 3D dispersal system, prone to additional advective mechanisms, in contrast to 2D estuaries of confined environments [45]. These constraints should be borne in mind for the following interpretations of mixing curves.

5.1
Nutrients

The salinity of the estuarine mixing zone along the turbid bulge of the coastal plume (Figs. 8 and 9) corresponds to about 5–15 (Fig. 7), which still represents the range where changes in salinity, pH, and also turbidity generally enhance particle–water reactions and thus the behavior of dissolved inorganic and organic constituents [4, 5, 9]. However, the small range of pH from about 7.8 to 8.0 along the mixing zone (Fig. 10a) is rather dampened, making it difficult to assess by the simple mixing curve approach whether it affects particle–water reactions. Other estuarine systems of the east coast of Brazil and elsewhere in the tropics operate over a wider pH range and are also more turbid [46].

The entire region is highly impoverished in dissolved inorganic nitrogen and phosphorous. The nutrient nitrate ($NO_3 - N$, Fig. 10b) and also orthophosphate (DIP) generally behave in a conservative fashion during average and above average river flow. Under conditions of below average river flow, as during the extreme drought event in 2001, minor trends towards a slight sink of nitrate were detected, either attributed to the longer flushing time of the internal estuarine portion (enabling some uptake by phytoplankton) or enhanced dilution by advection of water masses originating from the updrift side of the external mixing zone. The variability of the inclination of the nitrate mixing curves of Fig. 10b, are however largely driven by the input changes from the fresh water end-member.

The DIN : DIP molar ratios (Fig. 10c) vary considerably from 5 : 1 during lower river flow (i.e., the drought event of 2001) to over 150 : 1 at above average river flow. The river thus generally imposes a potential phosphorous limitation for the sustenance of primary production in the estuary, and a marine end-member nitrogen limitation at the outer premises of the coastal plume, when compared to the uptake demand by phytoplankton indicated by the Redfield ratio of 16 : 1 [47] or 20 : 1 for coastal and oligotrophic waters [48]. The more ideal mix in relation to the demand by phytoplankton

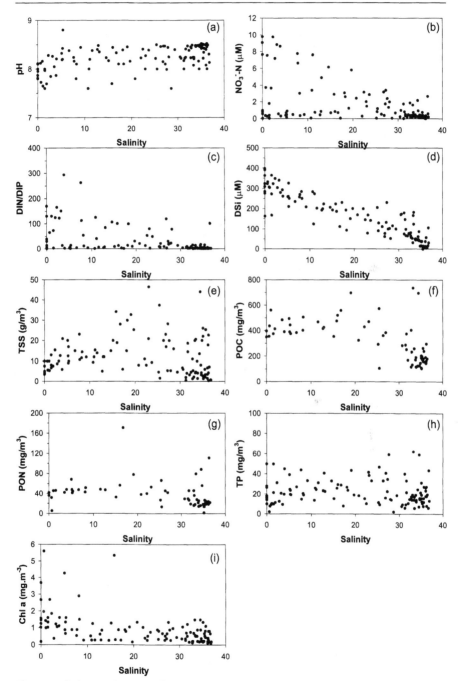

Fig. 10 Salinity mixing plots from the São Francisco estuarine region: pH (**a**), $NO_3^- - N$ (**b**), DIN:DIP ratio dissolved inorganic nitrogen and dissolved inorganic phosphorus (**c**), DSi dissolved silicate (**d**), TSS total suspended solids (**e**), POC particulate organic carbon (**f**), PON particulate organic nitrogen (**g**), TP total phosphorus (**h**), and Chl a chlorophyll a (**i**). Obtained from nine runs during the period Nov 2000 to Sept 2001 [34]

is found at the polyhaline premises of the plume. Similar trends in shifts of the nutrient mix also characterize other estuarine coastal waters of the east coast of Brazil, but are less dramatic than those found along the São Francisco plume [42].

Dissolved silicate (DSi), furnished in large quantities by the river, behaves entirely conservatively, proportional to the mixing of fresh and marine waters (Fig. 10d). DSi probable uptake by primary production in small quantities and input from the bottom from the relict deposits by dissolution are undetectable by the present approach. Due to the impoverishment of DIN since the construction of the dams, the DSi : DIN ratios shifted from about 5 : 1 to over 200 : 1, which now characterize the inner to middle portions of the plume. DSi : DIN uptake ratios of around 1 : 1 generally characterize siliceous phytoplankton [49]. The marked changes induced in the quality of the nutrient mix may be followed along the estuarine mixing zone up to the salinity limit of about 25. From then onwards the predominance of the marine endmember induces normalization of typical conditions encountered in coastal shelf waters of eastern and northeastern Brazil, impacted by western boundary currents [22, 34, 50].

5.2
Suspended Solids and Particulate Organic Matter

The bulge and the updrift and downdrift sides of the estuarine-coastal plume are more turbid than the fresh and marine end-members (Figs. 8 and 9), and two portions of the estuarine mixing zone, one over the shoals (3 < S < 10) and another (15 < S < 33) between about 15 and 20 km towards the SW, are enriched with suspended matter (TSS, Fig. 10e). This clearly indicates that other sources are furnishing materials to the coastal plume. The wind-wave driven erosion/resuspension of particulates from older relict muddy deposits beyond the sandy shoals and coastal erosion at the updrift side act in concert in feeding the first portion of the estuarine-plume with suspended matter. The second portion is henceforth thought to be enriched by lateral input of resuspended materials originating in nearshore deposits along the SW coast (Fig. 8). The changes in riverine TSS concentrations and loads between the predam and postdam periods thus induced a shift in the predominance of material sources feeding the plume, with the river being a major source in the predam period, and the pro-delta deposits and coastal erosion representing the main sources of today.

The behavior and composition of the particulate organic pool along the estuarine plume also reveals some unexpected features. Particulate organic carbon (POC, Fig. 10f) concentrations range from 350 to 500 mg m^{-3} between S = 0 and 30 and from 100 to 200 mg m^{-3} at the marine end-member, forming a convex plot against salinity. Particulate organic nitrogen (PON, Fig. 10g) exhibits a more dampened convex plot in comparison to POC. Total phosphorous (TP, Fig. 10h) shows an erratic scatter along the entire salinity spectrum. In contrast, however, the phytoplankton biomass indicator chlorophyll a (Fig. 10i) exhibits low concentrations and even shows a slight trend towards a sink along the mixing zone.

The gain of POC against the loss of chlorophyll a strengthens the fact that the coastal plume is being largely enriched by detrital carbon originating from the bottom deposits and/or other sources, but not from the river. However, the loss of chlorophyll also shows that an albeit minor contribution to the carbon pool may arise from detritus derived from phytoplankton senescence induced by several limiting factors, such as the low riverine nutrient inputs, the quality of the nutrient mix, and the higher turbidity and the physical vertical mixing of the lower salinity portion of the mixing zone, which impose stress upon the phytoplankton populations [49]. Both the $C:N_{wt}$ and POC : chlorophyll a ratios demonstrate a variability within the range of 8 to 12 : 1 and 100 to 700 : 1 along the entire mixing zone up to S = 30 and are indicative of the detrital nature of the organic pool [49, 51].

6
Export Processes over the Shelf Margin

The preferential direction of dispersal of the São Francisco plume is to the SW along the inner shelf. Nevertheless, the signal of the São Francisco has been detected over the shelf edge and deeper slope by deployment of sediment traps positioned at 500 m and 1550 m water depths 50 km off the river mouth during January to May 1995 [48]. The period still coincided with a higher water discharge of the river, as the Xingó postdam period was just at its onset. The measured total flux for the 4 months, Fig. 11, was 16.5 g m^{-2} in 1550 m. The flux for carbonate was 5.5 g m^{-2}, for biogenic opal 0.97 g m^{-2}, for lithogenics 8.8 g m^{-2}, and for organic matter 1.2 g m^{-2}. Despite the low productivity in the region [21] and the low sediment input from the river, the fluxes measured lie at the higher end of annual fluxes of other tropical regions of the world [19]. One of the explanations is that primary production enhanced by nutrient inputs (particularly DSi) from the river contributed during a short period to the considerable fluxes of biogenic opal and organic matter.

To understand the impact of the São Francisco dispersal system upon the shelf and beyond the margin, further studies have to address the link between nutrient bypassing and primary productivity at the outer boundary of the river's influence. The low chlorophyll a concentrations within the plume are not only maintained by low nutrient concentrations but also to some extent by the still fairly turbid conditions, at present generated by bottom and coastal erosion processes. To what extent the seasonal pulsation of material fluxes over the shelf margin detected in 1995 is still in operation remains to be

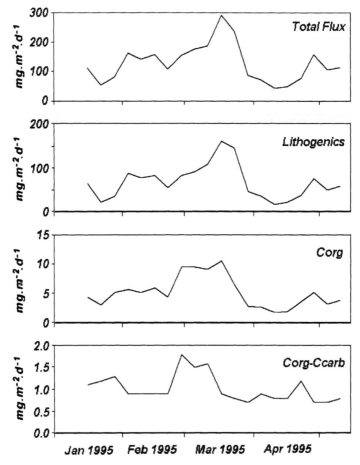

Fig. 11 Particle flux of the São Francisco river, captured by sediment traps 50 km offshore at the shelf slope in 1550 m depth [52]

verified. However, some seasonality in material fluxes should remain in view of the variability of the trade winds and easterly wave regime in spring, summer, and autumn (which control the resuspension processes along the coast and probably also control subsequent material cascading along the shelf bottom [21]) and the impact of SE polar fronts in winter. All of these processes have impact upon the nature of dispersal of the plume and the extension of material bypassing.

Acknowledgements This work was supported by CNPq grant no 476833/2001-9 to B. Knoppers, PRONEX-FAPERJ/CNPq grant E-26/171.175/2003, and a PhD scholarship from CAPES/PROPEP-UFAL to P.R.P. Medeiros. The authors wish to thank Manuel M. Messias (Labmar-UFAL) for the nutrient analysis and Matthias Birkicht (ZMT, Bremen) for the CHN analysis. Particular thanks are also due to Prof. Dr. J.A. Lorenzetti (INPE, São Paulo) for furnishing the LANDSAT 7 TM image and the corresponding TSS calibration.

References

1. Nixon SW (1988) Limnol Ocean 33:1005
2. Ketchum BH (1951) J Mar Res 10:18
3. Cameron WM, Pritchard DW (1965) Estuaries. In: Hill MN (ed) The Sea 2. Wiley, New York, p 306
4. Dyer KR (1973) Estuaries: a physical introduction. Wiley, London
5. Burton JD, Liss PS (1976) Estuarine chemistry. Academic, London
6. Moris AW (1985) Estuarine chemistry and general survey strategy. In: Head PC (ed) Practical estuarine chemistry: a handbook. Cambridge University, Cambridge, p 1
7. Day JW, Hall AS, Kemp WM, Yanez-Arancibia A (1989) Estuarine ecology. Wiley, New York
8. Smith SV, Atkinson MJ (1994) Mass balance of nutrient fluxes in coastal lagoons. In: Kjerfve B (ed) Coastal lagoon processes. Elsevier, Amsterdam, p 133
9. Turner A, Millward GE (2002) Estuar Coast Shelf Sci 55:857
10. Walsh JJ (1988) On the nature of continental shelves. Academic, San Diego
11. Mantoura RFC, Martin JM, Wollast R (eds) (1991) Ocean margin processes in global change. Wiley, Chichester
12. Wollast R, Mackenzie FT, Chou L (eds) Interactions of C, N, P and S biogeochemical cycles and global change. NATO ASI Series, London
13. Meybeck M (1993) C, N, P and S in rivers: from sources to global inputs. In: Wollast R, Mackenzie FT, Chou L (eds) Interactions of C, N, P and S biogeochemical cycles and global change. NATO ASI Series, London, p 163
14. Wright LD, Nittrouer CA (1995) Estuaries 18:494
15. Hay WW (1998) Chem Geo 145:287
16. Milliman JD, Syvitski JPM (1992) J Geo 100:525
17. Meybeck M (2003) Phil Trans R Soc Land B http://dx.doi.org/10.1098/rstb.2003.1379
18. Vörösmarty CJ, Sharma KP, Fekete BM, Copeland AH, Holden J, Marble J, Lough JA (1997) Ambio 26:210
19. Ittekkot V, Humborg C, Schäfer P (2000) BioScience 50:776
20. Richter C, Burbridge PR, Gätje C, Knoppers BA, Martins O, Ngoile MAK, O'Toole MJ, Ramachandran S, Salomons W, Talue-McManus, L (2001) Group report: integrated coastal management in developing countries. In: von Bodungen B, Turner RK (eds). Dahlem University, Berlin, p 253
21. Knoppers B, Ekau W, Figueredo AG (1999) Geo Mar Letters 19:171
22. Ekau W, Knoppers B (1999) Arch Fish Mar Res 47:113
23. Humborg C, Ittekkot V, Cociasu A, von Bodungen B (1997) Nature 386:385
24. Milliman J (1997) Nature 386:325
25. Marques M, Knoppers B, Machmann Oliveira A (2001) AMH First Int Symp Transbound Waters Man 1:1
26. ANA (2004) Agência Nacional de Águas http://www.ana.gov.br
27. Nimer E (1972) Rev Bras Geog 34:3
28. Guerra HT (1962) O litoral Atlântico: Paisagens do Brasil. IBGE, Rio de Janeiro
29. Dominguez JML (1996) The São Francisco strandplain: a paradigm for wave-dominated deltas? In: De Batist M, Jacobs P (eds) Geology of siliciclastic shelf seas. Geological Society, London, Special Publication 117:217
30. Martin L, Suguio K, Flexor J-M, Dominguez JML, Bittencourt ACSP (1996) An Acad BrasCiên 68:303
31. Suguio K, Martin L, Bittencourt ACSP, Dominguez JML (1985) Rev Bras Geociên 15:273

32. Angulo JA, Lessa GC (1997) Mar Geo 140:141
33. Komar PD (1973) Geo Soc Am Bull 84:2217
34. Medeiros PPR (2003) Aporte fluvial, transformação e dispersão da matéria em suspensão e nutrientes no estuário do rio São Francisco, após a construção da usina hidroelétrica do Xingó (AL/SE), PhD Thesis. Universidade Federal Fluminense, Niterói
35. Viana JB (1972) Estimativa do transporte litorâneo em torno da embocadura do rio Sergipe. Instituto de Pesquisas Hidroviárias, Belo Horizonte
36. Tintelnot M (1995) Transport and deposition of fine-grained sediments on the Brazilian continental shelf as revealed by clay mineral distribution, PhD thesis. University of Heidelberg, Heidelberg
37. Tundisi JG, Rocha O, Matsumura-Tundisi T, Braga B (1998) World Water 14:141
38. Kane RP (1998) Rev Bras Geofis 16:37
39. Stramma L, Ikeda Y, Peterson RG (1990) Deep-Sea Res 37:1875
40. Milliman JD (1975) Contr Sedimentol 4:151
41. Bessa MF, Paredes JF (1990) Geochimica Brasiliensis 4:17
42. Souza WFL (2002) A interface terra-mar leste do brasil: tipologia, aporte fluvial, águas costeiras e plataforma continental, PhD Thesis. Universidade Federal Fluminense, Niterói
43. Meybeck M, Ragu A (1995) GEMS/Water contribution to the global register of river inputs (GLORI), Provisional Final Report. UNEP/WHO/UNESCO, Geneva
44. Gordon Jr DC, Boudreau PR, Mann KH, Ong JE, Silvert WL, Smith SV, Wattayakorn G, Wulff F, Yanagi T (1996) LOICZ Biogeochemical modelling guidelines. LOICZ Reports & Studies 5. Texel
45. Cifuentes LA, Schemel LE, Sharp JH (1990) Estuar Coast Shelf Sci 30:411
46. Eyre BD (1998) Estuaries 21:540
47. Redfield AC (1958) Am Sci 46:205
48. Golterman HL, Oude NT (1991) Eutrophication of lakes, rivers and coastal seas. In: Hutzinger O (ed) Handbook of environmental chemistry, vol 5(A). Springer, Berlin Heidelberg New York, p 79
49. Harris GP (1986) Phytoplankton ecology: structure, function and fluctuation. Chapman and Hall, London
50. Knoppers B, Meyerhöfer M, Marone E, Dutz J, Lopes J, Leipe T, Camargo R (1999) Arch Fish Mar Res 47:285
51. Banse K (1974) Mar Biol 41:199
52. Jennerjahn T, Ittekkot V, Carvalho CEV (1996) Preliminary data on particle flux off the São Francisco river, Eastern Brazil. In: Ittekkot V, Schäfer P, Honjo S, Depetris PJ (eds) Particle flux in the ocean. Wiley, London, p 215

Hdb Env Chem Vol. 5, Part H (2006): 71–90
DOI 10.1007/698_5_029
© Springer-Verlag Berlin Heidelberg 2005
Published online: 20 October 2005

Geochemistry of the Amazon Estuary

Joseph M. Smoak[1] (✉) · James M. Krest[1] · Peter W. Swarzenski[2]

[1]Environmental Science, University of South Florida, St. Petersburg, Florida 33701, USA
smoak@stpt.usf.edu, krest@stpt.usf.edu

[2]Center for Coastal and Regional Marine Studies, US Geological Survey,
St. Petersburg, Florida 33701, USA
pswarzen@usgs.gov

Abstract The Amazon River supplies more freshwater to the ocean than any other river in the world. This enormous volume of freshwater forces the estuarine mixing out of the river channel and onto the continental shelf. On the continental shelf, the estuarine mixing occurs in a very dynamic environment unlike that of a typical estuary. The tides, the wind, and the boundary current that sweeps the continental shelf have a pronounced influence on the chemical and biological processes occurring within the estuary. The dynamic environment, along with the enormous supply of water, solutes and particles makes the Amazon estuary unique. This chapter describes the unique features of the Amazon estuary and how these features influence the processes occurring within the estuary. Examined are the supply and cycling of major and minor elements, and the use of naturally occurring radionuclides to trace processes including water movement, scavenging, sediment-water interaction, and sediment accumulation rates. The biogeochemical cycling of carbon, nitrogen, and phosphorus, and the significances of the Amazon estuary in the global mass balance of these elements are examined.

Keywords Major elements · Minor elements · Nutrients · Organic matter · Radionuclides

1
Introduction

The enormous supply of water, solutes and particles makes the Amazon estuary unique among the world's estuaries. The Amazon River is the largest river in the world in terms of freshwater discharge (5.8×10^{12} m^{3}year^{-1}; [1]), which is approximately 20% of global river volume discharge worldwide [2]. This magnitude of freshwater discharge forces the river and seawater mixing onto the continental shelf. Therefore the Amazon estuary does not conform to the common definition of an estuary as a semi-enclosed body of water [3]; instead the Amazon estuary must be defined as an estuary based solely on the mixing of river water and seawater.

The area of the Amazon estuary, for the purpose of this chapter, is defined as the southern limit of the Amazon River plume (i.e., northeast of the Para River) northward to the Brazilian/French Guiana border (4°N), and seaward to the shelf break, which starts at the 100-m isobath (Fig. 1). This is the study area defined by AMASSEDS (A Multidisciplinary Amazon Shelf SEDiment Study) [1]. This encompasses a total shelf area of 1.1×10^{11} m^2 and includes the area of modern sedimentation [4]. AMASSEDS was a large-scale investigation of the fate of the Amazon River discharge and much of what we know about the Amazon estuary comes from the AMASSEDS investigation. While the above is the defined area of the estuary, the influence of the Amazon is far beyond this area, with relatively low-salinity water reaching thousands of kilometers into the Caribbean and equatorial Atlantic [5].

1.1
Background setting

The tropical setting and the enormity of the drainage basin (6 million km^2) combine to produce the largest river in the world in terms of freshwater discharge (Table 1). The tropical setting and size of the drainage basin also minimize changes in discharge both within and between years [1]. The maximum sediment discharge is about six times greater than the minimum sediment discharge and the freshwater discharge varies by less than a factor of

Table 1 Background setting

Amazon estuary	
Drainage basin area	6×10^6 km^2
River water discharge	5.8×10^{12} m^{3}yr^{-1}
River sediment discharge	7.7×10^{14} g yr^{-1}
Shelf sediment accumulation	6.2×10^{14} g yr^{-1}
Offshore water entrainment	5.8×10^{13} m^{3}yr^{-1}

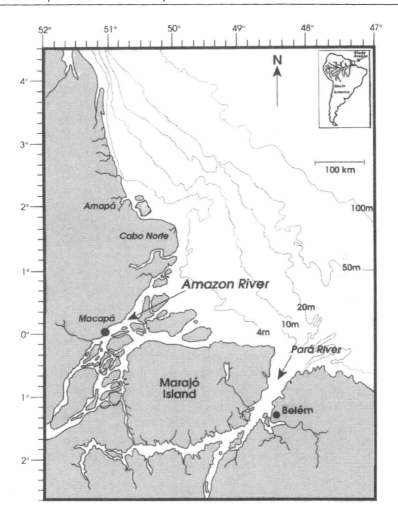

Fig. 1 The Amazon Estuary is defined as the southern limit of the Amazon River plume northward to the Brazilian/French Guiana border, and seaward to the shelf break, which starts at the 100-m isobath

four [6, 7]. Approximately 80% of the sediment discharged by the Amazon River is derived from the Andes Mountains [8] despite most of the drainage basin being occupied by rainforest. The sediment discharge is about 6% of the sediment discharge of rivers worldwide [2, 9] and is 85–95% silt and clay-sized particles [2, 8].

In addition to the enormous volume of water supplied by the river, an even greater volume is supplied by entrainment of offshore water into the mixing zone. This entrainment is the result of estuarine circulation that occurs on the continental shelf. The entrained volume of water onto the continental shelf is 10 times the river discharge [10] and supplies a substantial portion of nutrients to the estuary. DeMaster and Pope [11] estimated the nitrate, phosphate and ammonium that is supplied by offshore advection as compared to the river and found that 38, 52, and 80% of these nutrient totals is supplied by entrainment of offshore water, respectively.

The great physical energy on the shelf also contributes to the uniqueness of the Amazon estuary. The outer Amazon continental shelf is swept by the North Brazilian Current (NBC), a boundary current that can exceed $75 \, \text{cm s}^{-1}$ [12–14]. The Amazon estuary also is influenced by very energetic tides. The tides typically have a range of 2 m at the mouth of the river [15], and tidal currents can attain velocities in excess of $200 \, \text{cm s}^{-1}$ [16, 17].

Sediments are actively accumulating over an area of $7 \times 10^{10} \, \text{m}^2$, and the rest of the shelf area is composed of relict sands [4, 18]. Within the actively accumulating area, $6.2 \times 10^{14} \, \text{g year}^{-1}$ of sediment are accumulating [18]. Sediment accumulation on the shelf is deposited in a prograding subaqueous delta and is mostly silt and clay from the Amazon River [19]. Northward moving currents export approximately $1.2 \times 10^{14} \, \text{g year}^{-1}$ of sediment from the shelf area to the north mostly along the coast [20, 21]. Negligible sediment is exported seaward and southeastward of the Amazon shelf [19].

Shelf sediment can be physically reworked down to depths of several meters by physical energy [4, 18]. This reworking influences burial and export of chemical species in the Amazon estuary. The reworking often occurs in association with extremely high-concentration bottom layers called fluid muds (suspended sediment concentration $> 10 \, \text{g L}^{-1}$), which can be mobilized by changes in the fortnightly tidal cycles [22]. Fluid muds on the Amazon continental shelf can be found up to 7 m thick [22]. The areal extent, thickness, and occurrence of fluid muds make the Amazon shelf distinct from other regions of the world's continental shelves.

2
Chemical Processes

Chemical, physical, and biological processes occurring in the Amazon estuary control the fate of chemical species in this unique environment. The estuary may act as a source or sink for chemical species depending on these processes and the nature of the species. As described above, the physical processes operating within the Amazon estuary are extreme, and as a consequence, influence much of the chemical cycling in the Amazon estuary. The most obvious of these physical controls is the magnitude of water discharged from the river that forces the estuary onto the shelf. Forcing the estuary onto the shelf causes the chemical cycling to occur in a high-energy environment (e.g., NBC, tides, winds) unlike those of a typical estuary in a protected drowned river. A prime example of the influence of the physical processes on the biogeochemical cycles can be observed in the primary production. Despite abundant nutrients, primary production is inhibited over much of the inner shelf due to high suspended sediment concentration limiting light penetration [11]. The suspended sediment is supplied by the river and resuspension created by the dynamic energy on the shelf. The abundant nutrients are supplied by the river and by entrainment of offshore water into the mixing zone from the estuarine circulation that occurs on the continental shelf.

2.1
Major and Minor Elements

As with any major river, the elemental composition of the Amazon River (before it undergoes estuarine mixing) is controlled primarily by the geology of the different parts of the basin through which the Amazon and its tributaries flow and the weathering processes occurring in the sub-basins. One important difference between the Amazon and other major rivers like the Mississippi or the Chiang Jiang is that the Amazon basin is generally lacking in platform marine sediments and basic igneous rocks, and as a consequence, concentrations of many dissolved species are relatively low in comparison to the other rivers [23].

The majority of the anions and cations originate from weathering processes in the Andes. Downstream, anion and cation concentrations generally decrease, reflecting a dilution of the high-ionic strength Andean waters by the low-ionic strength water draining the lowlands [8, 23–25]. Not surprisingly, elemental distributions change with the hydroperiod—during the dry season (approx. June through September) when Andean source waters are at or are approaching their minimum discharge rates, concentrations of cations, anions, alkalinity and suspended sediments are all decreasing. Silicate concentrations are inversely proportional to the percent contribution of Andean waters, reflecting the weathering of silicates in floodplains and lowland tributaries. As a result, Si/Alk and K/Ca ratios are highest in the dry season [25].

Trace metal and major-ion transport in the Amazon is highly dependent upon suspended sediment transport. It is estimated that less than 1% of the Fe and Al in the water column is dissolved (i.e., passes through a membrane filter with nominal pore size of 0.45 or 0.2 μm [26, 27]. Similarly, V, Cr, Mn, Co, Ni, Zn, Cs, and Pb are almost entirely associated with the river particulate matter [28, 29]. Of the dissolved fraction for Fe and Al, 80% and 90%, respectively, is associated with colloids [27]. Of the particulate fractions, Gibbs [26] showed the distribution of Fe, Ni, Co, Cr, Cu and Mn in different phases using sequential extraction techniques.

Obviously, the composition and fluxes of particulate matter in the Amazon play important roles in the cycling of the trace metals. Although organic matter may be only a small percentage of the particulate matter in the mainstem Amazon, tributaries draining lowlands may have much higher fractions [30, 31]. This is especially true at the confluence of the Rio Negro and Rio Solimões Rivers near Manaus, where the two tributaries come together to form the Amazon River (Tables 2 and 3). The Rio Negro is a "black water" river, relatively acidic, rich in humics and dissolved organic carbon, but low in total cations and suspended particulate loads, for it flows predominately through "transport-limited" lowlands as described by Stallard and Edmond [23, 24]. Complexation with organic ligands leads to a relative enrichment of Fe, Al, Cs, Zn, and Pb in the particulate matter in the Rio Negro [28]. The Rio Solimões is a "white water" river with near neutral pH. It is rich in total cations and has a high suspended sediment load, typical of "weathering-limited" tributaries originating in the Andes. Suspended particulate matter in the Solimões exhibits an enrichment of the trace metals Fe, Al, V, Cr, Mn, Co, Ni, Cu, As, Cs and Pb, relative to concentrations seen in average rocks, most likely as a result of their ease of dissolution from the rock material [28]. At the confluence, some elements act conservatively (e.g., K, Ca, and Mg), but others show either gains or losses, including loss due to sedimentation processes [32]. Therefore, elemental concentrations should be sampled downstream of the confluence when estimating discharge fluxes to the coastal ocean (Table 3).

Downstream from the confluence of the Negro and the Solimões is Óbidos, the last gauging station before the river begins to show a marine influence. The distribution of trace elements between dissolved and particulate frac-

Table 2 Major-ion dissolved concentrations in the Amazon River and its major tributaries (concentrations in micromoles per liter)

	Negro near Manaus	Solimões near Manaus	Amazon below confluence
Cl	8–22 [a,c]	39.3–382 [a,b,c]	36–210 [a,c]
Ca	2.0–8.0 [a,c]	155–288 [a,b,c]	127–218 [a,c]
Mg	3.7–6.2 [a,c]	39–73 [a,c]	39–64 [a,c]
K	3.0–13 [a,c]	19.5–29 [a,b,c]	22–26.1 [a,c]
Na	5.0–30.9 [a,c]	81–357 [a,b,c]	82–199 [a,c]
Fe	2.2–3.4 [a]	0.34–0.95 [a]	0.82–0.98 [a]
Al	2.85–3.19 [a]	0.59–0.78 [a]	1.11–1.30 [a]
Mn	0.14–0.19 [a,c]	0.10–0.29 [a,c]	0.054–452 [a,c]
Si	–	132–184 [b]	–
NO$_3$	–	4.7–20.9 [b]	–
PO$_4$	–	0.47–1.48 [b]	–
SO$_4$	2.0 [c]	18–77 [b,c]	19–50 [c]

[a] Aucour at al. 2003
[b] Devol et al. 1995
[c] Seyler and Boaventura 2003

Table 3 Major-ion dissolved concentrations in the Amazon River and its major tributaries, and the estuary with export fluxes to the ocean (concentrations in micromoles per liter)

	Negro near Manaus	Solimões near Manaus	Amazon below confluence	Salinity 0–5	Salinity 5–15	Salinity 15–30	Salinity 30–36	Export Flux t per year	Export Flux % from particulates
V	9.54–10.16[a]	15.5–26.8[a]	12.4–21.2[a]	–	–	–	–	80.6[a]	93%[a]
Cr	9.69–17.0[a]	5.5–12.9[a]	4.1–13.8[a]	–	–	–	–	–	–
Mn	136–194[a]	101–291[a]	53.9–452[a]	–	–	–	–	416[a]	85[a]
Co	1.16–2.79[a]	0.95–1.85[a]	0.64–1.9[a]	–	–	–	–	8.99[a]	95[a]
Ni	2.63–19.1[a]	3.51–19.2[a]	2–20[a]	2.7–6.3[b]	2.7–6.9[b]	1.7–4.7[b]	2.7–2.9[b]	46.5[a]	86[a]
Cu	3.07–4.9[a,b]	22.4–25.3[a,b]	15–25.7[a,b]	20.8–25.8[b]	14.5–20.9[b]	4.1–18.7[b]	1.6–6.9[b]	38.9[a]	51[a]
Zn	10.5–15.2[a]	2.15–11.2[a]	6.9–20.6[a]	–	–	–	–	–	–
Rb	14.0–20.6[a]	20.7–21.4[a]	18.2–21.3[a]	–	–	–	–	50.2[a]	76[a]
Sr	31.4–49.3[a]	404–767[a]	290–613[a]	–	–	–	–	438[a]	53[a]
Mo	0.11–0.86[a]	1.95–3.67[a]	1.3–3.5[a]	–	–	–	–	–	–
Cd	0.125–2.235[a]	0.15–1.50[a]	0.12–1.6[a]	0.048–0.085[b]	0.053–0.123[b]	0.041–0.133[b]	0.037–0.063[b]	–	–
Sb	0.02–0.10[a]	0.54–1.07[a]	0.31–2.12[a]	–	–	–	–	–	–
Ba	45.4–46[a]	181–242[a]	128–205[a]	–	–	–	–	430[a]	71[a]
U	0.097–0.130[a]	0.155–0.197[a]	0.130–0.215[a]	–	–	–	–	1.69[a]	78[a]
Ti	–	–	5.2[c]	3.3–7.1[c]	0.15–0.31[c]	0.21–0.42[c]	0.10–0.41[c]	–	–

[a] Seyler and Boaventura 2003
[b] Boyle et al. 1982
[c] Skrabal 1995

tions is intermediate between the distributions observed in the Rio Negro and the Rio Solimões above the confluence. Being the most easily leached of the trace elements in this tropical climate, As, Cu, Rb, Sr, Ba, and U have the highest dissolved fractions while V, Cr, Mn, Co, Ni, Zn, Cs and Pb are, generally, more particle reactive and occur predominately in the particulate fraction.

As the Amazon discharges its dissolved and particulate load into the estuary, the major and trace elements undergo a variety of reactions—many typical of estuarine systems and some unique to the Amazon. Complexities are introduced by the large amounts of sediments and the physically dynamic conditions on the shelf which rework these sediments. These sediments are rich in Fe, Al and Mn-oxides, acting as "sub-oxic fluidized bed reactors", driving much of the elemental cycling of the Amazon estuary [33].

Of the trace metals, Cu and Ni are largely unreactive in the Amazon plume on time scales of a few days. However, as a micronutrient, Cu concentrations may be decreased in the plume by biological uptake [34]. Cd, as in other rivers, seems to exhibit desorption from suspended sediments between a salinity of 0 and 5, but the trends are not clear, partly as a result of poor precision of the data set [34].

For dissolved Fe, Ti, and rare earth elements (REE's), 86 to 97% is removed between a salinity of 0 and 6 [35, 36]. At high salinities, Ti and REE's are elevated over open ocean end-member concentrations, suggesting that a minor input of these elements is being contributed from suspended particles or bottom sediments. Bottom fluxes have also been proposed to explain elevated plume concentrations of U [37, 38] and Al [39].

Much of the elemental flux from the river gets deposited in the delta sediments, which are subjected to constant reworking (and remineralization, especially in the cases of REE's, Ti, U, Al, etc.). However, these sub-oxic sediments also may be responsible for authigenic clay formation. In an example of reverse weathering, Michalopoulos and Aller [40], demonstrated that as much as 22% of the total riverine silica flux (and 90% of the biogenic SiO_2) is stored on the shelf, and a large portion of this silica is being incorporated into authigenic aluminosilicates, rich in K and Fe. Biogenic silica apparently reacts with detrital $Al(OH)_3$, $Fe(OH)_3$, and dissolved cations in seawater (esp. K, Mg, F) to form these authigenic clays, which may mask the fate of some portion of the flux of other elements as they are stored in forms not detected by most analytical techniques [39, 40].

2.2
Radionuclide Tracers

Naturally occurring (e.g., U/Th series (Fig. 2)) radionuclides have been applied successfully as tracers for varied biogeochemical studies in the Amazon estuary. Because these radionuclides decay predictably with time, exponen-

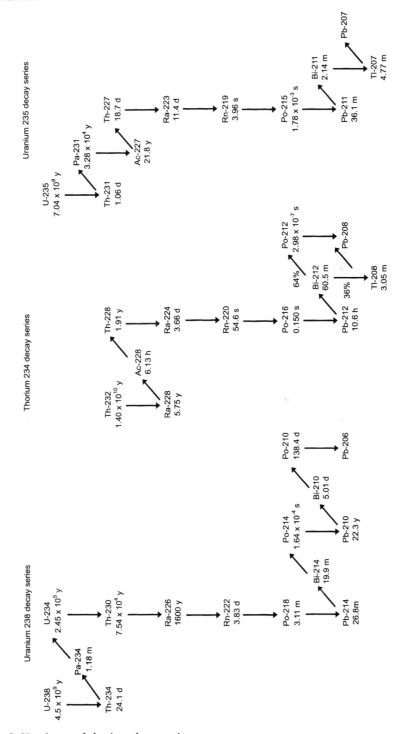

Fig. 2 Uranium and thorium decay series

tial decay laws and numerical models can be developed to derive rate constants applicable for scavenging, transport and removal processes, as well as more traditional sediment accumulation/deposition rates. The use of certain radionuclides as analogs for particle-reactive elements requires that these radionuclides behave similarly to the elements in question. This entails both similar removal mechanisms from the water column and similar phase associations in suspended particles and sediments. This section illustrates the application of several U/Th series radionuclides to coastal processes unique to the Amazon shelf, where the sheer magnitude, energy and abundance of fine-grained particles of this system control many of physicochemical reactions and processes.

2.2.1
Uranium

Although the earliest dissolved ^{238}U concentration for the mainstem Amazon River was measured by Moore [41], McKee et al. [37] were the first to systematically examine the fate of uranium within the Amazon estuary. In this work, the authors observed consistently elevated ^{238}U ($t_{1/2} = 4.5 \times 10^9$ years) concentrations relative to the well established, conservative U: salinity correlation (e.g., [42]), and concluded that widespread release of U from riverine particles is likely the dominant source of this enhanced uranium. The phase partitioning of U on the Amazon Shelf was subsequently investigated by Swarzenski et al. [38], who examined the estuarine behavior of "truly" dissolved (< 0.001 μm), dissolved (< 0.4 μm), particulate (> 0.4 μm) and colloidal-sized U (0.001–0.4 μm) (Table 4). In this detailed examination of U in the water column of the Amazon shelf during a low discharge river stage, ∼ 90% of the U proximal to the river mouth was found to be associated with particulates, and most (> 92%) of the dissolved U was bound up in a reactive colloidal pool. To further investigate the role of shelf sediments in the U budget on the Amazon shelf,

Table 4 Salinity, truly dissolved, dissolved, particulate and colloidal uranium. (based on Swarzenski et al., 1995)

Salinity (ppt)	"Truly" dissolved (μg l^{-1})	Dissolved (μg l^{-1})	Particulate (μg l^{-1})	Colloidal (μg l^{-1})
0.31	0.01	0.12	No data	0.10
9.74	0.31	0.44	3.54	0.15
19.72	1.20	1.68	2.53	0.47
28.20	2.22	2.67	5.40	0.43
33.82	2.78	3.26	1.31	0.52

Table 5 Source of ^{210}Pb supplied to the Amazon Estuary (based on Smoak et al. 1996)

Source	Percentage
Offshore water	67
River water	31
Atmospheric fallout	2.3
In situ production	0.1

porewaters from sediments adjacent to the Amazon River mouth were analyzed for U, Fe, and Mn. At varying river discharge stages, pore water U and Fe were closely co-varied to depths greater than 1 m, which confirms a strong dependence between the diagenetic transformations of Fe in the reducing sediments and U [58]. Such pore water profiles also suggest authigenic U-bearing mineral formation in the shallow sediments of the Amazon shelf.

2.2.2
Thorium

The short-lived, highly particle reactive tracer ^{234}Th ($t_{1/2} = 24.1$ days) was examined on the Amazon shelf to study rates of scavenging [10, 43]. The partitioning of thorium between soluble and particulate in the water column of the Amazon estuary is controlled by particle dynamics. That is, the particle residence times are shorter than the sorption times; therefore more thorium is found in the solution than would be predicted based on an equilibrium model [43]. The temporal and spatial variability in unsupported (excess) ^{234}Th inventories in sediments typically ranged from about 2 to 5 dpm cm^{-2} [10]. Such excess ^{234}Th inventories suggest that particle scavenging rates were controlled foremost by particle/colloid dynamics between the water column and the seabed. The presence of widespread fluid muds above the sediment–water interface effectively limited the short-term accumulation of ^{234}Th onto the seabed and provided a unique dispersal process for particle reactive species discharging from the Amazon without becoming incorporated into the seabed.

2.2.3
Lead

Lead-210 ($t_{1/2} = 22.3$ years) has been important in understanding the sedimentary processes on the Amazon Shelf. Kuehl et al. [4] used ^{210}Pb to examine sediment accumulation on the shelf and found some areas with an accumulation rate as high as 10 cm year^{-1}. Lead-210 can also be used to examine the potential input of particle reactive elements that are mixed from offshore waters onto the Amazon shelf. This entrained water is often difficult to quantify, especially on a century time scale that will be comparable with the other measurements, such as accumulation rates. Therefore ^{210}Pb, which integrates over a century time scale, is an excellent water mass tracer for this purpose. As ^{210}Pb-rich open ocean waters mix into Amazon estuarine waters, ^{210}Pb is rapidly scavenged by reactive particles/colloids. Smoak et al. [10] used total excess ^{210}Pb inventory on the Amazon continental shelf, which exceeded that supplied by other sources (Table 5), to determine the total excess ^{210}Pb inventory supplied from offshore water. Smoak et al. [10] then used the concentration of ^{210}Pb in offshore water to estimate that lateral advection of offshore water supplies 5.8×10^{13} m^3 year^{-1}. This oceanic water contribution is on the order of 10x more than what is typically discharged from the Amazon River. For trace elements that are derived mostly from atmospheric fallout, ^{210}Pb can thus be applied as a useful tracer for coastal uptake processes as oceanic waters mix into the sediment-rich waters of the Amazon River–ocean mixing zone.

2.2.4
Radium

Because of the strong sediment source function, several isotopes of Ra (^{226,228,224}Ra) were utilized to evaluate resuspension processes on the Amazon shelf over different time scales and discharge regimes [43]. In general, the upward flux of ^{226}Ra ($t_{1/2} = 1600$ years) from the shelf was balanced by the annual desorption of ^{226}Ra from Amazon River particles, although such budgets did reflect nuances as a function of variable discharge (i.e., supply vs resuspension). In order to balance a flux of ^{228}Ra ($t_{1/2} = 5.7$ years) from the Amazon shelf, approximately 10^{13} kg of sediment, equivalent to the top 35 cm of sediment across the entire shelf, must be resuspended annually. Energetic desorption of the shortest-lived Ra isotope, ^{224}Ra ($t_{1/2} = 3.7$ days), from suspended sediments (i.e., fluid muds) alone can support the inventory of ^{224}Ra on the Amazon shelf. Such results provide compelling evidence for the unparalleled energetics of the Amazon River–ocean mixing zone, where unusual particle dynamics (i.e., prevalence of fluid mud banks, colloidal-sized metal oxides) control the fate of most particle reactive constituents. Strong covariance between Ra and phosphate in these waters further suggests that Ra can also be a useful tracer for more biologically dominated geochemical cycles.

There are certain areas where the ^{210}Pb geochronology assumption of constant initial activity do not hold true. Dukat and Kuehl [44] developed a geochronometer utilizing the two long-lived isotopes of radium, ^{228}Ra/^{226}Ra on the Amazon shelf to augment the ^{210}Pb geochronology. This approach uses the ingrowth of ^{228}Ra into secular equilibrium with ^{232}Th and can be applied to determine sediment accumulation rates on an approximate 30-year time scale. Using the ^{228}Ra/^{226}Ra technique Dukat and Kuehl [44] found accumulation rates as high as 57 cm year^{-1} and discovered that the sediment accumulation in these areas is controlled by the input of fluid mud.

Radium isotope ratios (e.g., ^{228}Ra/^{226}Ra) have been utilized successfully to fingerprint and track the movement of the Amazon River plume northward into the Atlantic Ocean for more than 1500 km [43]. Moore et al. [43] found, based on ^{224}Ra, waters 380 km from the Amazon mouth had taken < 5 days to travel that distance. A sustained current > 80 cm s^{-1} is required to advect the water this distance. Drifter measurements agree well with the radium calculations.

2.3
Biogeochemical Cycling

Input of nutrients from rivers, entrainment of offshore water, and atmospheric deposition enhance production rates in estuaries, including the Amazon. As the Amazon estuary occurs on the continental shelf, it becomes an interesting site for investigation of biogeochemical cycling within the context of continental shelves. Continental shelves are of particular interest as global sources and sinks of carbon and other nutrients because of their high production rates, high sedimentation rates of terrigenous material, and their shallow water depths.

Remineralization on the Amazon shelf is enhanced by the physical energy reworking sediments and leads to more than 90% of the organic matter produced on the shelf being remineralized within the water column [45]. While primary production on the Amazon continental shelf is relatively high, only a few percent of the organic carbon and nitrogen produced in the estuary will eventually be buried in the sediment of the shelf. The high remineralization rates not only support much of the productivity in the estuary but also enhance the export of nutrients. Despite high remineralization rates within the Amazon estuary, a substantial quantity of organic C and N is buried on the shelf as a result of the very large initial supply.

2.3.1
Carbon

The Amazon River supplies an estimated 14×10^9 mol C d^{-1} to the estuary (Table 6). The total organic carbon (TOC) flux from the river is 6.5×10^9 mol C d^{-1} of which 2/3 is dissolved organic carbon (DOC) [46–48]. Dissolved inorganic carbon (DIC) is the dominant form of carbon supplied from the river with a flux of 7.4×10^9 mol C d^{-1}. The largest source of carbon, in the DIC form, 336×10^9 mol d^{-1}, is supplied by the entrainment of offshore water. The offshore water also supplies 13×10^9 mol d^{-1} of DOC. En-

Table 6 Carbon input and export

Estuary carbon	mol d^{-1}
River input	14×10^9
Total organic	6.5×10^9
Dissolved inorganic	7.4×10^9
Advection from offshore	349×10^9
Dissolved organic	13×10^9
Dissolved inorganic	336×10^9
Sediment deposition	1.1×10^9
Organic sediment	0.8×10^9
Sediment export	2.1×10^8
Ocean and atmosphere export	360×10^9

trained DOC and DIC were calculated from typical concentrations [49] and entrainment rates [10].

The total C supplied from the river and offshore water is 363×10^9 mol d^{-1}. Removal of C from the estuary includes deposition to the shelf sediments $(1.1 \times 10^9$ mol C d$^{-1})$ and sediment export out of the shelf area to the north $(2.1 \times 10^8$ mol C d$^{-1})$. Carbon deposition and export values were calculated based on the sediment deposition rate [18] and export rates [20], and sediment C concentration values [45]. Using the deposition rate and the organic C sediment concentration [45] total organic C burial rate is 0.8×10^9 mol C d^{-1}, which is 1.5% of the organic C buried globally on continental shelves. The Amazon continental shelf represents less than 0.5% of the total global shelf area. The global rate is estimated to be 5.0×10^{10} mol C d^{-1} [50]. Total C exported to the ocean (and atmosphere) is 360×10^9 mol d^{-1}, as calculated by mass balance of input and removal. The atmospheric exchange is unknown. Note that the C exported to the ocean (and atmosphere) is almost equivalent to that supplied from the ocean in the entrained water.

Some of the highest primary production rates in the marine environment are found on the Amazon shelf. Rates of primary production on the Amazon shelf average 0.22 mol C m^{-2}d^{-1} in the "optimal-growth zone" [51]. The "optimal-growth zone" occurs where suspended sediments are less than 10 mg l^{-1} and nutrients are high. Primary production is inhibited in the estuary over much of the inner shelf because the suspended sediment affects the supply of light [11]. Despite the inhibited production, shelf primary production is 18×10^9 mol C d^{-1}. Less than 2% of the organic carbon produced in the estuary is buried in the sediment of the continental shelf (Fig. 3). The remaining organic carbon is remineralized and exported offshore, exported along the coast to the north, or released into the atmosphere and out of the Amazon continental shelf area. Seventy three percent of the primary production on the shelf is estimated to be recycled in the water column and 10% advected off-

Marine Organic Carbon

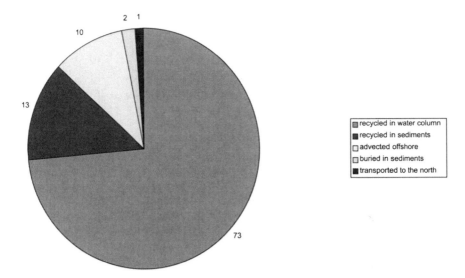

Fig. 3 Percentage of total particulate marine organic carbon produce on the shelf that is recycled in water column, recycled in sediments, advected offshore, buried in sediments and transported along the coast to the north

shore [21]. Eighty percent of the particulate organic carbon (POC) produced on the shelf that reaches the seafloor is estimated to be regenerated in the sediments and returned to the water column [21].

Approximately 25% of the riverine supply of POC is buried in the sediments on the shelf. Less than 5% of riverine POC is recycled in the water column, and 55% of the deposited riverine POC is regenerated in the sediment and returned to the water column [21]. More terrestrial organic carbon than marine organic carbon is buried on the shelf, despite the level of organic carbon produced on the shelf. Seventy percent of the total organic carbon buried on the shelf is terrestrial and 30% is marine [45, 52]. The selective burial of terrestrial over marine organic carbon is attributed to the refractory nature of the terrestrial material as compared to the marine organic material.

2.3.2
Nitrogen

Total nitrogen flux from the river is 6.8×10^8 mol N d^{-1}, 3.6×10^8 mol N d^{-1} of which occurs as total organic nitrogen (TON) [11, 21, 46–48] (Table 7). Sixty percent of the TON flux is dissolved organic nitrogen and an addi-

Table 7 Nitrogen input and export

Estuary nitrogen	mol d^{-1}
River input	6.8×10^8
Total organic	3.6×10^8
Dissolved inorganic	2.7×10^8
Advection from offshore	1.6×10^8
Sediment deposition	1.3×10^8
Organic sediment	0.85×10^8
Sediment export	0.3×10^8
Atmospheric input	0.04×10^8
Ocean and atmosphere export	6.8×10^8

tional 2.7×10^8 mol N d^{-1} is in the form of dissolved inorganic nitrogen. The entrainment of offshore water supplies DIN, including nitrate, nitrite and ammonium (1.6×10^8 mol N d^{-1}). The entrained DIN was calculated using DIN values [11], and entrainment rates [10]. The atmospheric input of 0.04×10^8 mol N d^{-1} [53] represents a very minor contribution. The input of total N supplied by the river, entrainment of offshore water and atmospheric deposition is 15.2×10^8 mol N d^{-1}. The removal of total N to shelf sediments was 1.3×10^8 mol N d^{-1}, and sediment export removed 0.3×10^8 mol N d^{-1} out of the system to the north. Nitrogen deposition and export values were calculated based on the sediment deposition rate [4] and export rates [20], and sediment N concentration values [54]. Using the deposition rate and the organic N sediment concentration [21] the total organic N burial rate is 0.85×10^8 mol d^{-1}, which is 1.9% of the organic N buried globally on continental shelves. The global rate is estimated to be 4.6×10^9 mol N d^{-1} [55]. According to mass balance calculation, the remaining 6.8×10^8 mol N d^{-1} is exported to the ocean (and atmosphere). Approximately 24% of this value was supplied by the entrainment of offshore water. This mass balance does not consider loss of molecular N from the marine environment or N fixation, which has not been measured on the Amazon continental shelf. Aller et al. [45] suggest loss of molecular N from the marine environment might be significant.

Using the carbon primary production and the Redfield C/N ratio (6.6) the marine production of organic N is estimated to be approximately 27×10^8 mol N d^{-1}. When comparing the input with the production on the shelf, it becomes evident that about 70% of the N required for shelf production must be recycled. Sixty five percent of the total N buried on the shelf is organic N, and of that total it is estimated that 43% is of marine origin and 57% is from terrestrial sources [21]. Thirty six percent of the PON supplied by the river is buried on the shelf. It is estimated that 16% of marine organic N produced in the surface waters is deposited in the seabed and less than 9% of marine organic N initially deposited on the seafloor is buried [21]. The burial is < 1.5% of marine organic N produced. About one third of the N buried on the shelf is inorganic and exists as fixed ammonium [21].

2.3.3
Phosphorus

Total phosphorus input from the river is 63×10^6 mol P d^{-1} and half is in dissolved inorganic form (DIP) (Table 8). Total organic phosphorus (TOP) flux from the river is 23×10^8 mol P d^{-1} [11, 47, 56] with approximately 35% as dissolved organic phosphorus (DOP). The total riverine supply of P is 63×10^6 mol P d^{-1} to the estuary, and the entrainment from offshore water is 24×10^6 mol P d^{-1} for a total supply of 87×10^6 mol P d^{-1}. The entrained DIP was calculated using DIP values [11], and entrainment rates [10]. P accumulation on the shelf totals 28×10^6 mol P d^{-1}, and 5.4×10^6 mol P d^{-1} is carried away by sediment export to the north. Phosphorus deposition and export values were calculated based on the sediment deposition rate [18] and export rates [20], and sediment P concentration values [56]. Total P exported to the ocean is 54×10^6 mol P d^{-1} based on the mass balance.

Marine production of organic P based on carbon primary production and applying the Redfield C/P ratio (106/1) is approximately 17×10^6 mol P d^{-1}. Comparing the external input with the production reveals that 45% of the P production in the shelf water must be recycled P. Approximately 3.5% (7.3×10^6 mol P d^{-1}) of the combined P produced in the water column and particulate organic P supplied by the river accumulates on the shelf. However, organic P can be converted to inorganic P during burial [57]. It is estimated that approximately 16% of the P fixed during primary production is deposited on the shelf as organic P [21]. A more detailed description of the C, N, and P biogeochemical cycles can be found in DeMaster and Aller [21].

Table 8 Phosphorus input and export

Estuary nitrogen	mol d^{-1}
River input	63×10^6
Total organic	23×10^8
Dissolved inorganic	32×10^8
Advection from offshore	24×10^6
Sediment deposition	28×10^6
Organic sediment	7.3×10^6
Sediment export	5.4×10^6
Ocean export	54×10^6

3
Conclusions

The enormous magnitude of water supplied by the Amazon River creates a unique setting in the Amazon estuary. Not only does the water supply large quantities of solutes and particles but it forces the Amazon estuary onto the shelf. As a result the estuarine mixing occurs in a very dynamic place, which influences many of the processes occurring in the estuary. The already high suspended sediment concentrations are increased through the dynamic energy resuspending and reworking sediments. These sediments influence everything from trace metal cycling to primary production. While anthropogenic activities are increasing in the Amazon drainage basin, these natural processes are of such magnitude that they have yet to be influenced. The basin is large, the population relatively low, and the magnitude of the river great.

References

1. Nittrouer CA, DeMaster DJ (1996) Cont Shelf Res 16:553
2. Meade RH, Dunne T, Richey JE, De M Santos U, Salati E (1985) Science 270:614
3. Pritchard DW (1967) What is an estuary: physical viewpoint. In: Lauff GH (ed) Estuaries. American Association for the Advancement of Science, Washington, DC, p 37
4. Kuehl SA, DeMaster DJ, Nittrouer CA (1986) Cont Shelf Res 6:209
5. Muller-Karger FE, McClain CR, Richardson PL (1988) Nature 133:56
6. Meade RH, Nordin CF, Curtis WR, Rodriques FMC, Do Vale CM, Edmond JM (1979) Nature 178:161
7. Nittrouer CA, DeMaster DJ, Figueiredo AG, Rine JM (1991) Oceanography 4:3
8. Gibbs RJ (1967) Geol Soc Am Bull 78:1203
9. Gibbs RJ (1970) J Mar Res 28:113
10. Smoak JM, DeMaster DJ, Kuehl SA, Pope RH, McKee BA (1996) Geochim Cosmochim Acta 60:2123
11. DeMaster DJ, Pope RH (1996) Cont Shelf Res 16:263
12. Cochrane JD (1963) Science 142:669
13. Metcalf WG, Stalcup MC (1967) J Geophys Res 72:4959
14. Metcalf WG (1968) J Mar Res 26:232
15. Gibbs RJ (1982) Est Coast Shelf Sci 14:283
16. Beardsley RC, Candela J, Limeburner R, Geyer WR, Lentz SJ, Castro BM, Cacchione D, Carneiro N (1995) J Geophys Res 100:2283
17. Geyer WR, Beardsley RC, Candela J, Castro BM, Legeckis RV, Lentz SJ, Limeburner R, Miranda LB, Trowbridge JH (1991) Oceanography 4:8
18. Kuehl SA, Nittrouer CA, Allison MA, Faria LE, Dukat DA, Jaeger JM, Pacioni TD, Figueiredo AG, Underkoffler EC (1996) Cont Shelf Res 16:787
19. Nittrouer CA, DeMaster DJ (1986) Cont Shelf Res 6:5
20. Allison MA, Nittrouer CA, Faria LEC Jr (1995) Mar Geol 125:373
21. DeMaster DJ, Aller RC (2001) Biogeochemical process on the Amazon Shelf: changes in dissolved and particulate fluxes during river/ocean mixing. In: McClain ME, Victoria RL, Richey JE (eds) The biogeochmistry of the Amazon Basin. Oxford Press, New York, p 328

22. Kineke GC, Sternberg RW, Trowbridge JH, Geyer WR (1996) Cont Shelf Res 16:667
23. Stallard RF, Edmond JM (1983) J Geophys Res-Oceans 88:9671
24. Stallard RF, Edmond JM (1987) J Geophys Res-Oceans 92:8293
25. Devol AH, Forsberg BR, Richey JE, Pimentel TP (1995) Global Biogeochem Cycles 9:307
26. Gibbs RJ (1973) Science 180:71
27. Benedetti M, Ranville JF, Ponthieu M, Pinheiro JP (2002) Org Geochem 33:269
28. Seyler PT, Boaventura BR (2001) Trace elements in the mainstem Amazon River. In: McClain ME, Victoria RL, Richey JE (eds) The biogeochmistry of the Amazon Basin. Oxford Press, New York, p 307
29. Seyler PT, Boaventura GR (2003) Hydrol Process 17:1345
30. Leenheer JA (1970) Acta Amazon 10:513
31. Hedges JI, Clark W, Quay PD, Richey JE, Devol AH, Ribeiro N (1986) Limnol Oceanogr 31:717
32. Aucour A-M, Tao F-X, Moreira-Turcq P, Seyler P, Sheppard S, Benedetti MF (2003) Chem Geol 197:271
33. Aller RC (1998) Mar Chem 61:143
34. Boyle E, Huested SS, Grant B (1982) Deep-Sea Res 29:1355
35. Sholkovitz ER (1993) Geochim Cosmochim Acta 57:2181
36. Skrabal SA (1995) Geochim Cosmochim Acta 59:2449
37. McKee BA, Demaster DJ, Nittrouer CA (1987) Geochim Cosmochim Acta 51:2779
38. Swarzenski PW, McKee BA, Booth JG (1995) Geochim Cosmochim Acta 59:7
39. Mackin JE, Aller RC (1986) Cont Shelf Res 6:245
40. Michalopoulos P, Aller RC (2004) Geochim Cosmochim Acta 68:1061
41. Moore WS (1967) Earth Planet Sci Lett 2:231
42. Ku TL, Knauss KG, Mathieu GG (1977) Deep-Sea Res 24:1005
43. Moore WS, DeMaster DJ, Smoak JM, McKee BA, Swarzenski PW (1996) Cont Shelf Res 16:645
44. Dukat DA, Kuehl SA (1995) Mar Geol 125:329
45. Aller RC, Blair NE, Xia Q, Rude PD (1996) Cont Shelf Res 16:753
46. Richey JE, Hedges JI, Devol AH, Quay PD (1990) Limnol Oceanogr 35:352
47. Richey JE, Victoria RL, Salati E, Forsberg BR (1991) The biogeochemistry of a major river system: the Amazon case study. In: Degens ET, Kempe S, Richey JE (eds) Biogeochemistry of major work rivers. Wiley, New York, p 57
48. Hedges JI, Ertel JR, Quay PD, Grootes PM, Richey JE, Devol AH, Farwell GW, Schmidt FW, Salati E (1986) Science 231:1129
49. Drever JI (1988) The geochemistry of natural waters, 2nd edn. Prentice Hall, New Jersey
50. Wollast R (1991) The coastal organic carbon cycle: fluxes, sources and sinks. In: Mantoura RFC, Martin J-M, Wollast R (eds) Ocean Margin processes in global change. Wiley, p 365
51. DeMaster DJ, Smith WO Jr, Nelson DM, Aller JY (1996) Cont Shelf Res 16:617
52. Showers WJ, Angle DG (1986) Cont Shelf Res 6:227
53. Nixon SW, Ammerman JW, Atkinson P, Berounsky VM, Billen G, Boicount WC, Boynton WR, Church TM, Ditoro DM, Elmgren R, Garber JH, Giblin AE, Jahnke RA, Owens NJP, Pilson MEQ, Seitzinger SP (1996) The fate of nitrogen and phosphorus at the land-sea margin of the North Atlantic Ocean. In: Howarth RW (ed) Nitrogen cycling in the North Atlantic Ocean and its watersheds. Kluwer, Dordrecht, p 141
54. Aller JY, Aller RC (1986) Cont Shelf Res 6:291

55. Wollast R (1993) Interactions of carbon and nitrogen cycles in the coastal zone. In: Wollast R, Mackenzie FT, Chou L (eds) Interactions of C, N, P and S biogeochemical cycles and global change. Springer, Berlin Heidelberg New York, p 195
56. Berner RA, Rao JL (1994) Geochim Cosmochim Acta 58:2333
57. Ruttenberg KC, Berner RA (1993) Geochim Cosmochim Acta 57:991
58. Swarzenski PW, Campbell PL, Porcelli D, McKee BA (2004) The estuarine chemistry and isotope systematics of 234,238U in the Amazon and Fly Rivers. Cont Shelf Res 24:2357–2372)

Hdb Env Chem Vol. 5, Part (2006): 91–120
DOI 10.1007/698_5_027
© Springer-Verlag Berlin Heidelberg 2005
Published online: 6 October 2005

The Mackenzie Estuary of the Arctic Ocean

Robie W. Macdonald[1] (✉) · Yanling Yu[2]

[1]Department of Fisheries and Oceans, Institute of Ocean Sciences, PO Box 6000,
Sidney, BC V8L 4B2, Canada
macdonaldrob@pac.dfo-mpo.gc.ca

[2]Polar Science Center, Applied Physics Laboratory, University of Washington,
1013 NE 40th St., Seattle, WA 98105-6698, USA
yanling@apl.washington.edu

Abstract The Mackenzie Estuary is a seasonally ice covered, deltaic estuary. It receives over 300 km^3 of freshwater and 125×10^6 t of sediment annually in a strongly modulated seasonal cycle. Ice cover plays a crucial role in the physical setting by limiting air–sea interaction (energy and gas exchange), reducing mixing, and withdrawing freshwater from the estuary while leaving behind the bulk of the dissolved components. Few studies have been conducted on estuarine processes occurring in this estuary and, although we can project from temperate estuaries what the important conservative and nonconservative processes are likely to be, the winter encroachment by ice sufficiently alters the physical, chemical, and biological processes that projections from other estuaries will likely be wrong. Here we discuss how the estuary evolves through the seasonal cycles of temperature, ice cover, river inflow, particle loadings, and winds, and review what is known of the biogeochemical cycling within the estuary. Given that the Arctic is exceptionally vulnerable to change, especially in the marginal seas, it is safe to predict that remote, pristine estuaries of the Arctic are as much at risk in the future as estuaries more directly impacted by human encroachment.

Keywords Climate change · Ice · Mackenzie Estuary · Nutrients · Organic carbon

1
Introduction

The Mackenzie River terminates in the Arctic Ocean at one of the world's great estuaries (Fig. 1). With a freshwater discharge of about $330\,\mathrm{km}^3\mathrm{year}^{-1}$ [1], the Mackenzie ranks fourth within the Arctic and 18th globally. More remarkable is the delivery of almost 130×10^6 t year^{-1} of sediments to the Beaufort Shelf [2], which by far exceeds that of any other arctic river and places it within the top 12 rivers of the world [3, 4]. Most of the sediments transported by the lower reaches of the river come from the west side of the drainage basin, which is mountainous [5, 6], in contrast to the eastern side, which has far less topography and contains large lakes that remove most of the sediment mobilized in the drainage basin above the lakes. The Mackenzie's drainage basin extends as far south as 53°N, spanning arctic and temperate ecosystems including tundra and boreal forest, and regions that have no permafrost and intermittent or fully developed

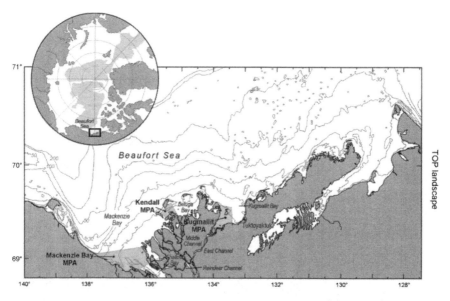

Fig. 1 Map showing the location of the Mackenzie River and delta, and the bathymetry of its estuary. The three outlined regions in the estuary are proposed to become Marine Protected Areas

permafrost [5, 7]. Therefore, unlike its estuary, the Mackenzie River cannot be considered truly arctic and we may expect its hydrology and organic and inorganic chemistry to reflect its mixed arctic and temperate character [8–12].

Due to the high sediment loads, the river has built a large postglacial delta ($\sim 13\,000\,\mathrm{km}^2$), within which there is a complex of multiple channels, lakes, and thermokarst embayments [5, 13–15] and a shallow, muddy inner shelf [16] onto which the river's components impinge (Fig. 1). According to Brunskill [5], river flow in the delta partitions with about two thirds following the Middle Channel and one sixth following the West Channel into Mackenzie Bay, and the remaining sixth following the East Channel into Kugmallit Bay. With freezing in October and ice jamming in spring, we may expect considerable variation in the distribution of channel flow, although there appears to be no quantitative information on this topic. Ice jams in spring can result in the flooding of up to 95% of the delta, but despite the large flows and often violent jamming of those flows, the main channels of the Mackenzie Delta have changed little since the first maps were drawn in 1826 [5].

Like any large estuary, mixing and entrainment are displayed over the inner shelf as multiple plumes and fronts [17–19]. The Mackenzie River exhibits a strong seasonal variation in its flow (Fig. 2a), ranging from 3000–$5000\,\mathrm{m}^3\mathrm{s}^{-1}$ during winter to as much as $40\,000\,\mathrm{m}^3\mathrm{s}^{-1}$ during freshet in early June. The variation in sediment load is more poorly documented but certainly exhibits an even larger temporal variance (Fig. 2b). Wide seasonal variation, forced predominantly by the large range in air temperature (average daily of -30 to $+16\,^\circ\mathrm{C}$) and water temperature (0–$20\,^\circ\mathrm{C}$), is a prominent characteristic of an arctic river. Many arctic rivers freeze completely to the bottom in winter, especially if they are small and the drainage basin is contained wholly within the Arctic. Because the Mackenzie River basin includes large headwater lakes (Great Slave, Great Bear, Athabaska, and Williston Lakes), it maintains substantial winter inflow rivaling that of the Lena River [20]. Nevertheless, lying north of the Arctic Circle, the delta, estuary, and adjacent shelf experience a large variation in solar radiation (from none to 24 hours a day) and are covered by ice for 8–9 months of the year (Fig. 3). The ice cover grows throughout winter, achieving thicknesses of ~ 2 m by spring (May).

The characteristics that make the Mackenzie Estuary exceptional include large freshwater discharge with a wide seasonal variation, a substantial winter inflow, inordinately large amounts of particulates in transport during spring and summer, an ice cover for most of the year, and a very strong variation in light climate. Despite the size of this estuary and its importance to the adjacent shelf, and despite the interesting geochemical processes that likely occur under the ice in winter, few studies have focused on the functioning of this estuary. This is partly because logistics are difficult during the period of 24-hour darkness and the period of breakup toward the end of May, when ice conditions become dangerous both to sampling platforms and moorings,

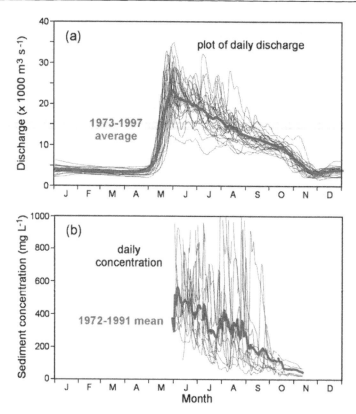

Fig. 2 a The annual cycle of the Mackenzie River discharge showing variation and mean flow for the period from 1973 to 1997 (Water Survey of Canada, Environment Canada). **b** The sediment load carried by the Mackenzie River during the period between 1972 and 1991

and partly because the estuary is exceptionally large and complex, containing three major channels only one of which is within reasonable distance from a base of operations at Tuktoyaktuk.

Despite a poor database for the geochemical processes associated with this estuary, it is clear that the delta and nearshore areas provide a critical habitat for fish, birds, and mammals and that natives have long depended on this region for its wildlife resources [21]. In recognition of the importance of this estuary, especially to belugas, three areas adjacent to the mouths of major river channels are proposed to become Marine Protected Areas (Fig. 1). Given the turbidity of the Mackenzie River, its remoteness, and the long period of ice cover, many gaps remain in our knowledge of how migratory and resident animals use the estuary. The Mackenzie River and its estuary are pristine. Perhaps one of the greatest threats to this estuary from human activities is offshore oil [22]. Forty years of exploration have established reservoirs containing 1.1×10^9 m^3 of recoverable oil [23], with promising areas in the

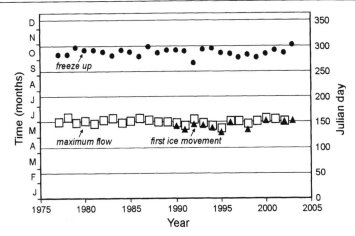

Fig. 3 Dates of freezeup, breakup (as indicated by first ice movement), and maximum flow in the Mackenzie River's East Channel at Inuvik ($\sim$ 130 km above Kugmallit Bay) (Water Survey of Canada, Environment Canada)

offshore Mackenzie Delta (the largest offshore field is Amauligak, which lies off Kugmallit Bay in 32 m of water, with estimated recoverable oil reserves of 37×10^6 m^3).

Climate change looms large over the cryosphere, and the marginal seas of the Arctic Ocean are projected to be among the world's most sensitive regions to warming [24–26]. Clearly we must understand how arctic estuaries presently function before speculating on the sorts of changes they likely will face. In particular, the Mackenzie Estuary's shoreline, which consists of poorly bonded material of low topography, must be viewed as the present manifestation of a coast that has been progressively inundated since the last ice age [27]. This estuary remains especially vulnerable to sea level rise, an increased incidence of large-fetch storms, and the destruction of permafrost.

Here we discuss what is known about how the Mackenzie Estuary processes freshwater, dissolved components, and sediments, focusing particularly on the role of ice in constraining the pathways and mixing of Mackenzie River water in the coastal region of the shelf.

2
Physical Setting

The bathymetry of the lower reaches of the Mackenzie River and the Beaufort Sea nearshore control the manner in which river water enters the sea and how ice growth in the estuary affects that entry. For example, from depths of up to 30 m in the lower reaches of the East Channel, Kugmallit Bay shoals to < 2 m. The region offshore of the Mackenzie Delta presents a broad transverse bar of less than 2 m water depth covering an area of about 3500 km^2 and containing a volume during open water of ~ 3.5 km^3 (Fig. 1). Water less than 5 m in depth is found over ~ 7000 km^2 and comprises ~ 18 km^3 of water. During freshet (~ 2.5 km^3 day^{-1}), these nearshore regions can be flushed in a day to a few days. At winter-minimum flow (~ 0.4 km^3 day^{-1}), flushing time for the < 2 m isobath might be as much as a week but is probably far less considering that ice encroaches on the upper 2 m of the water column throughout winter. The uniformly shallow water off the mouths of the Mackenzie River, with depths of < 2 m found 10 or 20 km offshore, suggests that the sediment gradient is relatively low and is maintained that way by the enormous sediment supply and by frequent storms to resuspend sediments and rapidly fill gouges produced by ice or other processes [1, 16].

Reimnitz [28] emphasizes that the shallow nearshore region at the mouth of many arctic rivers, which he terms "the 2-m ramp" (Fig. 4), is unique to arctic estuaries: the processes maintaining these arctic ramps, however, remain unclear but seem to relate to ice. In October, a relatively smooth cover of landfast ice begins to form over the estuary, reaching seaward beyond the 2-m

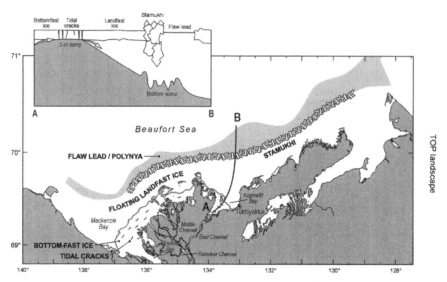

Fig. 4 Schematic diagram of the Mackenzie Estuary during the ice-covered season showing features important to the transport of water and associated properties within the delta. Noted is the bottomfast ice which may extend as far as the 2-m isobath at the end of winter, tidal cracks which develop at the boundary between bottomfast ice and floating landfast ice, the landfast ice zone predominantly within the 20-m isobath, the zone of accumulating rubble ice (stamukhi) at the outer edge of the landfast ice, and the flaw lead (intermittently open water) beyond the stamukhi. The *inset* is a schematic section (marked A–B) showing how the river passes across the bottomfast ice covering the 2-m ramp to accumulate under the floating landfast ice. The stamukhi can exhibit sufficient topography to ground out in 20 m of water, scouring the bottom

ramp out to as far as the 20-m isobath [20, 29]. To the east, landfast ice is generally stable as a solid sheet, remaining in place throughout winter, whereas to the west leads and ridging are more common due to winds and currents [30].

As the thickness of the ice progressively increases, attaining about 2 m by the end of winter, it grounds in the shallows, potentially forming a cover of bottomfast ice over much of the nearshore with the outer limit corresponding approximately to the outer edge of the 2-m ramp. Because bottomfast ice does not move up and down with tides in the same way as floating landfast ice, tidal cracks form between the two types of ice (Fig. 4) and, as ice thickens through winter, the region of tidal crack formation migrates seaward. This process of forming tidal cracks is probably relatively subdued in the Mackenzie Estuary given that tides are weak [17, 31]. The winter inflow of the Mackenzie River ($3000-5000 \ m^3s^{-1}$) must make its way across the 2-m ramp during winter through channels as described for the Colville River in Alaska or the Lena River in the Laptev Sea [28, 32], but little is known about their locations or whether they remain in the same place from year to year.

As the shorefast ice evolves and grows through winter, the outer edge of the landfast ice (~ 20 m isobath) forms a region of broken, rafted blocky ice—termed the stamukhi zone (Fig. 4)—in response to westward motion of pack ice over the central shelf, together with intermittent winds that sometimes force ice northward, opening up flaw leads, and sometimes southward, producing more ice jams, shoves, and keels [29]. Gradually through winter, the stamukhi zone may expand in area and volume, becoming partially grounded and thus helping to anchor the outer edge of the landfast ice. There is likely considerable variation from year to year in the intensity and location of stamukhi, and it has been reported that a well-developed stamukhi may not form in some years [33]. The intense region of seabed ice scour located approximately at the 20-m isobath [17, 34] suggests that ice buildup through convergence to produce stamukhi and bottom grounding is a fairly regular occurrence.

Breakup commences in the Mackenzie River's headwaters in late April and progresses northward until it reaches the delta in late May. Peak flow, therefore, impinges on the estuarine shallows while they are still covered with bottomfast ice, the consequences of which are relatively violent as river water over- and underflows the ice [28, 35, 36]. Although the processes of overflow, drainage through the ice, and strudel scour of sediments have been documented by aerial reconnaissance, and in some cases by field studies [28], the hazard associated with this process means that it remains relatively poorly described for the Mackenzie Estuary. From the above discussion, it is clear that sediments in the 2-m ramp must be exposed to very dynamic conditions that include (1) reworking by storms during late summer [16], (2) a complete ice cover that stops oxygen exchange with the atmosphere in winter and possibly produces brines that flush porewater, and (3) bottomfast ice which freezes to the sediment surface and promotes sediment freezing. What these processes mean for sedimentary geochemistry and the carbon cycle is completely unknown. Macdonald et al. [1] suggest that the delta sequesters about 65×10^6 t year^{-1} of sediments and receives over 1×10^6 t year^{-1} of terrestrial and marine organic carbon, implying a vigorous sedimentary redox cycle with implications for the small volume of water maintained between the ice and sediments for nine months of the year.

The Mackenzie River's hydrology undergoes considerable intra- and inter-annual variability (Fig. 2) with components of seasonal changes reflecting the strength of flow within the Mackenzie's various subdrainage basins [37]. In particular, the Rocky and Mackenzie Mountain subbasins, which supply 60% of the flow [37], are carbonate- and SO_4^{2-}-rich [11]. On the other hand, the basins of the interior plains are organic-, silicate-, and Na$^+$-rich while rivers in the Canadian Shield are relatively dilute in solutes and dominated by silicate [11]. As shown by the stable isotopic composition ($\delta^{18}O$), the Mackenzie River below the junction between the Liard River and the Mackenzie main-stem maintains a distinct cross-channel signature for over 200 km below the confluence with the isotopically lighter Liard water ($\sim$ – 22‰) on the west side and isotopically heavier Mackenzie mainstem water ($\sim$ – 17‰) on the east side [38]. Clearly, these isotopic gradients will be matched by gradients in, for example, carbonate and silicate. Although complete cross-channel mix-ing occurs before the river water reaches the sea, variation in the relative strength of flow from the Liard or the Mackenzie as documented by Woo and Thorne [37] will likewise produce seasonal variation in the chemical com-position of the water arriving at the estuary, with more of the winter water coming from the mainstem Mackenzie (silicate) and more of the spring and summer water coming from the Liard (carbonate).

3
Seasons in the Estuary

The Mackenzie Estuary exhibits four seasons that reflect the annual air-temperature cycle and the Mackenzie's hydrography [17, 29].

3.1
Winter

Commencing in October, as freezing-degree days accumulate to cool the sea to the point where ice starts to form, the entire inner shelf will become ice covered. The Mackenzie River inflow declines to seasonal lows and ice pro-gressively thickens until the end of winter in late May of the following year (Fig. 5a). Given the physical constraints described above, the Mackenzie River winter inflow passes across the 2-m ramp, probably using channels that be-come progressively longer and possibly smaller as ice thickens and eventually grounds out in much of the shallows (Fig. 4 inset). As the ice encroaches on

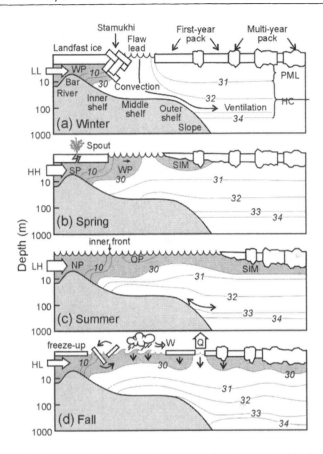

Fig. 5 Schematic diagram [17] showing the seasonal patterns exhibited in the Mackenzie Estuary as representative sections. **a** Winter is characterized by the formation of a large impoundment of Mackenzie River inflow under the floating landfast ice. The isolation of this winter inflow allows salinification and mixing in the flaw lead due to ice formation/brine rejection. **b** Spring is characterized by a large pulse of turbid freshwater from the Mackenzie River hitting the estuary before the landfast ice has melted. At freshet, river inflow, water impounded over winter, and rapid ice melt all combine to produce an exceptionally large inventory of freshwater in the nearshore. **c** Summer is characterized by open water, fronts, and plumes. **d** Fall is characterized by mixing of properties due to storms over open water and the export of freshwater from the shelf. Abbreviations: HH = higher high river discharge; LH = lower high river discharge; LL = lower low river discharge; NP = new plume water; OP = old plume water; Q = surface heat flux; SIM = sea ice melt; SP = spring inflow (warm and turbid); W = wind; WP = winter inflow (cold and clear)

the 2-m ramp, tides and storm surges may provide important modulating mechanisms for the leakage of freshwater from the river out under the floating ice, but almost nothing is known about how such processes might work in the Mackenzie Estuary. Throughout winter, the inflow accumulates under the floating landfast ice, slowly spreading along the shore and into the interior of the landfast zone [20, 29, 39]. Eventually at the end of winter, the estuary is manifested as a large lake under the landfast ice ($\sim 12\,000\,\text{km}^2$), estimated to contain $70\,\text{km}^3$ of freshwater, which is most of the Mackenzie's winter inflow [20]. A further $12\,\text{km}^3$ of the river water freezes into the landfast ice cover. Mixing in the estuary is severely limited because tidal currents and current shears are weak and the ice cover dampens air–sea interaction, with the result that a very sharp pycnocline (going from a salinity of < 5 to over 30 in a few centimeters) may be found at a depth of ~ 5–$10\,\text{m}$ for much of the nearshore region (Fig. 6, left-hand panels). Because the entire water column is at its freezing point, the strong gradient in salinity is paralleled by a strong temperature gradient (Fig. 6, right-hand panels) with the saline water below the pycnocline exhibiting temperatures down to $-1.5\,°\text{C}$. As heat diffuses faster than salt, these circumstances favor the formation of frazil ice at the interface between saline water and freshwater, and ice platelets formed in this manner would scavenge inorganic and biological particles out of the pycnocline and transport them to the bottom of the landfast ice.

Toward the end of winter, the horizonal gradient in the salinity of the fresh surface layer under the ice is relatively uniform out to the stamukhi, at about $0.3\,\text{km}^{-1}$. The brackish water or freshwater in the under-ice plume tends to be accompanied by only a small particle load. Macdonald et al. [20] estimated plume spreading rates to be 1–$2\,\text{cm s}^{-1}$ along the coast and perhaps $0.2\,\text{cm s}^{-1}$ perpendicular to the coast. Whether the stamukhi zone provides an inverted dam that acts as a barrier to surface flow in winter or whether the plume

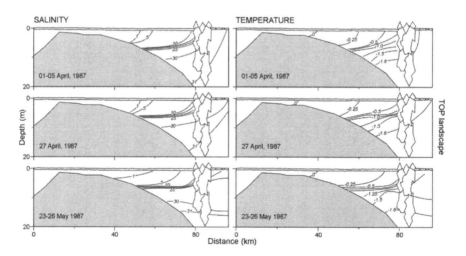

Fig. 6 The distribution of salinity (*left*) and temperature (*right*) along a section approximated by the A–B line in Fig. 4. During late winter (April–May), the strong salt stratification of water at the freezing point implies downward heat diffusion leading to frazil ice formation where temperature gradients are strong

spreading rates are simply too slow to reach the stamukhi before the end of winter, freshwater from winter inflow appears unable to reach the flaw lead over the central shelf during winter. Because there is no solar radiation at this time of year, nutrients supplied by the river or entrained from saline water beneath the spreading plume accumulate in the surface water under the landfast ice throughout winter.

The disposition of river inflow, landfast ice, stamukhi at the end of the landfast ice, and flaw lead beyond the stamukhi appears to be a common feature of many arctic shelves [40, 41]. Despite these similar features, differing river inflows, shelf volumes, shelf residence times, and even the presence or absence of stamukhi zones mean that the evolution of winter under the landfast ice will vary between estuaries. We presently lack comparative studies to assess how arctic estuaries vary between shelves.

3.2
Spring

Spring in the estuary is initiated by air temperatures rising above freezing and the Mackenzie River going into freshet toward the end of May (Fig. 5b; Fig. 2a). The peak flow closely follows the first movement of ice in the river (Fig. 3) and, because landfast ice and bottomfast ice are still in place when freshet arrives, the estuary becomes a very turbulent and turbid environment. Despite ample incident radiation, the ice cover and turbid water together limit primary production within the advancing plume of water associated with freshet. Sensible heat from the Mackenzie River advances the early meltout of coastal landfast ice by anywhere from two weeks to as much as a month or more [29, 36, 42, 43] and, accordingly, the large pulse of river water entering from freshet is augmented by river and ocean water released from the melting estuarine ice and by water stored in the under-ice lake. However, the chemical composition and particle load of the river water entering with the freshet differ from those of river water frozen in ice, because most solutes are excluded from the ice during the freezing process whereas particles may become entrained by the ice. Similarly, the water stored under the ice through winter has different properties (dissolved and particulate) based on the source of inflow during winter (mostly from the large headwater lakes and the eastern part of the Mackenzie Basin), and on processes that may have occurred over winter to alter the water's composition (diffusion of regenerated products from shallow sediments, flocculation and settling of particles, injection of brine from growing ice). We can assume that the rejection of solutes from the ice and their entry into stratified water in the estuary below the ice is, in the grand scheme, a conservative process because melted ice gets remixed into the estuary during the following summer, but the distillation during ice formation offers the possibility of altering the distribution of properties within the system. Nothing has been published on this aspect of arctic estuaries.

Observations suggest that all of the landfast ice accumulated in winter melts close to shore and reenters the estuary [28, 36]. Because landfast ice can trap or carry a considerable amount of sediment during fall freeze-up or spring flooding, the in situ melting will deposit much of this sediment back into the estuary. Landfast ice can thus be an effective means for redistributing and transporting sediments within the estuary. Heat supplied by the Mackenzie River together with an increased air temperature melt much of the sea ice both within the landfast zone and over the central shelf, adding an additional source of freshwater or brackish water to the shelf. The Mackenzie River's yield (volume of inflow divided by shelf area) is about 5 m compared to perhaps 2 m of brackish water (salinity ~ 5) for sea-ice melt. The distribution of these two sources of freshwater over the shelf differs vastly, with runoff entering from the land while ice melt is produced widely in the ocean [20, 44, 45]. This simple comparison suggests that care needs to be exercised in using salinity–property plots to evaluate conservative or nonconservative behavior in arctic estuaries.

3.3
Summer

Beginning in about late July or early August, summer is the season when the Mackenzie River flows out onto an open shelf, behaving much the same as one might expect for a large temperate river impinging on a wide, shallow shelf. The Mackenzie River inflow remains strong (Fig. 2) and is evident in the nearshore as turbid, warm plumes and fronts [46] (Fig. 5c) where the chemical composition of the surface water can be mapped on the plume structure [45]. The plume invades the nearshore and its distribution is very much affected by winds and the extent of open water [18, 19], as well as by the Coriolis force which tends to bend the plume toward the east along the Tuktoyaktuk Peninsula. Under favorable open-water conditions, the plume can extend hundreds of kilometers across the shelf and into the ocean interior [45, 47]. The Mackenzie River inflow remains fresh until it passes over the 2-m ramp (termed transverse bar by Carmack and Macdonald [17]), where it then produces a partially mixed estuary with surface brackish water entraining saltier bottom water with a strong pycnocline at 10–20 m (Fig. 5c, [17, 29]). Carmack and Macdonald [17] describe three oceanographic domains associated with the shelf in summer (inner, middle, and outer shelf), among which the estuary comprises the inner shelf. Seaward of the 2-m ramp, the inner shelf exhibits a fresh, 2–4-m turbid layer extending outward by about 5 to 10 km. Salinities go from 0 to 25 within this very turbid zone and thus many of the processess associated with non-conservative behavior will occur here, including flocculation and sedimentation. Due to the shallowness of this region, mixing by winds associated with storms can produce strong vertical mixing which enhances bottom resuspension; therefore, sediments within the 10-m isobath may provide only a temporary repository for riverine particulates and sedimented material [16]. Winds and ice cover shape the structure of the Mackenzie Plume and insert a great deal of intra- and interannual variability into the nearshore, as discussed in greater detail below.

3.4
Fall

By mid-September, the shelf has usually become completely clear of ice and the Mackenzie River inflow has weakened considerably (Fig. 2). Although a reduced plume or turbid estuary is maintained by the $10\,000\,\text{m}^3\,\text{s}^{-1}$ inflow, storms help to mix the freshwater components (runoff and ice melt) into the top 10–20 m of the water column and weaken frontal structures (Fig. 5d). Freshwater accumulated on the shelf during the summer may also be effectively exported by winds over open water; this period of freshwater removal "conditions" the salinity over the shelf with important consequences for the formation of dense water by brine rejection from ice production in the winter to come [20, 48]. Essentially, the fall period is setting up the ocean conditions upon which ice formation will operate and over which winter inflow will spread and mix. Intense fall storms from the east, more common during the 1960s, not only aid in the removal of fresh surface water but also promote shelf-edge upwelling [49], which potentially provides the source of nutrients entrained by the estuary in the following year. On the other hand, storms from the northwest, which were more common in the 1970s and 1990s, produce storm surges in the estuary of as much as 2.4 m (100 years) that inundate the low-gradient coastal regions and promote retreat of the coastline by rates typically over 1 m year^{-1} and in some places by over 10 m year^{-1} [17, 50].

As fall progresses, the ocean cools until about mid-October when sufficient heat has been lost to allow the formation of ice. Once the sea is cooled to its freezing point, storms over open water promote suspension freezing, an exceptionally effective means of transporting bottom sediments into ice [51]. Although this process is known to remove large quantities of sediment from the Laptev Shelf [52], almost nothing is known of its importance to the Mackenzie Estuary. In late fall, the Mackenzie inflow has declined to its winter low (Fig. 2) and as ice covers the sea, energy from the winds declines and winter progresses once again as an estuary defined by a plume spreading under the landfast ice.

4
The Role of Ice and Winds in Plume Spreading

As described above in sectional view (Figs. 5 and 6), the Mackenzie River discharge enters the estuary as buoyant plumes which then spread outward from the three main river mouths. Here we discuss variation and what controls it during the horizontal spreading of the Mackenzie Plume and its dispersion into the offshore. During winter, the landfast ice isolates the plume from winds, slowing the spreading and mixing and leading to a well-stratified upper layer (e.g., see [53, 54]). Likely due to the Coriolis effect, freshwater under the fast ice spreads eastward parallel to the coast (1–2 cm s^{-1}), with a much slower spreading rate into the interior of the landfast ice ($\sim$ 0.2 cm s^{-1}), although Mackenzie River water also spreads along the western boundary of Mackenzie Bay as far out as Herschel Island [20]. As winter progresses, the plumes emanating from the various river mouths eventually amalgamate into a large, slowly expanding lake under the ice $\sim$ 5–10 m deep extending from the coast to the stamukhi boundary [17]. The isolation from winds and the weak tides suggest that most of the under-ice plume spreading is forced by the buoyancy difference between the fresh, lighter river water in the estuary and the saline, denser shelf water farther offshore. Because the topography of the landfast ice is small (few ridges, small relief), winter plume spreading is likely predictable from year to year, perhaps with a smaller lake/plume produced in years with low winter inflow from the Mackenzie River.

When landfast ice cover melts away in summer, the spreading of the river plumes becomes considerably more variable and complex due to the winds, variation in the river discharge (Fig. 2), tides, and coastal circulation. The tides on the Mackenzie Shelf, which are semidiurnal with a range typically between 0.3 and 0.5 m, produce currents of only 2–3 cm s^{-1}, which are far weaker than those produced by other forces [31]. The impact of tides on plume spreading and vertical mixing, therefore, is relatively small and will not be further discussed.

In contrast to tides, winds are of paramount importance in this region on timescales of days to weeks, as emphasized by both observations and models (e.g., see [18, 55]). Since the Mackenzie Plume is confined to an Ekman layer, it responds quickly to wind stress and change in wind direction. Wind conditions on the Mackenzie Shelf are variable during the open-water season (June to October) with predominant winds from the northeast or southeast and from the northwest. The wind-forced coastal surface currents and the distribution of the river plume differ remarkably under these different wind regimes.

Winds from the northwest lead to the development of strong southeast currents offshore with a longshore current trapped to the coast (Fig. 7a). These same winds favor downwelling, piling turbid freshwater in the plume onto the coast and moving the freshwater eastward and off the shelf toward Amundsen Gulf [18, 19]. When an ambient longshore current is aligned with the coastal buoyant current, the plume thickens and tends to concentrate in a band closely attached to the coast. Such plumes are less susceptible to horizontal mixing [56, 57]. A SeaWiFS image from 27 August 2002 (Fig. 7b) shows the river plume during a period when winds had been mostly from the

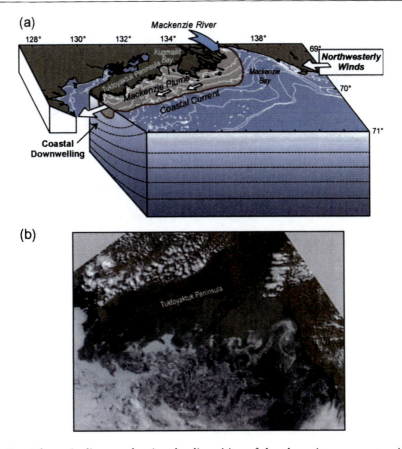

Fig. 7 **a** Schematic diagram showing the disposition of the plume in response to winds from the northwest. These winds favor downwelling and pile turbid freshwater onto the coast to the east of the river. **b** The disposition of Mackenzie Plume on 27 August 2002, during a period when winds were mostly from the southwest to northwest (courtesy SeaWiFS Project, NASA/Goddard Space Flight Center and ORBIMAGE)

southwest to northwest directions. Shelf water and eddies were forced toward the coast, causing the river plume to spread northeastward, confined mostly to the estuary.

Winds from the east, on the other hand, tend to push the surface water offshore under Ekman transport [17, 55], thus bringing higher-salinity shelf water up to the surface due to coastal upwelling (Fig. 8a). When these conditions occur, the longshore current can be retarded, causing the buoyant river plume to thin and spread offshore. Strong offshore transport like this has been modeled in a jetlike region within the plume under upwelling winds [58]. If the upwelling-favorable winds are particularly strong, the plume can become detached from the bottom with the consequence that wind can more easily mix the plume with ambient shelf water [57]. Persistent winds from the east in

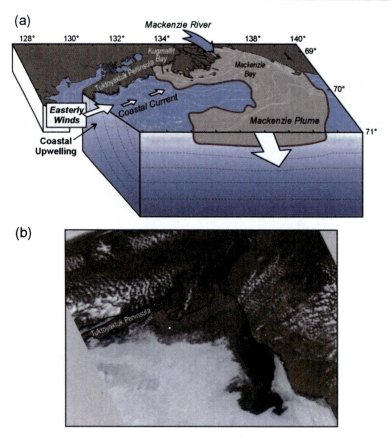

Fig. 8 a Schematic diagram showing the disposition of the Mackenzie Plume in response to winds from the east. These winds promote Ekman transport and coastal upwelling along the Tuktoyaktuk Peninsula and strong offshore transport of the plume to the northwest. **b** Persistent northeast to southeast winds in the Beaufort Sea during July 2001 produced this classical example of offshore spreading of the Mackenzie River (courtesy SeaWiFS Project, NASA/Goddard Space Flight Center and ORBIMAGE)

the Beaufort Sea during July 2001 produced a classical example of offshore spreading by the Mackenzie River (Fig. 8b). Starting about July 13 and lasting for at least a week, the plume extended northwestward from Mackenzie Bay and out past Herschel Island where eddies revealed instability along the plume front. Instabilities provide the means to move river water and suspended matter quickly offshore. According to Weingartner et al. [59] the currents in these meanders can be up to 50 cm s^{-1} with meander amplitudes of about 30 km. Thus, the cross-shelf flows associated with these instabilities can be vigorous even in the absence of any significant wind forcing. Instabilities and meanders are common on arctic shelves extending well downstream from river mouths.

A feature that distinguishes arctic estuaries from temperate or tropical estuaries is the role played by surface gravity waves. Because the wave field depends on open-water fetch, the Arctic can display two very different sediment transport and resuspension responses to a given wind condition depending on the ice cover. Large open-water areas with a fully developed wave field enhance vertical mixing in shallow water and resuspend sediments, whereas ice cover effectively dampens waves and sediment resuspension.

4.1
Ice Cover and the Transition from Winter to Summer

A group of SeaWiFS images has been assembled from four different years to illustrate how the Mackenzie Estuary typically progresses from ice cover in late winter to open water in late summer. The estuary is covered by snow and ice before June (Fig. 9a–c), but the positions of the intermittently open flaw lead and the landfast ice are easily identified in the sequence of images for 2003. As winds push the drifting ice pack to the northwest between mid-April and early June, the large interconnected lead system, called the Bathurst Polynya, can be seen opening across the entire shelf into Amundsen Gulf and northward along the Banks Island coast (not shown) for a total length exceeding 1200 km. The polynyas and flaw-lead system may open or close during this period, but the landfast ice remains firmly in place and there is usually no evidence of turbid Mackenzie River inflow invading the flaw lead.

As air temperatures rise above freezing in early June due to the increase in solar radiation [17], offshore pack ice gradually retreats northward and landfast ice starts to break up along the coast (Fig. 9d). At the same time, the freshet of the Mackenzie River surges into the estuary, rising rapidly to its annual maximum discharge. This sudden surge of river runoff imports a large amount of heat to the estuary [29], contributing to an intense meltout and breakup of the ice around the river mouths. The flooding during freshet normally starts during the first two weeks of June and lasts for about two weeks [36]. The large decrease in surface albedo that accompanies turbid Mackenzie River ice overflow enhances the absorption of the solar radiation and further contributes to deterioration of the landfast ice. Coastal ice can be eliminated within two weeks, depending on how widely spread the flooding is [43]. By late June (Fig. 9e) the Mackenzie Plume is able to spread freely over the estuary, depending especially on the winds that control the direction in which the plume spreads (Figs. 7 and 8) and the location of the ice edge. In years when the ice clears from the shelf in early summer, as it did in 1975, the plume can spread well out onto the shelf, whereas in years when the pack ice encroaches on the inshore, as it did in 1974, the plume remains confined in the nearshore [60].

During June and July, the landfast ice in the estuary gradually disappears from the coastline, bays, and inlets (Fig. 9d–f). By late July to perhaps early

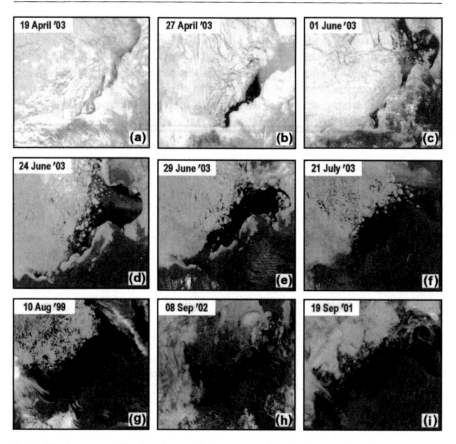

Fig. 9 The sequence of ice breakup and plume spreading in the Mackenzie Estuary as illustrated by series of SeaWiFS satellite images collected in 1999, 2001, and 2003. In April, the ocean is completely ice covered but a flaw lead over the middle shelf can open or close at the outer edge of the landfast ice (**a–c**). In early June (**d**) the estuary remains completely ice covered even though the Mackenzie River has gone into freshet (Fig. 2a). Through June to early July (**e–h**), ice withdraws over the middle shelf while landfast ice becomes melted out, assisted by sensible heat from the Mackenzie River and absorbed solar radiation due to turbid water. By late July, the landfast ice has melted and the Mackenzie River now spreads into the ocean as plumes from the three major river mouths (**i**) (Courtesy SeaWiFS Project, NASA/Goddard Space Flight Center and ORBIMAGE)

August, the estuary becomes mainly ice free, with the greatest open-water fetch usually observed in late September or early October (Fig. 9g–i). As noted above, winds dominate the disposition of the plume during this period and there is ample evidence of the plume's bimodal behavior (Figs. 7 and 8) within and between years (cf. Fig. 9e–i).

4.2
Freshwater Processing in the Estuary

Arctic estuaries have two sources of freshwater manipulation: river runoff and sea ice formation/melting [29]. The supply of buoyancy to arctic shelves and the interior ocean, and its export into the North Atlantic convecting sites, provides a sensitive pivot upon which climate change can operate [61, 62]. The manner in which shelves process freshwater is a crucial upstream component of global thermohaline circulation, considering that shelves can store freshwater and, through ice formation, produce exportable freshwater [29]. Because polar shelves produce ice in winter, especially over the mid-shelf flaw leads where ice diverges, the shelves also produce brine, which can either destabilize shelf surface water, producing a polar mixed layer (30–50 m), or produce convecting brines which are exported from the shelf after transiting the bottom [63, 64]. The ability of the Mackenzie Shelf to produce dense water in winter within the flaw lead despite the continued inflow of the Mackenzie River led Macdonald [29] to propose that in winter the Mackenzie Estuary exhibits both a positive and a negative estuarine behavior, and that frozen estuaries do not fit into standard classifications such as salt wedge, partially mixed, or well-mixed. Within the landfast ice, freshwater inventories and stratification increase throughout winter, producing an exceptionally well-defined salt wedge, whereas within the flaw lead salt inventories increase, producing a completely mixed water column (Fig. 5a; [44]). The distance between these two domains has been observed to be less than 500 m. In this scheme, the ice is far more than a passive, unevenly distributed barrier to air–sea exchange. Instead, it provides the crucial structure in the winter estuary that defines where stratification occurs and where mixing occurs for the entire inner half of the shelf. The separation of domains in the Mackenzie Estuary could represent a feature unique to shelves that form stamukhi zones. Other shelves that have a weak or absent stamukhi, as observed in the Laptev Sea [35], may exhibit different exchange rates in winter between water in the mid-shelf flaw lead and water under the landfast ice. Clearly, further studies comparing shelf behavior in the Arctic are warranted.

The manner in which winter inflow is stored behind the stamukhi zone under the fast ice and the consequences for producing convecting water over the middle shelf have been well described, as has the manner in which climate change might operate on this couple [29]. However, there are other consequences to biogeochemical properties which have not previously been discussed or studied. The convecting water and the wind-driven upwelling in the flaw lead provides an efficient way to mix surface- and deep-water components together, thus bringing up nutrients while distributing riverine components deeply (up to 50 m) into the water column. Thus, the stage is set over the middle shelf in winter for productivity in the following spring. Within the floating landfast ice zone, the slowly spreading plume invades the region as a moving, buoyant front. Buoyancy in the spreading plume is capable of capturing solutes excluded from the ice and transporting them toward the flaw lead, including nutrients which would be available in late spring to support ice algal growth or pelagic production at the outer edge of the landfast zone in spring. Although evidence for this process is clear [20, 29], nothing is known about how important it is to the distribution of primary production in the estuary.

5
The Biogeochemical Setting

5.1
Particulate Components

The enormous supply of organic and inorganic material by the Mackenzie River (e.g., [1, 65–68]) implies that these materials should dominate the adjacent shelf's geochemistry, especially for particulates and sediments, and this conclusion has been corroborated by numerous studies of water and sediment properties over the shelf [9, 10, 69–73]. However, there remains far less understanding of how particulate substances are processed within the estuary. Lisitzin [74] suggests that the bulk of riverine particulates is removed early in estuaries, and thus described the Arctic's inner shelves as a "marginal filter" past which most particles do not transport. The removal of particles would definitely be facilitated by the strong salinity gradients, observed in both winter and summer (Figs. 5 and 6), through flocculation due to increased ionic strength. The Mackenzie Estuary is no different—the annual particle delivery (124×10^6 t year^{-1}) divided by the inflow (330 km^3 year^{-1}) suggests an average suspended load of 375 mg l^{-1}. This value is far larger than any observed particulate concentrations in the estuary or offshore (e.g., see [69, 75–77]), suggesting losses of sediments and terrigenous organic carbon in the delta and nearshore [1]. Suspended particulate matter (SPM) and particulate organic carbon (POC) plotted against salinity for samples collected on cross-shelf sections (Fig. 10a, b) [69, 77] show high variability and nonconservative behavior reflected in curvature. Although these samples and those collected by others suggest large losses of particles from the inner shelf's water column, the situation is far more complicated than simple sedimentation fallout from the Mackenzie Plume. Resuspension of bottom sediments provides a second large, sporadic source of particles that originally derive for the most part from the Mackenzie River. Bottom turbid layers have frequently been observed over the shelf's nearshore [17]; these indicate an especially important particulate transport process in summer and fall, when storms from the northwest occur over a large fetch. Although water profiles (transmissometers, bottle samples) over the inner shelf reveal almost ubiquitous bottom turbidity, these layers are usually hidden

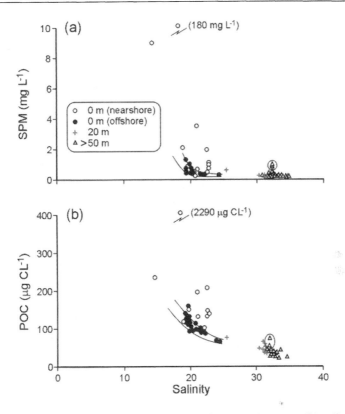

Fig. 10 Suspended particulate matter (**a**) and particulate organic matter (**b**) collected over the Beaufort Shelf in 1986 [77]. At high salinities (> 30), particulates and POC are generally low. Curvature in the plots, with much higher values observed in lower-salinity water, implies nonconservative loss of suspended sediment and organic carbon in the estuary

from satellites and it is likely their importance within the estuary is completely underestimated. The sediments underlying most of the estuary are, accordingly, in a dynamic state of accumulation, resuspension, and transport.

5.2
Dissolved Components

The Mackenzie River also delivers dissolved or colloidal inorganic and organic material [11, 12, 69, 78]. Telang et al. [66] suggest that dissolved organic carbon (DOC) supplied by the river ($\sim 1.3 \times 10^6$ t year^{-1}) is almost equivalent to the POC, but little is known of its chemical composition or whether much of it becomes metabolized or photolyzed. It has been assumed that most of it simply passes across the shelf conservatively [1]. Many chemical components of the DOC likely provide useful conservative tracers of terrigenous carbon and perhaps some of these tracers might be specific to the Mackenzie River (cf., [71, 79, 80]). In the case of inorganic dissolved chemicals, Ba clearly fingerprints Mackenzie River water far out into the Beaufort Sea [47, 81] and it is reported that alkalinity does likewise. These tracers are important adjuncts to stable isotopic composition (δ^{18}O) because they provide added specificity, with the potential to distinguish, for example, between river runoff and direct precipitation onto the ocean.

From the perspective of the estuary and shelf, dissolved nutrients (silicate, phosphate, and nitrate) provide great interest due to their potential importance for shelf primary production [60]. Silicate and phosphate, which have been measured over the shelf during a number of summer cruises [45, 60], exhibit mixing lines during late summer (Fig. 11a, b). As mentioned earlier, sea-ice melt, which has a low salinity ($\sim$ 5) and low silicate ($\sim$ 2), complicates mixing behavior over arctic shelves. In the cases shown here, the effect is relatively small in the nearshore during summer, probably due to rapid flushing of ice and ice melt out of the estuary. Silicate shows an enriched river water end-member mixing with saline shelf water that is silicate depleted, implying that the mixing line is produced predominantly by turbid, low-productivity, silicate-rich plumes in the nearshore mixing into the shelf's interior where clear water has had its silicate removed through spring and summer diatom blooms (e.g., [82]). In contrast to silicate, almost no dissolved phosphate is supplied by the river (Fig. 11b); indeed, the phosphate mixing line goes to zero at a salinity of about 12 [60]. This nonzero intercept implies that the Mackenzie Estuary removes reactive phosphate, possibly by reactions with iron, which might explain phosphorus enrichments in shelf sediments [70]. The only data set to include both winter and summer samples over the inner Mackenzie shelf (Fig. 11) surprisingly reveals two mixing lines for silicate (Fig. 11a), one for winter (empty circles) and one for summer (filled circles). The summer mixing line implies a riverine silicate concentration of about 50–60 mmol m^{-3}, similar to values established previously in the river [45, 60]. Water below the euphotic zone over the middle shelf derives from the Pacific, which contains relatively high silicate concentrations (up to 35 mmol m^{-3}; filled circles on the right-hand side of Fig. 11a). In winter, the saline end-member suggests that resupply of silicate to the estuarine water spreading under the ice occurs partly through entrainment of this nutrient-rich Pacific water (salinity 32–33, silicate 15–25 mmol m^{-3}). However, the positive offset between summer and winter at low salinities reveals an additional silicate source within the estuary. The most obvious mechanism to supply the extra silicate to the plume spreading under the ice is the rejection of dissolved silicate from the accumulating ice. Accordingly, the clear water under the ice collects much of its nutrient inventory from river water that has been frozen, leaving the landfast ice nutrient-impoverished. Phosphate (Fig. 11b) shows a similar offset in the mixing line but with less evidence of phosphate exclusion from the ice, which is reasonable considering that river water con-

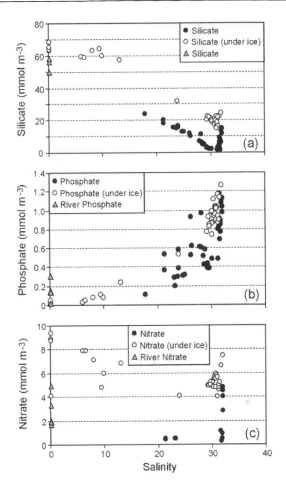

Fig. 11 Plots of silicate (**a**), phosphate (**b**), and nitrate (**c**) for samples collected at the end of winter prior to freshet (April–June) and the following summer (August–September) in 1987 along a transect approximately following A–B marked in Fig. 4. In the silicate plot there are clearly two apparent mixing lines, one each for winter and summer. In summer, mixing between silicate-depleted shelf surface water and silicate-rich Mackenzie River water produces the lower line (*filled circles*). The upper line (*empty circles*) is produced by mixing nutrient-enriched shelf water (upwelled and mixed over the outer shelf) and Mackenzie River water with the added, nonconservative input of silicate by rejection from ice growing in the estuary. The phosphate distribution is a product of similar processes except that the Mackenzie River carries almost no dissolved phosphate and may even provide the means to remove phosphate when mixed with salt water

tains little phosphate. The nitrate data are more equivocal simply due to fewer data points (Fig. 11c), but the evidence seems to show that nitrate does accumulate within the plume during winter (4–8 mmol m^{-3}), partly through entrainment and partly through rejection from the ice. These data and literature previously cited suggest that the Mackenzie River is an important source of silicate and nitrate to the estuary, but that it may actually lead to a removal of phosphate from the sea. In the case of nitrogen and phosphorus, dissolved organic forms are likely to be as important as inorganic forms but there are no data with which to evaluate this notion. In summary, Fig. 11 implies that the nonconservative behavior of nutrients associated with the rejection of ionic material from ice is a distinguishing feature of arctic estuaries.

5.3
Primary Production in the Estuary

The Mackenzie Estuary is an important component of shelf productivity in terms of accumulating and releasing nutrients, but much of that productivity may occur beyond the estuary and out over the shelf [82]. In winter there is insufficient light everywhere for phytoplankton productivity. As spring progresses, the landfast ice and snow cover limit light such that a small amount of production may occur on the bottom of the ice as ice algae [83], but pelagic production cannot occur to any great degree [82]. With freshet and for much of the spring and summer, the landfast ice disappears but is replaced by a turbid plume which is equally—or more—effective at stalling primary production through light limitation. Nutrient budgets, ^{14}C measurements, and modeling all imply that the limiting nutrient, nitrate, becomes depleted for much of the shelf by the end of summer suggesting that the estuarine nutrients get used before surface waters are exported to the interior ocean and, indeed, the strong supply of silicate from the river likely promotes diatom growth farther offshore [82, 84]. Although autotrophic production may be limited by light in much of the Mackenzie Estuary, there is an ample supply of terrigenous organic carbon to support heterotrophic production. Parsons and coworkers [84, 85] suggest that the Mackenzie Shelf exhibits two plankton communities, one community near the river mouth characterized by bacteria and amphipods and a second community offshore characterized by phytoplankton, copepods, hydromedusae, and ctenophores. Furthermore, stable isotope composition within the foodweb (δ^{13}C) led these authors to propose two parallel food chains, one offshore with an autotrophic base and a second in the estuary with a heterotrophic base. The strength and distribution of heterotrophy would be controlled by the plume dynamics forced by winds and the warmth of the inflowing water from the Mackenzie River.

6
The Mackenzie Estuary and Change

As a component of the cryosphere, the Arctic is sensitive to change and it is likely that such change will feed back into the global climate system either through the radiation balance (albedo), the ice cover (changes in thickness or volume), or the thermohaline circulation (freshwater processing) [86–88]. Change will be felt first, and probably most severely, in the Arctic's marginal seas [25]. For the Mackenzie Estuary, we project two primary components of change, the first in the hydrological cycle and the second in the ice climate. Within Canada, the Mackenzie Basin has experienced the greatest warming over the past 100 years [89] but there is as yet little evidence that temperature rise has been accompanied by any trend in the mean annual flow of the Mackenzie River, at least since about 1972 when reliable data were collected [90, 91]. Given the large interannual variability in mean flow ($\pm25\%$) [1], data will have to be collected for perhaps several more decades before significant trends will emerge [2]. Even if the mean flow does not vary, the timing of breakup, freshet, and freeze-up offer considerable leverage for change within the estuary. While data collected during the past two decades do not as yet reveal significant trends (Fig. 3), the future will likely witness more extreme flood events than are already observed (Fig. 2a). Extreme events would certainly impact the estuary in terms of connectivity with no-closure lakes [14] and with sediment deposition within channels and delta lakes [92]. Furthermore, the loss of permafrost, especially in the discontinuous zone of the Mackenzie Basin [25, 93], will alter hydrological coupling between basin and river, with consequences for water, sediment, organic carbon, and nutrient transport [94, 95]. Perhaps equally important are the possible consequences of damming for power or irrigation, something that has already occurred in the Liard system (Williston Lake Reservoir). Diversion of water will decrease the mean flow entering the Mackenzie Estuary, with ancillary effects on sediment transport and an alteration of the freshwater balance over the Mackenzie Shelf, with possible impact on ice cover and the effects of brine production. Upstream storage of water for power would increase winter inflow, enlarge the freshwater storage under the winter ice cover, and thereby reduce the capacity of the shelf to produce dense water by brine rejection in winter [29, 55].

The second component of change, the ice climate, will have dramatic consequences for estuaries like the Mackenzie's which have low topographical relief and poorly bonded soil. Continued sea-level rise by perhaps 0.5–1 m over the next century [25] together with larger open-water fetch extending later in fall mean a vigorous assault on the deltaic coastline. This will provide an added source of sediment and turbidity to the nearshore, a rearrangement of the estuary's coastline—breaching of some lakes in the delta—and almost certainly critical impacts on humans, fish, belugas, and other animals that use the estuary. As noted by Reimnitz [28], a 10–20-km-wide 2-m ramp is unique to arctic deltas. If, as seems likely, these ramps depend on ice cover and especially bottomfast ice, then a change in ice cover will put the estuary's offshore out of dynamic balance. As a consequence, bottom erosion will occur to produce a more normal gradient with the outer edge of the 2-m ramp being deepened to perhaps 10 m. Another possible change could occur in the average position of the stamukhi zone due to changes in landfast ice thickness and seaward extension. Under a warmer climate, for example, a shorter ice growing season would lead to a retreat of the stamukhi toward the coast leaving a smaller volume to contain the Mackenzie's winter inflow.

Finally, the clearance of ice across the shelf/basin boundary will likely alter the intensity of shelf-edge upwelling particularly in late summer–early fall (see, for example, [96]). More intensive nutrient regeneration would then occur over the middle shelf which, by enhancing the nutrient content of the winter saline end member (see Fig. 11), would increase late-winter nutrient inventories in the estuary during spring, promoting greater spring productivity.

7
Future Studies

Clearly the threats of climate change and the fact that the Mackenzie Basin has experienced exceptional warming suggest that we need to continue to collect time series on the Mackenzie River's hydrology (flow, ice cover, biogeochemistry) and augment these with better records of breakup, ice cover, channel disposition, and coastlines—which can be assisted by satellite imagery. As should be clear from this review, for its size and importance the Mackenzie Estuary is very much understudied and further efforts to determine biogeochemical cycling are warranted. Some of these studies should focus especially on the interaction between sediments and water in a seasonally ice-covered environment, and the sources and mechanisms of storage of organic carbon.

An ice-covered estuary with low winter flow and small water volumes between sediment and ice also offers unique opportunities to determine process rates at the regional scale. For example, measuring the ingrowth of methane (or CO_2) into the estuary during winter might be an elegant way to estimate methane production (or metabolism) in sediments [97]. Likewise, denitrification is likely to be an exceptionally important process in the Arctic Ocean due to the enormous shelves [98] and yet we have only a few spot measurements [99]. Could the ingrowth of N_2 under the landfast ice in winter provide a valid new estimate? Winter ice cover and freezing punctuated by a large supply of organic carbon in summer imply large, but as yet unknown, variances in sediment or water redox conditions in winter. The Mackenzie Estuary, with a wide range in temperature, solar radiation, and ice cover, offers a laboratory to study organic carbon metabolism and elemental cycling under extremes—processes that will be crucial to project the Arctic's response to changes likely to occur within this century.

Acknowledgements We thank Tom Weingartner and Mike Schmidt for providing the Sea-WiFS satellite images and Patricia Kimber for producing the drawings. Tom Weingartner contributed numerous helpful suggestions to improve an earlier version of the paper. This work was supported by the Department of Fisheries and Oceans Strategic Science Fund (RWM) and the National Foundation through Grant OPP-0229473 (YU). We are grateful to numerous colleagues, many of whom are cited here, who have spent hard-fought time collecting data and who have produced the foundation for writing this paper.

References

1. Macdonald RW, Solomon SM, Cranston RE, Welch HE, Yunker MB, Gobeil C (1998) Mar Geol 144:255
2. Carson MM, Jasper JN, Conly FM (1998) Arctic 51:116
3. Stein R, Macdonald RW (eds) (2003) The organic carbon cycle in the Arctic Ocean. Springer, Berlin Heidelberg New York
4. Milliman JD, Meade RH (1983) J Geol 91:1
5. Brunskill GJ (1986) In: Davies BR, Walker KF (eds) The ecology of river systems. Junk, Dordrecht, p 435
6. Milliman JD, Syvitski JPM (1992) J Geol 100:525
7. Walker HJ (1998) J Coastal Res 14:718
8. Yunker MB, Backus SM, Graf Pannatier E, Jeffries DS, Macdonald RW (2002) Estuar Coastal Shelf Sci 55:1
9. Yunker MB, Macdonald RW, Veltkamp DJ, Cretney WJ (1995) Mar Chem 49:1
10. Yunker MB, Macdonald RW (1995) Arctic 48:118
11. Millot R, Gaillardet J, Dupré B, Allègre CJ (2003) Geochim Cosmochim Acta 67:1305
12. Reeder SW, Hitchon B, Levinson AA (1972) Geochim Cosmochim Acta 36:825
13. Dallimore A, Schroder-Adams CJ, Dallimore SR (2000) J Paleolimnol 23:261
14. Burn CR (1995) Can J Earth Sci 32:926
15. Solomon S, Mudie PJ, Cranston R, Hamilton T, Thibaudeau SA, Collins ES (2000) Int J Earth Sci 89:503
16. Hill PR, Nadeau OC (1989) J Sediment Petrol 59:455
17. Carmack EC, Macdonald RW (2002) Arctic 55:29
18. MacNeill MR, Garrett JF (1975) Department of the Environment, Beaufort Sea Technical Report #17. Sidney, BC, p 113
19. Giovando LF, Herlinveaux RH (1981) Pacific Marine Science Report 81–4. Institute of Ocean Sciences, Sidney, BC p 198
20. Macdonald RW, Paton DW, Carmack EC, Omstedt A (1995) J Geophys Res 100:895
21. North/South Consultants (2002) Freshwater Institute, Winnipeg, MB, p 76
22. Stirling I (1990) Sea mammals and oil: confronting the risks. Academic, San Diego, p 223
23. Dietrich J, Dixon J (2000) Abstract from GeoCanada 2000 http://www.bmmda.nt.ca/background.htm
24. Manabe S, Spelman MM, Stouffer RJ (1992) J Clim 5:105
25. Macdonald RW, Harner T, Fyfe J, Loeng H, Weingartner T (2003) In: Macdonald RW (ed) AMAP assessment 2002. Arctic Monitoring and Assessment Programme, Oslo, p 65
26. SEARCH (2001) Polar Science Center, Applied Physics Laboratory, University of Washington, Seattle, WA, p 91

27. Héquette A, Ruz M-H, Hill PR (1995) J Coastal Res 11:494
28. Reimnitz E (2002) Polarforschung 70:123
29. Macdonald RW (2000) In: Lewis EL (ed) The freshwater budget of the Arctic Ocean. NATO, ASI Series, p 383
30. Cooper PF Jr (1974) In: Reed JC, Sater JE, Gunn WW (eds) Proceedings of a symposium on Beaufort Sea coast and shelf research. Arctic Institute of North America, Arlington p 235
31. Huggett WS, Woodward MJ, Stephenson F, Hermiston FV, Douglas A (1975) Department of the Environment, Beaufort Sea Technical Report #16, Victoria, p 38
32. Reimnitz E (1996) The plumbing system of an Arctic river delta and related sedimentary processes, Proposal for research in the Laptev Sea; unpublished manuscript
33. Dickens Associates Ltd (1987) Bedford Institute of Oceanography. Atlantic Geoscience Centre, Dartmouth, NS, p 90
34. Blasco SM, Shearer JM, Myers R (1998) 13th international symposium on Okhotsk Sea & sea ice. Proceedings of ice scour and arctic marine pipelines workshop, Mombetsu, Hokkaido, p 53
35. Reimnitz E, Toimil L, Barnes P (1978) Mar Geol 28:179
36. Dean KG, Stringer WJ, Ahlnäs K, Searcy C, Weingartner T (1994) Polar Res 13:83
37. Woo M-K, Thorne R (2003) Arctic 56:328
38. Krouse HR, Mackay JR (1971) Can J Earth Sci 8:1107
39. Macdonald RW, Carmack EC (1991) Atmos Ocean 29:37
40. Reimnitz E, Dethleff D, Nürnberg D (1994) Mar Geol 119:215
41. Rigor I, Colony R (1997) Sci Total Environ 202:89
42. Antonov VS (1978) Polar Geogr 2:223
43. Searcy C, Dean K, Stringer W (1996) J Geophys Res 101:8885
44. Macdonald RW, Carmack EC, Paton DW (1999) Mar Chem 65:3
45. Macdonald RW, Carmack EC, McLaughlin FA, Iseki K, Macdonald DM, O'Brien MO (1989) J Geophys Res 94:18
46. Carmack EC, Macdonald RW, Papadakis JE (1989) J Geophys Res 94:18
47. Macdonald RW, Carmack EC, McLaughlin FA, Falkner KK, Swift JH (1999) Geophys Res Lett 26:2223
48. Melling H (1996) In: Lemke P, Anderson L, Barry R, Vuglinsky V (eds) Proceedings of the ACSYS conference on the dynamics of the arctic climate system, vol WCRP-94, WMO/TD No 760, Goteborg, p 78
49. Kulikov EA, Carmack EC, Macdonald RW (1998) J Geophys Res 103:12725
50. Harper JR, Henry RF, Stewart GG (1988) Arctic 41:48
51. Lange MA, Pfirman SL (1998) In: Lepparanta M (ed) Physics of ice-covered seas, vol 2. Helsinki University Printing House, Helsinki, p 651
52. Eicken H, Kolatschek J, Lindemann F, Dmitrenko I, Freitag J, Kassens H (2000) Geophys Res Lett 27:1919
53. Ingram RG, Larouche P (1987) J Geophys Res 92:9541
54. Ingram RG, Larouche P (1987) Atmos Ocean 25:242
55. Omstedt A, Carmack EC, Macdonald RW (1994) J Geophys Res 99:10011
56. O'Donnell J (1990) J Phys Oceanogr 20:551
57. Münchow A, Garvine RW (1993) J Mar Res 51:293
58. Kourafalou VH, Oey L-Y, Wang JD, Lee TN (1996) J Geophys Res 101:3415
59. Weingartner TJ, Danielson S, Sasaki Y, Pavlov V, Kulakov M (1999) J Geophys Res 104:29697
60. Macdonald RW, Wong CS, Erickson PE (1987) J Geophys Res 92:2939
61. Aagaard K, Carmack EC (1989) J Geophys Res 94:14

62. Rahmsdorf S (2002) Nature 419:207
63. Melling H, Lewis EL (1982) Deep Sea Res 29:967
64. Melling H (1993) Cont Shelf Res 13:1123
65. Thomas DJ, Macdonald RW, Cornford AB (1986) Rapports et proces-verbaux des reunion conseil international pour l'exploration de la mer 186:165
66. Telang SA, Pocklington R, Naidu AS, Romankevich EA, Gitleson II, Gladyshev MI (1991) In: Degens ET, Kempe S, Richey JE (eds) Biogeochemistry of major world rivers. Wiley, New York, p 75
67. Yunker MB, Macdonald RW, Fowler BR, Cretney WJ, Dallimore SR, McLaughlin FA (1991) Geochim Cosmochim Acta 55:255
68. Peake E, Baker BL, Hodgson GW (1972) Geochim Cosmochim Acta 36:867
69. Whitehouse BG, Macdonald RW, Iseki K, Yunker MB, McLaughlin FA (1989) Mar Chem 26:371
70. Ruttenberg K, Goñi MA (1996) Mar Geol 139:123
71. Goñi M, Yunker MB, Macdonald RW, Eglinton TI (2000) Mar Chem 71:23
72. Macdonald RW, Naidu AS, Yunker MB, Gobeil C (2003) In: Stein R, Macdonald RW (eds) The organic carbon cycle in the Arctic Ocean, Chapt 7.2. Springer, Berlin Heidelberg New York, p 177
73. Macdonald RW, Thomas DJ (1991) Cont Shelf Res 11:843
74. Lisitzin AP (1995) Oceanology 34:671
75. Matsumoto E, Wong CS (1977) J Oceanogr Soc Japan 33:227
76. Bornhold BD (1975) Department of the Environment, Victoria, BC, p 23
77. Iseki K, Macdonald RW, Carmack EC (1987) NIPR symposium on polar biology, vol 1. National Institute of Polar Research, Tokyo p 35
78. Yunker MB, Cretney WJ, Fowler BR, Macdonald RW, McLaughlin FA, Whitehouse BG (1991) Org Geochem 17:301
79. Lara RJ, Rachold V, Kattner G, Hubberten HW, Guggenberger G, Skoog A, Thomas DN (1998) Mar Chem 59:301
80. Opsahl S, Benner R, Amon RMW (1999) Limnol Oceanogr 44:2017
81. Falkner KK, Macdonald RW, Carmack EC, Weingartner T (1994) The polar oceans and their role in shaping the global environment, Geophysical Monograph, vol 85. American Geophysical Union, Washington, DC p 63
82. Carmack EC, Macdonald RW, Jasper S (2004) Mar Ecol Prog Ser 277:37
83. Horner R, Schrader GC (1982) Arctic 35:485
84. Parsons TR, Webb DG, Dovey H, Haigh R, Lawrence M, Hopky G (1988) Polar Biol 8:235
85. Parsons TR, Webb DG, Rokeby BE, Lawrence M, Hopky GE, Chiperzak DB (1989) Polar Biol 9:261
86. Perovich DK, Grenfell TC, Richter-Menge JA, Light B, Tucker WB, Eicken H (2003) J Geophys Res 108(C3) doi: 10.1029/2001JC001079
87. Rothrock DA, Yu Y, Maykut GA (1999) Geophys Res Lett 26:3469
88. Macdonald RW, McLaughlin FA, Carmack EC (2002) Deep Sea Res I 49:1769
89. Gullett DW, Skinner WR (1992) Atmospheric Environment Service, Environment Canada
90. Stewart RE, Burford JE, Crawford RW (2000) Cont Atmos Phys 9:103
91. Stewart RE (2000) In: Lewis EL, Jones EP, Lemke P, Prowse TD, Wadhams P (eds) The freshwater budget of the Arctic Ocean. Kluwer, Dordrecht, p 367
92. Graf Pannatier E (1997) Sediment accumulation and historical deposition of trace metals and trace organic compounds in the Mackenzie Delta (Northwest Territories, Canada). Terre & Environment, vol 10. PhD thesis, Université de Genève

93. Vörösmarty CJ, Hinzman LD, Peterson BJ, Bromwich DH, Hamilton LC, Morison J, Romanovsky VE, Sturm M, Webb RS (2001) Arctic Research Consortium of the US, Fairbanks, AK, p 84
94. Schindler DW (1997) Hydrol Process 11:1043
95. Benner R, Benitez-Nelson B, Kaiser K, Amon RMW (2004) Geophys Res Lett 31:L05305
96. Carmack E, Chapman DC (2003) Geophys Res Lett 30:1778
97. Kvenvolden KA, Lilley MD, Lorenson TD (1993) Geophys Res Lett 20:2459
98. Chen C-TA, Liu KK, Macdonald RW (2002) In: Fasham MJR (ed) Ocean biogeochemistry: a JGOFS synthesis. Springer, Berlin Heidelberg New York, p 53
99. Devol AH, Codispoti LA, Christensen JP (1997) Cont Shelf Res 17:1029

Hdb Env Chem Vol. 5, Part H (2006): 121–147
DOI 10.1007/698_5_023
© Springer-Verlag Berlin Heidelberg 2005
Published online: 8 November 2005

Biogeochemistry and Chemical Contamination in the St. Lawrence Estuary

C. Gobeil

INRS-ETE, Université du Québec, 490, de la Couronne, Québec, QC, G1K 9A9, Canada
charles_gobeil@ete.inrs.ca

Abstract The many contrasting environments one finds in the St. Lawrence estuary make it attractive for the study of estuarine biogeochemical processes. The landward part of the estuary is relatively shallow with a partially mixed poorly productive water column and high turbidity. The seaward portion of the estuary is deep with a permanently stratified water column. It is substantially more productive and the turbidity is low. In recent years, research on the St. Lawrence estuary has focussed on quantification of sedimentation by means of radionuclides; speciation and reactivity of trace metals in the water column; sources and composition of organic matter; sediment redox chemistry and early diagenesis of trace elements; contamination by persistent organochlorine pollutants; and stable lead isotope ratios as tracers of anthropogenic inputs.

Keywords Biogeochemistry · Contaminants · Sediment diagenesis · St. Lawrence estuary · Water column

1
Introduction

The estuary of the St. Lawrence River is the largest in northern America and perhaps the largest in the world. Downstream from the Great Lakes, the 400 km long, 20–50 km wide and 350 m deep St. Lawrence estuary (SLE) receives more than 1% of all the freshwater discharged to the world ocean [1]. Located in eastern Canada at the lower limit of the subarctic region, it ices over for 2 months every year, and navigation is virtually brought to a halt. The St. Lawrence estuary is an important biotope, conspicuous for the hundreds of whales that migrate every summer from the northern Atlantic to feed on krill. Like many other coastal areas in the world, the SLE has not escaped the heavy pressures of industrialization and urbanization.

Most studies of the chemistry of the SLE have been carried out during the last 30 years, although some surveys of dissolved oxygen and nutrients did take place as early as the 1930s [2]. The objective of these studies was to identify the biogeochemical processes involved in controlling the distribution of chemical constituents within the water column and at the sediment–water interface. So far, the pertinent peer-reviewed literature totals more than 50 titles; this chapter aims at assessing the present state of knowledge with emphasis on findings not reviewed in previous syntheses [1, 3, 4]. Recent studies of the burden of persistent organic pollutants and contaminant metals in the St. Lawrence estuary will also be reviewed.

2
Environmental Setting

The SLE is customarily viewed as the upper estuary, located between Ile d'Orleans and the mouth of the Saguenay Fjord, and the lower estuary, extending from there to Pointe-des-Monts (Fig. 1). Each region exhibits its own characteristics in terms of hydrography, turbidity, biological productivity and sedimentation. With a surface area of 3500 km^2, the upper estuary occupies one quarter of the total area. The depth is mostly less than 30 m, except in two basins along the north shore, downstream from Ile aux Coudres and further to the east, where the depth is greater than 50 m and 120 m, respectively. Topographically distinct, the morphology of the lower estuary is dominated by the Laurentian Trough, which occupies about half of the area and reaches 300–350 m depth. About 1200 km long overall, the Laurentian Trough stretches from the mouth of the Saguenay Fjord through the Gulf of St. Lawrence and the Cabot Strait to the edge of the continental shelf. In the lower estuary, the trough is flanked on both its sides by shelves that are relatively shallow and no wider than 15 km.

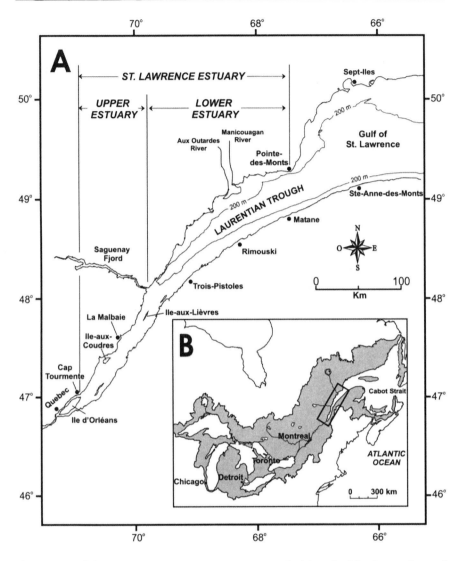

Fig. 1 Maps of the St. Lawrence estuary in eastern Canada (**a**), and of the Great Lakes and St. Lawrence drainage basin (**b**)

2.1
Hydrography

The discharge of the St. Lawrence River varies between 10 and 25×10^3 m^3 s^{-1} with an annual average of 13×10^3 m^3 s^{-1} [5,6]. The river is the only significant source of freshwater to the upper estuary. In periods of spring tides and when the run-off is low in winter, brackish waters may penetrate as far landward as Quebec City [7]. At other times, however, salt intrusion stops just downstream of Ile d'Orleans (Fig. 2). Strong salinity gradients characterize the region between Ile d'Orleans and Ile aux Coudres. Due to the strength of tidal currents, the water column is well mixed. This is also true along the south shore where many small channels and shallow plateaus may be found. However, along the north shore, the water column becomes more and more stratified as one moves seaward, and one finds salinities above 30 in the deepest parts of the upper estuary downstream of Ile aux Coudres [8].

Besides the discharge of the St. Lawrence River, the lower estuary receives the freshwater of the Saguenay, the Manicouagan, and the Aux Outardes rivers. The accumulated discharge of these rivers is roughly 20% of the St. Lawrence discharge [9]. The water column in the lower estuary is stratified. In winter, a surface layer of relatively cold and fresh water lies on top of a warmer and saltier layer. With the onset of spring, however, the surface layer warms up faster than the water below it, and a three-layer system gradually develops. The surface layer is relatively warm and fresh, the intermediate layer is cold, and the deep layer is both warmer and saltier than the intermediate layer [10] (Fig. 3). Between May and September, the thickness of the cold intermediate layer (defined as having a temperature of less than 3 °C)

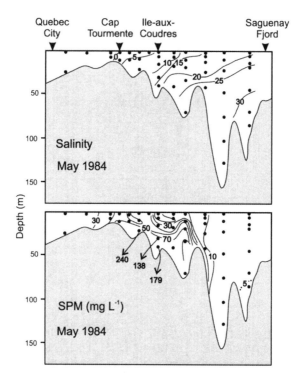

Fig. 2 Spatial distribution of salinity and suspended particulate matter in the upper estuary in May 1984 (modified from Takayanagi and Gobeil [18])

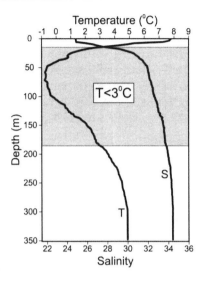

Fig. 3 Vertical profiles of temperature and salinity in the water column at a station off Rimouski in June 1992, showing the cold intermediate layer ($T < 3\,°C$)

averages 134 m in the lower estuary [11]. The circulation in the lower estuary is characterized by a seaward surface flow, counterbalanced by a deep landward transport of denser water from the Atlantic Ocean [12]. The approximate mean velocity of the deep-water flow between the Cabot Strait and the head of the Laurentian Trough is $0.5\ \mathrm{cm\ s^{-1}}$, which implies a transfer time of the order of 8 years between those two locations [13]. Driven by internal waves of great amplitude, upwelling of intermediate and deep waters enriched with nutrients occurs at the head of the Laurentian Trough [14, 15]. During its voyage from the Cabot Strait to the head of the Laurentian Trough, the deep water does not mix much, if at all, with the surface water.

2.2
Turbidity

A zone of maximum turbidity is a prominent feature of the upper estuary [16]. The 100 km long turbidity zone extends from near the salt intrusion limit to La Malbaie and exhibits concentrations of suspended particulate matter ranging between 20 and several hundred $\mathrm{mg\ L^{-1}}$ [17, 18] (Fig. 2). Such high turbidity is maintained by a complex density-driven circulation combined with a resuspension of the bottom sediments near the head of the estuary, which may be intensified by breaking internal waves [17]. Considerable quantities of sediments (5×10^5 t) deposit on the intertidal flats and marshes near Cap Tourmente in summer. This sediment is eroded in the fall with the unconscious aid of the Greater Snow Geese. These migrating birds destroy the vegetation cover during their stay over in the area. This cyclical deposition and erosion of sediments may exert a strong influence on distribution and storage time of suspended particles in the upper estuary [19]. By contrast, the concentrations of particulate matter are much lower in the lower estuary [20]. The suspended matter concentration is at the most $3\,mg\,L^{-1}$ in the 50 m thick surface layer, $0.4\,mg\,L^{-1}$ in a 50 m thick bottom nepheloid layer, and 0.05–$0.1\,mg\,L^{-1}$ in the rest of the water column.

2.3
Primary Production

The instability and the high turbidity of the water column in the upper estuary limit the production of phytoplankton. In the area of maximum turbidity, primary production is virtually insignificant [21]. In the lower estuary, the average yearly primary production of carbon is of the order of $100\,g\,m^{-2}\,year^{-1}$ with very high peaks in summer [22]. The most productive region of the lower estuary ($\sim 130\,g\,m^{-2}\,year^{-1}$) is in the central part because of the stabilizing influence of the freshwater plume that flows from the Manicouagan and Aux Outardes rivers. The less productive region lies ($\sim 30\,g\,m^{-2}\,year^{-1}$) along the south shore, where brackish waters arrive from the upper estuary and the Saguenay Fjord and turbidity is high. The levels of phytoplankton production in the SLE are considered to be low compared to other coastal water bodies at similar latitudes. The reason for this is primarily the brevity of the productive season [22].

2.4
Sedimentation

Analysis of the bottom sediments in the upper estuary revealed the presence of mostly sand and gravel, indicating an almost complete absence of net sedimentation [17, 23]. Less than 10% of the total surface area of the upper estuary could be covered with fine sediments deposited during the present sediment regime [24]. The suspended particulate material carried by the St. Lawrence River, which amounts to $6.5 \times 10^6\,t\,year^{-1}$ [25], is transported into the lower estuary and gulf of the St. Lawrence and is deposited in the Laurentian Trough, the only important sedimentation basin downstream from the Great Lakes. The sedimentation rate, determined by applying a two-layer biodiffusion model to the vertical distribution of ^{210}Pb ($t_{1/2} = 22.3$ years) in seven sediment cores from the lower estuary, decreases seaward along the Laurentian Trough from 0.47 to $0.12\,g\,cm^{-2}\,year^{-1}$ (Fig. 4) [26]. That assessment concords well with previous measurements obtained with a similar method on a smaller number of cores [27] and with measurements performed many times over one year with a sediment trap left drifting at a depth of 150 m near Rimouski [28]. Radiocarbon dates of the shells of benthic mol-

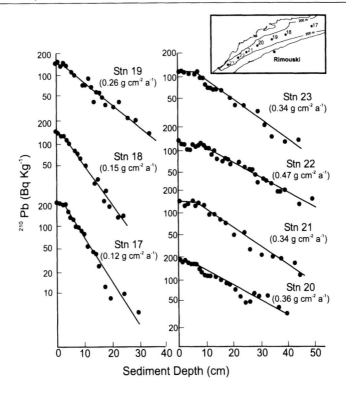

Fig. 4 Vertical profiles of excess ^{210}Pb in Laurentian Trough sediments in the lower estuary. Sedimentation rates given in parentheses were estimated through the application of a two-layer biodiffusion model to the profiles (modified from Smith and Schafer [26])

luscs found in two long piston cores give similar results [29]. According to these rates, the total accumulation of sediments reaches 10.9×10^6 t year^{-1} in the 5000 km^2 depositional area in the lower estuary that is delineated by the 200 m isobath. This is superior to the volume of sediment injected annually into the estuary by the St. Lawrence River. Other sources of sediments are therefore required to balance the annual sediment budget in the SLE. It has been suggested that biogenic material produced at the head of the lower estuary, as well as material resuspended along the walls of the Laurentian Trough and on the plateaus adjacent to it could contribute to the sediments accumulating in the trough [26].

3
Water Column Chemistry

The existing database on the chemical composition of the SLE waters suffers from a shortage of winter data and long time series data. Nonetheless, the available data show that biological activity, as well as constraints imposed by the estuarine circulation, influence the water composition in the estuary with regard to dissolved oxygen, nutrients and organic carbon.

3.1
Dissolved Oxygen

The concentration of dissolved oxygen in the lower estuary portion of the Laurentian Trough decreases with depth. Within the deep waters the oxygen concentration is lower than in the deep waters of the Gulf of St. Lawrence [30]. These trends are due to the consumption of oxygen by respiration and by remineralization of organic matter, and to the permanent pycnocline that effectively isolates the deep waters from the atmosphere. As the deep waters move landward from Cabot Strait to the head of the Laurentian Trough, dissolved oxygen is gradually depleted. The time it takes for the bottom water to traverse the trough is 5–10 years [13]. While the concentrations of dissolved oxygen in deep waters are as high as 185 μM in the Cabot Strait [31], they are currently less than 65 μM in the lowermost 50 m of the water column throughout an area of about 1000 km^2 in the central lower estuary [32]. Consumption of dissolved oxygen is not limited to the water column. Oxygen uptake rates by the sediment, estimated from microelectrodes profiles, are of the order of 0.25 μmol cm^{-2} day^{-1} [33]. Values of about 0.7 μmol cm^{-2} day^{-1} were measured by core incubation [34]. Over a 5 year period (corresponding to the minimum water transit time from Cabot Strait to the head of the trough), a flux value of 0.25 μmol cm^{-2} day^{-1} translates into an oxygen consumption by the sediments that is the equivalent to 25% of the total dissolved oxygen in a 100 m thick water column containing 185 μM, as in Cabot Strait.

Although we have a relatively good grasp of the spatial distribution of dissolved oxygen in the lower estuary, our knowledge of its temporal evolution is fragmentary. However, the deep water oxygen concentrations currently measured are lower than in the 1970s [32]. That decline may be attributed, among other factors, to an increase of the load of nutrients originating in the tributary basin, to a change in the flux of terrigenous organic matter, and to a change in the composition of oceanic waters entering the Laurentian Trough [32]. Considering that dissolved oxygen concentrations of less than 65 μM are considered to be inadequate to sustain biodiversity [35], research is urgently required to better understand the factors that are likely to impact on dissolved oxygen in the SLE deep waters.

3.2
Nutrients

As one might expect, the distributions of nitrate, phosphate and silicate in the lower estuary are negatively correlated with the distribution of dissolved oxygen. Nutrient concentrations increase with depth as a result of the mineralization of organic matter settling from surface waters [36]. Thus, in June 1992, the nitrate, phosphate and silicate concentrations in the lower estuary deep waters respectively reached 23 µM, 2.2 µM and 58 µM, which is 2–3 times more than in the surface waters (Fig. 5). The nutrient concentrations in the deep water of the Laurentian Trough increase during the voyage between Cabot Strait and the lower estuary [31]. However, the deep water nitrogen to phosphorus ratio (N : P) of about 10 : 1 deviates significantly from the classical 16 : 1 Redfield ratio observed in the oceanic environment [36]. It has been hypothesized that the deviation might reflect either an incomplete regeneration of nitrate or depletion of phosphorus in mineralized organic matter [30]. However, biogeochemical processes at the sediment–water interface such as denitrification can also influence the N : P ratio in bottom waters. A sharp decrease in nitrate in sediment porewater within 1 cm of the sediment–water interface shows that denitrification in Laurentian Trough sediments is important (Fig. 6). Sediment core incubation experiments indicate that nitrate is taken up by sediments at a rate of $0.01–0.05$ µmol cm^{-2} day^{-1} [34]. As above, a flux of 0.01 µmol cm^{-2} day^{-1} translates into a nitrate consumption over a 5 year period by the sediments of about 10% of the nitrate in a 100 m water column containing 17 µM nitrate (as in the deep waters in Cabot Strait [31]).

In the upper estuary, where productivity is low, mixing processes appear to control the distribution of nutrients in the water column. Thus, silicate and phosphate concentrations have occasionally been reported as being linearly

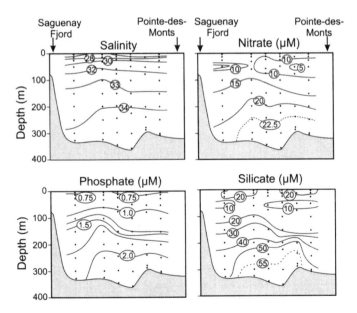

Fig. 5 Spatial distribution of salinity and nutrients in lower estuary in June 1992

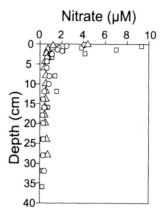

Fig. 6 Vertical distribution of nitrate in the sediment porewater in three cores collected at the same station (325 m) off Rimouski in May 1985

related to salinity [19, 30]. Such linear relationships, however, are not always observed, possibly because of the time-varying composition of the St. Lawrence River water. For example, silicate concentration at the mouth of the St. Lawrence River was 35 μM in May, 74.2 μM in September, 74.5 μM in May 1986 and 13 μM in June 1986 [18, 30]. Based on higher nitrate concentrations in waters with salinity between 5 and 20 than in higher salinity waters and in St. Lawrence River waters, it was suggested that there is internal production of dissolved nitrate in the upper estuary [30]. However, a source has not been identified. Considering the presence of ammonia in concentrations of 200–1000 μM in the interstitial waters of a sediment near Ile aux Coudres (Tremblay GH, personal communication) the high nitrate concentrations in the water column might be attributable to oxidation of ammonia released from the sediments during periodic resuspension/resedimentation events likely to occur in that region [37]. However, lack of data for the nutrient inputs into the St. Lawrence limits our understanding of nutrient dynamics in the upper estuary.

3.3
Organic Carbon

The yearly variations of particulate organic carbon (POC) in the lower estuary have previously been summarized [38]. Surface water POC concentrations reach their maximum of 200–400 mg m^{-3} during periods of intense biological activity, but they do not exceed 40 mg m^{-3} during the winter. Stable carbon isotope ratios (δ^{13}C) of suspended particulate matter in surface waters of the lower estuary also show a clear seasonal pattern indicating a strong terrigenous character in spring and a dominant contribution of autochthonous marine carbon in summer [39, 40]. On the other hand, the particulate organic carbon in deep waters does not vary much within an annual cycle and is significantly lower and has C : N ratios higher than the shallow water counterparts. This suggests a greater proportion of terrigenous organic matter in deep than in shallow waters.

We do not know of any database on dissolved organic carbon (DOC) in the SLE waters, but data have been obtained using the modified high-temperature catalytic oxidation method [41] for the Laurentian Trough in the Gulf of St. Lawrence [31]. Those results showed that, as for nutrients, DOC increases landward along the Laurentian Trough in the intermediate and deep waters, from 34 μM at Cabot Strait to 75 μM at a site located about 100 km seaward of the lower estuary. It was suggested that refractory forms of DOC accumulate in the Laurentian Trough deep waters and concluded that the DOC is a dynamic component of the carbon cycle within that system. The latter conclusion should also apply to the SLE where, according to the observed trend, one should expect higher levels of DOC than in the Gulf.

3.4
Trace Metals

The bulk of trace metal studies in the water column date from before the 1990s [42, 43]. However, new results on mercury and aluminium have recently been published.

Mercury concentrations in the water column of the lower St. Lawrence estuary range from 1.8 to 7.8 pM, with the highest values occurring in surface water [44]. Those values are very close to values in other northern Atlantic coastal waters such as the North Sea and the Baltic Sea. They are also close to Sargasso Sea values. Some 20% of total Hg in the water column belongs to easily reducible Hg forms, notably elementary gaseous mercury. Methylmercury, the main form of Hg found in fish muscles, was undetectable (< 0.2 pM).

The reactivity of dissolved aluminium during estuarine mixing in the upper estuary was also examined [18]. Dissolved Al concentrations vary significantly over time at the mouth of the St. Lawrence River, but decrease exponentially as salinity increases in the upper estuary. This implies removal of dissolved Al during the estuarine mixing (Fig. 7). In May 1984, an almost complete removal of dissolved Al was observed in water with salinity up to 10, which is found in the most turbid region of the estuary. The variations in dissolved Al in the St Lawrence River water were attributed to changing proportions of Great Lakes water, whose drainage basin is chiefly made up of sedimentary rock, and of water from tributaries that drain the igneous and metamorphic rocks of the Canadian Shield. The two most important tributaries are the Ottawa and the St. Maurice rivers. The discharge of these tributaries varies widely over the year, amounting to as much as 60% of the total St. Lawrence discharge in spring, but as little as 10% at other times [45, 46].

In order to better understand the geochemical cycle of Al in the upper estuary, laboratory experiments were conducted to estimate removal

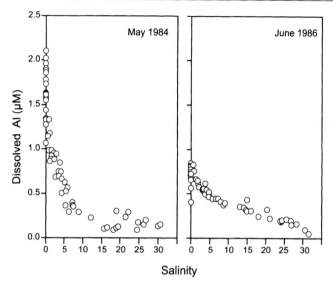

Fig. 7 Dissolved aluminium as a function of salinity in May 1984 and June 1986 in the upper estuary (modified from Takayanagi and Gobeil [18])

due to flocculation induced by increased ionic strength and adsorption on suspended particulate matter [18]. The results were that: (i) flocculation was an important removal mechanisms in May 1984 but not in June 1986, which is most likely caused by differences in the speciation of dissolved Al between the two sampling periods, and (ii) adsorption on suspended solid was a key mechanism at both periods. The formation of an authigenic aluminosilicate phase may also constitute a mechanism for removal of dissolved Al.

4
Sediment Chemistry

The Laurentian Trough is particularly well suited for studying the early diagenesis at the sediment–water interface because of the stability of its bottom conditions due to the great depth [47]. Most recent investigations have been focussing on the organic matter, and the redox reactions and elemental cycles, at the vicinity of the sediment–water interface.

4.1
Source and Composition of Sedimentary Organic Matter

The organic matter content in deep lower estuary sediments varies between 1.3 and 2.4%, decreasing progressively from the sediment surface to a depth of 40 cm [48]. Many observations point to the terrigenous origin of a sizable portion of the organic carbon preserved in the sediments. First, the C : N weight ratio is high ($\sim$ 10–12). Ratios of 13–15 are indicative of a terrigenous origin, whereas ratios of about 6 indicate a marine origin. Second, the stable carbon isotope ratio of the surface sediment is – 22.4‰ on average [40]. This has been interpreted as evidence that about 60% of total organic carbon is allochthonous, the isotopic signatures of the terrigenous and marine carbons being respectively of – 25.6 $\pm$ 0.6‰ and – 20.0 $\pm$ 0.3‰ in the SLE [39, 49]. Third, the Laurentian Trough sediments in the lower estuary contain between 0.1 and 0.5 mg g^{-1} of lignin. This is one of the most abundant biopolymers in vascular plants, representing about 3% of the total organic carbon in the surface sediments [50, 51]. Studies on lignin lead to the conclusion that industrial discharges of solid organic waste from the pulp and paper industry have significantly raised the inputs of terrigenous organic matter following the expansion of that industry in and after the 1920s [51].

The variations with depth of the main biochemical components of organic matter were determined on sediment cores collected at two lower estuary sites in the Laurentian Trough. Similar analyses were performed on settling particles sampled with a sediment trap placed at about mid-depth in the water column at the same sites [48, 52]. In the sediments, carbohydrates are the predominant constituent with 15–22% of the total organic carbon; hydrolysable amino acids, lipids and labile proteins come next with 7–13%, 1–5% and 0.3–1%, respectively. The non-characterized fraction represents 62–74%. In the settling particles (whose total organic carbon content varies between 2.6 and 6.7%), lipids predominate (17–37%), followed by carbohydrates (7.9–16%), hydrolysable amino acids (8.4–16%), and labile proteins (0.3–2.6%). This leaves a non-characterized fraction of 40–64%, likely consisting of both refractory organic matter and humic compounds.

The above findings made it possible to apply three approaches to determine the relative reactivity of the main biochemical components of the sedimentary organic matter during early diagenesis [48]:

1. The average concentration in the samples from the sediment traps were compared to those in the surface sediments (0–3 cm)
2. The annual fluxes of organic components in the water column based on the sediment trap results were compared to the inventories of each category of components in the surface sediments (0–3 cm) normalized to represent one year
3. The rates of loss of biochemical components over the whole length of the cores (35 cm)were examined

Regardless of the approach, the result was invariable: lipids appear as the dominant substrate near the sediment–water interface and carbohydrates and amino acids are the main source of energy in the deeper sediments.

4.2
Sediment Redox Reactions

Measurements with a gold amalgam voltammetric microelectrode have proved helpful in advancing our knowledge of the redox reactions involved in the microbial degradation of sedimentary organic matter [34]. The use of the voltammetric microelectrode allows the simultaneous determination, with a spatial resolution on the order of a millimetre, of O_2, Mn(II), Fe(II), HS(-I), and I(-I). Examples of profiles obtained with that electrode in a sediment core from the Laurentian Trough are given in Fig. 8. They show that oxygen decreases markedly below the sediment–water interface and becomes undetectable at a depth of 4 mm. They also reveal the occurrence of detectable levels of dissolved manganese (Mn(II)) and dissolved iron (Fe(II)), below depths of 14 mm and 45 mm, respectively, in the sediments, which bear witness to the reduction of manganese and iron oxyhydroxydes. Even more interesting, they point to the existence of a sharp peak of maximum I(-I) (iodide) concentration just at the depth where upward diffusing Mn(II) was removed, several mm below the depth where oxygen became undetectable.

To explain the above findings, the production in the interstitial waters of iodide, as a result of oxidation of Mn(II) by iodate (IO_3^-), was proposed. For a sharp peak of iodide to occur, iodide must be removed both from above and below. The produced iodide is then re-oxidized into iodate in the surface oxic layer and into I_2 in the deeper sediments, possibly through a reaction with nitrate. The produced I_2 is ultimately trapped through a reaction with organic matter. It has been demonstrated that the calculated production rate of io-

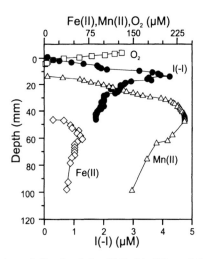

Fig. 8 Vertical distribution of dissolved O_2, I(-I), Mn(II), and Fe(II) in sediment pore-water at a station off Rimouski (325 m). The profiles were acquired with a solid-state gold/mercury amalgam microelectrode (modified from Anschutz et al. [34])

dide is high enough to account for the oxidation of all of the upward diffusing Mn(II) by IO_3^- [34]. Thus, the high spatial resolution profiles obtained with gold amalgam voltammetric microelectrode revealed the existence of a redox reaction, which is seldom envisaged in studies on diagenesis. Other reactions rarely envisaged may occur in the Laurentian Trough sediments:

1. The production of N_2 as a result of the reaction between ammonia and manganese oxides in presence of oxygen
2. The production of nitrate as a result of a reaction between ammonia and manganese oxides in presence of oxygen
3. The production of iodine as a result of a reaction between nitrate and iodide [34]

In bioturbated sediments such as in the Laurentian Trough, the relative importance of the above reactions varies over time implying that the steady state condition is never perfectly reached.

4.3
Sulfur Chemistry

The Laurentian Trough sediments in the lower estuary exhibit a greater abundance of iron monosulfides than pyrite (Fe_2S) [53]. This is in contrast to what is generally observed in coastal sediments, notably in those of the Laurentian Trough in the Gulf of St. Lawrence. One particularly interesting aspect of pyrite formation in coastal sediments is that pyrite represents a significant and stable sink for a number of trace elements, including arsenic and molybdenum [54–56]. Assuming that the conversion of monosulfide precursor phases to pyrite requires the formation of elemental sulfur or polysulfides [57], it was hypothesized that the factor causing the non-conversion of iron monosulfides in pyrite was the large excess of reactive iron in the lower estuary sediments, which inhibits the build-up of reduced sulfur in the porewater as a result of metal sulfide precipitation.

4.4
Phosphorus and Trace Metal Diagenesis

The available data on phosphorus in the sediments as well as in the interstitial waters in the Laurentian Trough have thrown light on the cycle of that element in coastal environments [58, 59]. It has been demonstrated that the surface sediments are enriched in phosphorus and that the dissolved phosphate concentrations in the interstitial waters do not vary much ($6 \pm 3 \mu M$) down to a depth of 5–15 cm, and then increase with depth up to values over $50 \mu M$ at a depth of 20 cm (Fig. 9). In the water overlying the sediments, the phosphate concentrations are of the order of $2 \mu M$. On the basis of the previous observations, it was proposed that an important portion of the organic phosphorus deposited at the sediment–water interface in the Laurentian Trough gets mineralized in the oxic surficial layer and the phosphate released is partitioned between the porewater and the adsorption sites of the solid phase, in particular those sites on the iron oxides. Part of the adsorbed phosphorus gets desorbed to replace the phosphorus escaping to the water column. Most of it, however, is remobilized during the reduction of iron oxides deeper in the sediments. The phosphorus that is not remobilized probably accumulates with stable minerals like apatite [58, 60]. By comparing the phosphorus concentration in sediment trap material, on the one hand, and in the deep sediments ($\sim$ 40 cm depth), on the other, it was concluded that about half of the phosphorus deposited at the sediment–water interface is recycled in the water column.

The diagenesis of mercury in the Laurentian Trough sediments has also been the subject of several studies. The sediment interstitial water contains up

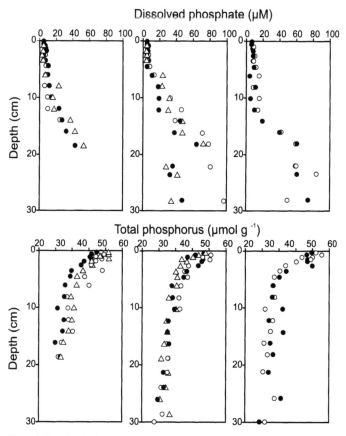

Fig. 9 Profiles of dissolved phosphate in the porewater and total phosphorus in replicate sediment cores taken in August 1984 (*left*), May 1985 (*centre*) and July 1985 (*right*) at the same station in the centre of lower estuary (modified from Sundby et al. [58])

to 20 times more Hg than the overlying bottom water [61, 62]. There is a coincidental increase in Hg and Fe into the porewater below the depth of oxygen penetration, which suggests remobilization of Hg as result of the reduction of iron oxides consequent to anaerobic oxidation of organic matter. However, diagenetic redistribution of remobilized Hg within the sediment column subsequent to deposition seems insufficient to induce important variations with depth of the Hg concentration in the solid phase. Finally, methylmercury accounts for a significant part of dissolved Hg in the anoxic sediments but is below detection limit in porewater recovered from the surficial oxic layers [63]. The oxic surficial sediments seem therefore to act as a barrier to the diffusion of Hg from the sediments to the overlying water column.

Research has very recently focussed on the early diagenesis of redox sensitive trace metals such as rhenium, molybdenum, uranium and cadmium in the Laurentian Trough sediments [56, 64] (Fig. 10). These elements deserve attention because they are potential tracers of paleo-redox conditions in the marine environment, although their diagenesis is as yet incompletely understood. The distributions of Re, U, Cd and Mo in marine sediments were found to be consistent with a mechanism whereby those elements diffuse downward and precipitate under reducing conditions. The Laurentian Trough results

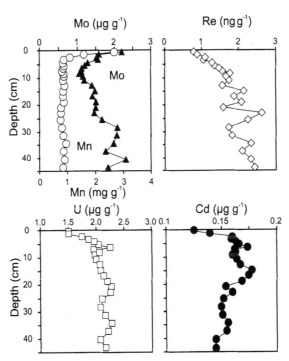

Fig. 10 Examples of manganese, molybdenum, rhenium, cadmium, and uranium profiles in the Laurentian Trough sediments (modified from Sundby et al. [56])

suggest that the kinetics of the reactions causing Cd, Re, U, and Mo to pre-cipitate in the sediments are slow comparatively to the diffusion of the metals in the interstitial water [56] and that they influence the accumulation rates and distribution of those elements. Furthermore, the Cd and Re accumulation rates are not sensitive to variations in sedimentation rates and sulfide con-centrations. The accumulation rates of U may depend on the abundance of reactive organic carbon, which controls the kinetics of the microbially medi-ated U reduction. The Mo accumulation rate, part of which is co-precipitated with pyrite, increases with the sulfide concentration.

5
Anthropogenic Pressure

The St. Lawrence estuary receives runoff from the Great Lakes and St. Law-rence River drainage basin, home to 35 million people. Concern about the impacts of human activity on the SLE, especially as regards the influx of con-taminants, is often voiced by the general public and the scientific community. High concentrations of organochlorine compounds and mercury in the tissue of top predators in the marine food chain, as well as elevated inventories of contaminants in the Laurentian Trough sediments, amply justify these appre-hensions.

5.1
Organochlorine Pollutants

An important database is now available on a wide range of organochlorines in both biota and sediment cores from the SLE. Organochlorines are toxic liposoluble compounds, very resistant to biotransformation and tending to biomagnification through the food web. Quite ubiquitous, it is usually as-sumed that they have no natural sources.

Polychlorinated biphenyls (PCBs) and several other organochlorine con-taminants in the blubber of beluga whales (*Delphinapterus Leucas*) from the SLE have repeatedly been analysed over the past two decades [65–69]. The samples were taken on carcasses found on the shores of the SLE, where a population of beluga whales is permanently established. The measured concentrations of many organochlorines are elevated. For example, ΣPCB concentrations were 25 times higher in animals from the SLE than from the Canadian Arctic. Among the 209 PCB congeners, the three with a planar con-figuration, also called non-*ortho*-substituted PCBs, which are known to be the most toxic, are easily detected with modern analytical tools. The most abun-dant was CB126; its concentration ranged from 600 to 11 800 ng Kg^{-1} in the ten samples that were analysed [66]. The concentrations of pesticides such as DDTs, chlordanes and mirex were also considered high relative to other belugas in Canadian waters. Polychlorinated dibenzo-*p*-dioxins and dibenzo-furans (PCDD/Fs) were also detected in the SLE beluga blubber at the low level of ng Kg^{-1}.

Other SLE organisms were investigated for their content in organochlorines. The levels of PCBs and of several pesticides, including DDT- and chlordane-related compounds, in the harbour seal (the other marine mammal permanently residing in the estuary) are higher than in seal species that come only seasonally to the estuary [70]. American eels, which migrate through the St. Lawrence estuary, are highly contaminated with PCBs, mirex and pesticides compared to eels of the same species from an uncontaminated reference tributary [71]. They may constitute an important source of contaminants to the beluga whales [72]. PCBs and organochlorine pesticides have been detected in muscle and liver of the Atlantic cod, American plaice and Greenland halibut although at levels well below the critical thresholds for human consumption [73]. Finally, low PCDD/F concentrations of ng Kg^{-1} were measured in the snow crab, northern shrimp, and in the gastropod whelk in the SLE [74]. In all cases the tissue concentrations were below international guidelines for fishery products.

The distribution of organochlorine contaminants in sediment cores from the lower estuary portion of the Laurentian Trough has also been determined. Sedimentary PCB concentrations reach at the most 30–50 µg Kg^{-1} [75, 76]. As for the three planar congeners mentioned above, their relative abundance is different in the sediments than in the beluga blubbers; congener CB77 exhibits the highest concentration, 600 ng Kg^{-1} at the most [77]. The highest sedimentary concentrations of PCDD/Fs, and mirex, reach a few hundred ng Kg^{-1} [78, 79]. The total burdens in the lower estuary sediments below 200 m were reported as 550 Kg for dioxins and furans and 184 Kg for mirex [79, 80]. Besides, sedimentary records of most organochlorines in the Laurentian Trough exhibit their maximum values at a few cm below surface, which suggests a recent decrease in the inputs, including PCBs, DDT-related compounds, mirex, as well as PCDD/Fs [75–79]. That temporal trend is consistent with a decline in PCBs and DDT-related compounds in male belugas, eels, harp seals and seabirds [68]. However, not all organochlorines present in the SLE beluga whale blubber show signs of decrease.

The relative proportions of the various congeners of dioxins and furans in the lower estuary surface sediments closely recall those observed in the sediments from a remote lake where the atmosphere is the sole source of contamination [80]. However, the presence of highly chlorinated furan congeners in the deeper sediment layers has been ascribed to sources located in the Great Lakes region [80]. Mass balance calculations also suggested that about 20% of the 2700 Kg discharges of mirex into Lake Ontario prior to 1976 have finally deposited in the SLE portion of the Laurentian Trough sediments [79]. Indeed, while many contaminants, including highly hydrophobic organic pollutants originating from the heartland of the northern American continent, are retained in the sediments of the Great Lakes, a portion is exported to the SLE after transiting through the St. Lawrence River.

5.2
Contaminant Metals

The presence of excess trace metals in the SLE is well documented for at least three elements: mercury, silver and lead.

The discovery, in the 1970s, at the head of the lower estuary portion of the Laurentian Trough, of Hg concentrations in surface sediments ten times superior to the natural level for the area was the first hard evidence of chemical contamination [81]. Later, Hg concentrations in the liver, kidney and muscle of beluga whales were found to be 2–4 times higher in the SLE than in the Canadian Arctic and pointed to the widespread contamination of the food chain [82]. The vertical distribution of Hg in sediment and sediment porewater from the trough were also established [26, 61]. From there, the total quantity of anthropogenic Hg deposited in the lower estuary portion of the trough since the beginning of industrialization was shown to exceed 100 t, that is about five times the natural accumulation of Hg for the same period. The inputs of anthropogenic Hg into the estuary have considerably diminished in recent decades. That decrease was attributed to implementation of government regulations in the early 1970s regarding the reduction of industrial Hg discharges from chlor-alkali plants and to the subsequent 1976 closure of such a plant on the Saguenay River. However encouraging the regression of Hg inputs into the SLE, the sediments are likely to remain a durable source of bioavailable Hg as demonstrated by the positive correlation between the Hg concentrations in the epibenthic northern shrimps and in the sediments [44].

Anthropogenic silver, also, is present. In sediment cores from the Laurentian Trough portion of the lower estuary, the Ag content increases from values close to crustal abundance ($0.07 \, \mu g \, g^{-1}$) at the bottom of the cores, to values 2–4 times the background at a depth of 5–15 cm, and then decreases by 10–25% at the surface [83]. The above background sedimentary Ag concentrations in the estuary – as well as in the St. Lawrence River – were ascribed to the widespread dispersion of anthropogenic Ag. The burden of anthropogenic Ag was estimated as about 43 t in the lower estuary sediments. As in other coastal environments [84, 85], the discharges of urban effluents have been identified as the most likely pathway for the introduction of anthropogenic Ag into the St. Lawrence system. The recent reduction of Ag deposition observed in the sediments is consistent with the introduction in the early 1980s of wastewater treatment facilities in the city of Montreal and in some other 300 riverine municipalities.

The presence of anthropogenic lead in the SLE has been demonstrated using stable Pb isotope ratios in sediments [86]. Geological systems differ according to their age and primeval characteristics. Furthermore, among the four stable lead isotopes (^{204}Pb, ^{206}Pb, ^{207}Pb and ^{208}Pb), the three heaviest are produced by radioactive decay of uranium (^{238}U $\rightarrow$ ^{206}Pb, ^{235}U $\rightarrow$ ^{207}Pb) and thorium (^{232}Th $\rightarrow$ ^{208}Pb). The interplay of all such factors in any system gives rise to an isotopic signature, i.e. the stable lead isotope ratios (^{206}Pb : ^{207}Pb, ^{206}Pb : ^{208}Pb, ^{206}Pb : ^{204}Pb) corresponding to that system [87]. Such signatures may then be used to discriminate the sources of the Pb found in the environment and to trace its transport mechanisms at the regional and global scales [88].

Lead concentrations in the Laurentian Trough sediments are negatively correlated with the ^{206}Pb : ^{207}Pb ratio at a station located in the centre of the lower estuary (Fig. 11). While the sedimentary Pb concentrations increase from a constant background, below 50 cm depth, to a maximum value, at 8 cm, and then diminish toward the sediment surface, the ^{206}Pb : ^{207}Pb ratio decreases from a constant value, below 50 cm, to a minimum value, at 8 cm, and then increases toward the sediment surface. The variations in the lead isotopic composition as a function of depth in the cores demonstrate the existence of at least one Pb source that had been "inactive" in pre-industrial time and which is most likely anthropogenic. If we assume a sedimentation rate of a few mm year^{-1} [26], the sedimentary record shown on Fig. 11 becomes consistent with the chronology of environmental Pb contamination over the last century. The diminishing Pb concentrations in the shallowest portion of the core may then be attributed to the phasing out of leaded gasoline in north-

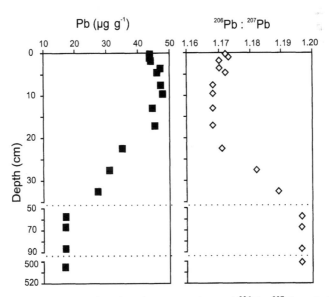

Fig. 11 Vertical distribution of total lead concentration and ^{206}Pb : ^{207}Pb ratio in the sediments collected at 325 m in the Laurentian Trough off Rimouski (modified from Gobeil et al. [86])

142

C. Gobeil

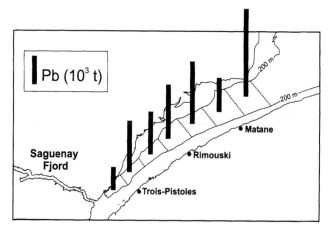

Fig. 12 Total burden of anthropogenic lead in sediments below 200 m in various sections of the Laurentian Trough in the lower estuary

ern America. This is also supported by the fact that the ^{206}Pb : ^{207}Pb ratio (1.151 ± 0.005) of the anthropogenic lead in the Laurentian Trough sediments, as determined through modelling [86], is in the middle of the range of atmospheric lead (^{206}Pb : ^{207}Pb ~ 1.141–1.159) reported in Canadian urban areas during the 1980s [89]. That atmospheric lead was ascribed mainly to the combustion of Canadian gasoline containing Pb from Bathurst, New Brunswick, which has a ^{206}Pb : ^{207}Pb ratio of 1.163 [90].

As is the case for Hg and Ag, the burden of anthropogenic Pb in the lower estuary portion of the Laurentian Trough was assessed by extrapolating the results obtained for sediment cores collected along the trough to the whole area below the 200 m isobath. The inventory of anthropogenic Pb in each core has been calculated with a sediment density of $2.65 \, \text{g cm}^{-3}$, using measured porosity values and assuming that all Pb in excess of $16 \, \mu\text{g g}^{-1}$ is anthropogenic. The total amount of industrial Pb deposited in the estuary since the onset of industrialization is thus estimated to be 13 000 t (Fig. 12), which amounts to an average of 140 t annually since the beginning of the last century. Considering that such an input represents 0.1% of the global discharge of anthropogenic Pb to the aquatic realm [91], the Laurentian Trough in the SLE constitutes a significant basin for anthropogenic metal deposition.

6
Summary and Concluding Remarks

While there is little or no accumulation of sediments in the upper estuary, sediments do accumulate in the lower estuary. The average yearly accumulation of sediments ($10.9 \times 10^6 \, \text{t year}^{-1}$) in the area delineated by the 200 m isobath in the lower estuary, an area of almost $5000 \, km^2$, exceeds the quantity of suspended solid injected annually into the estuary by the St. Lawrence River ($6.5 \times 10^6 \, t \, year^{-1}$). The other sources of sediments that are required to balance the annual sediment mass budget in the SLE have not been clearly identified.

Recent surveys indicate that dissolved oxygen concentrations, currently less than 65 µM in the bottom water in an area covering $1000 \, km^2$ in the central region of the lower estuary, are lower than recorded prior to the 1990s. The factors, whether anthropogenic or natural, affecting dissolved oxygen concentrations in the lower St. Lawrence estuary deep waters should be further investigated.

The spatial distribution of nutrients in the lower estuary is governed by regeneration of nutrients during the mineralization of the organic matter settling from the surface waters. Although not fully understood, the deviation of the N : P ratio in the deep waters, from the classical Redfield ratio in the ocean, may be closely linked to sediment diagenesis. Our as yet insufficient knowledge of how the riverine nutrient input varies, interferes with the understanding of nutrient dynamics in the upper estuary.

Recent efforts to better understand trace metal chemistry in the SLE water column have been limited to mercury and aluminium. The mercury concentrations (1.8–7.8 pM) are very close to those observed in other northern Atlantic coastal waters; 20% of the total mercury belongs to easily reducible mercury forms. Salt-induced flocculation and adsorption onto suspended particulate matter in the zone of maximum turbidity are important mechanisms governing the distribution of dissolved aluminium in the upper estuary.

A considerable fraction of the organic matter in the lower estuary sediments is terrigenous and in part anthropogenic. Analysis of the sediments and particles collected with a sediment trap indicates that lipids are the predominant substrate near the sediment–water interface while carbohydrate and amino acids are the main source of energy in the anoxic sediments.

Sediment diagenesis has been extensively investigated in the Laurentian Trough in the lower estuary. Redox reactions not frequently investigated have been brought to light by means of measurements with a gold amalgam voltammetric microelectrode at millimetre spatial resolution. The conversion of iron monosulfides to pyrite is inhibited by a large excess of reactive iron in the sediments.

The flux of phosphate to the overlying water is regulated by adsorption-desorption equilibria at the surface of the oxic surface sediments. Furthermore, the accumulation rates of rhenium, molybdenum, uranium and cadmium in the Laurentian Trough sediments are strongly influenced by the slow kinetics of the reactions causing those metals to precipitate in the anoxic sediments.

The high levels of organochlorines and mercury in top predators point to their wide occurrence throughout the SLE food web. The Laurentian Trough is a significant basin for contaminant deposition. The sedimentary record of organochlorines and metals show maximum values a few centimetres below surface, suggesting a recent decrease of the influx of contaminants, including PCBs, DDT-related compounds, mirex, dioxins and furans, mercury, silver and lead. Measurements of stable lead isotope ratios in trough sediments suggest that the combustion of Canadian leaded gasoline was an important source of contaminant lead in the SLE over the last hundred years.

References

1. El-Sabh MI, Silverberg N (eds) (1990) Oceanography of a large-scale estuarine system: The St. Lawrence. Springer, Berlin Heidelberg New York
2. Dugal L-P (1934) Nat Can 5:165
3. Strain PM (ed) (1988) Chemical oceanography in the Gulf of St. Lawrence. Can Bull Fish Aquat Sci 220
4. Thérriault J-C (ed) (1991) The Gulf of St. Lawrence: Small ocean or big estuary? Can Spec Publ Fish Aquat Sci 113
5. Pocklingtion R, Tan FC (1987) Geochim Cosmochim Acta 51:2579
6. Hélie J-F, Hillaire-Marcel C, Rondeau B (2002) Chem Geol 186:117
7. Ingram GI, El-Sabh MI (1990) Fronts and mesoscale features in the St. Lawrence estuary. In: El-Sabh MI, Silverberg N (eds) Oceanography of a large-scale estuarine system: The St. Lawrence. Springer, Berlin Heidelberg New York, p 71
8. Bewers JM, Yeats PA (1978) Estuar Coast Mar Sci 7:147
9. El-Sabh MI (1988) Physical oceanography of the St. Lawrence estuary. In: Kjerfve B (ed) Hydrodynamics of estuaries, vol II. CRC, Boca Raton Florida, p 61
10. Koutitonsky VG, Bugden GL (1991) The physical oceanography of the Gulf of St. Lawrence: A review with emphasis on the synoptic variability of the motion. In: Thérriault J-C (ed) The Gulf of St. Lawrence: small ocean or big estuary? Can Spec Publ Fish Aquat Sci 113:57
11. Gibert D, Pettigrew B (1997) Can J Fish Aquat Sci 54(1):57
12. Dickie L, Trites RW (1983) The Gulf of St. Lawrence. In: Ketchum BH (ed) Estuaries and enclosed seas. Elsevier, Amsterdam, p 403
13. Bugden GL (1991) Changes in the temperature-salinity characteristics of the deeper waters of the Gulf of St. Lawrence over the past several decades. In: Thérriault J-C (ed) The Gulf of St. Lawrence: small ocean or big estuary? Can Spec Publ Fish Aquat Sci 113:139
14. Thérriault J-C, Lacroix G (1976) J Fish Res Board Can 33:2747
15. Ingram RG (1983) Coast Shelf Sci 16:333
16. d'Anglejan B, Smith EC (1973) Can J Earth Sci 10:1380
17. Silverberg N, Sundby B (1979) Can J Earth Sci 16:939
18. Takayanagi K, Gobeil C (2000) J Oceanogr 56:517
19. Lucotte M, d'Anglejan B (1986) Estuaries 9:84
20. Sundby B (1974) Can J Earth Sci 11:1517
21. Thérriault J-C, Legendre L, Demers S (1990) Oceanography and ecology of phytoplankton in the St. Lawrence estuary. In: El-Sabh MI, Silverberg N (eds) Oceanography of a large-scale estuarine system: The St. Lawrence. Springer, Berlin Heidelberg New York, p 269

22. Thérriault J-C, Levasseur M (1985) Nat Can 112:77
23. d'Anglejan B, Brisebois M (1978) J Sediment Petrol 48:951
24. d'Anglejan B (1990) Recent sediments and sediment transport processes in the St. Lawrence estuary. In: El-Sabh MI, Silverberg N (eds) Oceanography of a large-scale estuarine system: The St. Lawrence. Springer, Berlin Heidelberg New York, p 109
25. Rondeau B, Cossa D, Gagnon P, Bilodeau L (2000) Hydrol Processes 14:21
26. Smith JN, Schafer CT (1999) Limnol Oceanogr 44:207
27. Silverberg N, Nguyen HV, Delibrias G, Koide M, Sundby B, Yokoyama Y, Chesselet R (1986) Oceanol Acta 9:285
28. Silverberg N, Edenborn HM, Belzile N (1985) Sediment response to seasonal variations in organic matter input. In: Sigleo AC, Hattori A (eds) Marine and estuarine geochemistry. Lewis, Chelsea, Michigan, p 69
29. St-Onge G, Stoner JS, Hillaire-Marcel C (2003) Earth Plan Sci Lett 209:113
30. Yeats PA (1988) Nutrients. In: Strain PM (ed) Chemical oceanography in the Gulf of St. Lawrence. Can Bull Fish Aquat Sci 220:29
31. Savenkoff C, Vézina A, Packard TT, Silverberg N, Thérriault J-C, Chen W, Bérubé C, Mucci A, Klein B, Mesplé F, Tremblay J-E, Legendre L, Wesson J, Ingram RG (1996) Can J Fish Aquat Sci 53:2451
32. Sundby B, Mucci A, Gobeil C, Gilbert D, Gratton Y, Archambault P (2002) Ocean Sciences Meeting, Honolulu, Hawai
33. Silverberg N, Bakker J, Edenborn HM, Sundby B (1987) Netherl J Sea Res 21:95
34. Anschutz P, Sundby B, Lefrançois L, Luther III GW, Mucci A (2000) Geochim Cosmochim Acta 64:2751
35. Sagasti A, Schauffner LC, Duffy JE (2000) Estuaries 23:474
36. Coote AR, Yeats PA (1979) J Fish Res Board Can 36:122
37. Gobeil C, Sundby B, Silverberg N (1981) Mar Chem 10:123
38. Gearing JN, Pocklington R (1990) Organic geochemical studies in the St. Lawrence estuary. In: El-Sabh MI, Silverberg N (eds) Oceanography of a large-scale estuarine system: The St. Lawrence. Springer, Berlin Heidelberg New York, p 170
39. Tan FC, Strain PM (1983) Geochim Cosmochim Acta 47:125
40. Lucotte M, Hillaire-Marcel C, Louchouarn P (1991) Estuar Coast Shelf Sci 32:297
41. Chen W, Wangersky PJ (1993) Mar Chem 42:95
42. Yeats PA (1988) Trace metals in the water column. In: Strain PM (ed) Chemical oceanography in the Gulf of St. Lawrence. Can Bull Fish Aquat Sci 220:79
43. Yeats PA (1990) Reactivity and transport of nutrients and metals in the St. Lawrence estuary. In: El-Sabh MI, Silverberg N (eds) Oceanography of a large-scale estuarine system: The St. Lawrence. Springer, Berlin Heidelberg New York, p 155
44. Cossa D, Gobeil C (2000) Can J Fish Aquat Sci 57 (Suppl 1):138
45. Cossa D, Tremblay GH, Gobeil C (1990) Sci Tot Environ 97/98:185
46. Cossa D, Tremblay G (1983) Major ions composition of the St. Lawrence River: seasonal variability and fluxes. In: Degens E, Kempe S, Soliman H (eds) Transport of carbon and minerals in major world rivers, Part II. Mitt Geol–Palaont Inst Uni Hamburg, SCOPE/UNEP Sonderband, p 253
47. Silverberg N, Sundby B (1990) Sediment–water interaction and early diagenesis in the Laurentian Trough. In: El-Sabh MI, Silverberg N (eds) Oceanography of a large-scale estuarine system: The St. Lawrence. Springer, Berlin Heidelberg New York, p 202
48. Colombo JC, Silverberg N, Gearing JN (1996) Mar Chem 51:295
49. Lucotte M (1989) Estuar Coast Shelf Sci 29:293
50. Louchouarn P, Lucotte M, Canuel R, Gagné J-P, Richard L-F (1997) Mar Chem 58:3
51. Louchouarn P, Lucotte M, Farella N (1999) Org Geochem 30:675

52. Colombo JC, Silverberg N, Gearing JN (1996a) Mar Chem 51:277
53. Gagnon C, Mucci A, Pelletier E (1995) Geochim Cosmochim Acta 59:2663
54. Huerta-Diaz MA, Morse JW (1990) Geochim Cosmochim Acta 56:2681
55. Belzile N (1988) Geochim Cosmochim Acta 52:2293
56. Sundby B, Martinez P, Gobeil C (2004) Geochim Cosmochim Acta 68:2485
57. Berner RA (1980) Early diagenesis: a theoretical approach. Princeton University Press, Princeton
58. Sundby B, Gobeil C, Silverberg N, Mucci A (1992) Limnol Oceanogr 37:1129
59. Anschutz P, Zhong S, Sundby B, Mucci A, Gobeil C (1998) Limnol Oceanogr 43:53
60. Louchouarn P, Lucotte M, Duchemin E, de Vernal A (1997) Mar Geol 139:181
61. Gobeil C, Cossa D (1993) Can J Fish Aquat Sci 50:1794
62. Gagnon C, Pelletier E, Mucci A (1997) Mar Chem 59:159
63. Gagnon C, Pelletier E, Mucci A, Fitzgerald WF (1996) Limnol Oceanogr 41:428
64. Gobeil C, Macdonald RW, Sundby B (1997) Geochim Cosmochim Acta 61:4647
65. Martineau DA, Béland P, Desjardins C, Lagacé A (1987) Arch Environ Contam Toxicol 16:137
66. Muir DCG, Ford CA, Rosenberg B, Norstrom RJ, Simon M, Béland P (1996) Environ Pollut 93:219
67. Muir DCG, Ford CA, Stewart REA, Smith TG, Addison RF, Zinck ME, Béland P (1990) Can Bull Fish Aquat Sci 224:165
68. Muir DCG, Koczanski K, Rosenberg B, Béland P (1996) Environ Pollut 93:235
69. Lebeuf M, Bernt KE, Trottier S, Noël M, Hammill MO, Measures L (2001) Environ Pollut 111:29
70. Bernt KE, Hammill MO, Lebeuf M, Kovacs KM (1999) Sci Tot Environ 243/244:243
71. Hodson PV, Castonguay M, Couillard CM, Desjardins C, Pelletier, McLeod R (1994) Can J Fish Aquat Sci 51:464
72. Hickie BE, Kingsley MCS, Hodson PV, Muir DCG, Béland P, Mackay D (2000) Can J Fish Aquat Sci 57:101
73. Lebeuf M, St-Pierre I, Clermont Y, Gobeil C (1999) Concentrations de biphényles polychlorés (BPC) et de pesticides organochlorés chez trois espèces de poissons de fond de l'estuaire et du golfe du Saint-Laurent et du fjord du Saguenay. Rapport statistique canadien des sciences halieutiques et aquatiques 1059, Ministère des pêches et des océans, Gouvernement du Canada
74. Brochu C, Moore S, Pelletier E (1995) Mar Pollut Bull 30:515
75. Cossa D (1990) Chemical contaminants in the St. Lawrence estuary and Saguenay Fjord. In: El-Sabh MI, Silverberg N (eds) Oceanography of a large-scale estuarine system: The St. Lawrence. Springer, Berlin Heidelberg New York, p 239
76. Gobeil C, Lebeuf M (1992) Inventaire de la contamination des sédiments du chenal Laurentien: les biphényles polychlorés. Rapport technique canadien des sciences halieutiques et aquatiques 1851, Ministère des pêches et des océans, Gouvernement du Canada
77. Lebeuf M, Gobeil C, Clermont Y, Brochu C, Moore S (1995a) Organohalogene Comp 24:293
78. Lebeuf M, Gobeil C, Clermont Y, Brochu C, Moore S (1995b) Organohalogene Comp 26:421
79. Comba ME, Norstrom RJ, Macdonald CR, Kaiser KLE (1993) Environ Sci Technol 27:2198
80. Lebeuf M, Gobeil C, Brochu C, Moore S (1996) Organohalogen Comp 28:20
81. Loring DH (1975) Can J Earth Sci 12:1219
82. Wagemann R, Stewart REA, Béland P, Desjardins C (1990) Can Bull Fish Aquat Sci 224:191

83. Gobeil C (1999) Environ Sci Technol 33:2953
84. Sanudo-Wilhelmy SA, Flegal AR (1992) Environ Sci Technol 33:848
85. Ravizza GE, Bothner MH (1996) Geochim Cosmochim Acta 60:2753
86. Gobeil C, Johnson WK, Macdonald RW, Wong CS (1995) Environ Sci Technol 29:193
87. Doe BR (1979) Lead isotopes. Springer, Berlin Heidelberg New York
88. Gobeil C, Macdonald RW, Smith JN, Beaudin L (2001) Science 293:1301
89. Sturges WT, Barrie LA (1987) Nature 329:144
90. Cumming GL, Richards JR (1975) Earth Plant Sci Lett 28:155
91. Nriagu JO, Pacyna JM (1988) Nature 333:134

Hdb Env Chem Vol. 5, Part H (2006): 149–173
DOI 10.1007/698_5_025
© Springer-Verlag Berlin Heidelberg 2005
Published online: 25 October 2005

The Nile Estuary

Waleed Hamza[1,2]

[1]Biology Department, Faculty of Science, United Arab Emirates University,
P.O. Box 17551, Al-Ain, UAE
w.hamza@uaeu.ac.ae

[2]Environmental Science Department, Faculty of Science, Alexandria University,
21511 Alexandria, Egypt
w.hamza@uaeu.ac.ae

Abstract The River Nile, the most famous river of the ancient world, is the dominant geographic feature of northeastern Africa and the longest river on Earth. At the point of discharge of the Nile into the Mediterranean, the great Nile delta has formed and furnishes the most fertile area for cultivation in the Egyptian territory. The delta is embraced by two large branches of the Nile (the Rosetta and Damietta branches and their promontories), as the northward flowing river bifurcates near the city of Cairo. Both the Rosetta and Damietta branches discharge freshwater directly and indirectly into the Mediterranean Sea to form the Nile estuary (also known as the Nile delta coastal area).

Fluctuations in both quantity and quality of the Nile water reaching the Mediterranean, especially as a result of the Aswan High Dam (AHD) construction in 1965, have profoundly influenced the morphometry and hydrology of the Nile, and the ecological characteristics of the river and the surrounding marine environment.

This chapter intends to highlight the range of characteristics of the Nile estuary and the main factors influencing them since the AHD construction. To this effect, the geography, hydrology, and ecology of this river-delta-estuary-coastal marine system will be described and illustrated, and recent numerical simulations of its hydrodynamics and ecosystem features will be discussed. The concluding remarks forecast future trends in the development of the Nile estuary and its vital role in the ecology of the Mediterranean Sea.

Keywords Egyptian coast ecosystem · Estuary · Hydrochemistry · Mediterranean Sea ·
River Nile

Abbreviations
AHD   Aswan High Dam
MFSPP   Mediterranean Forecasting System Pilot Project
FinEst   Finnish–Estonian
PAR   Photosynthetic Available Radiation
CE   Christian Era

1
Nile Estuary Development

The Nile estuary is the classical example of a transitional environment between the river and the sea. The geographical position and morphometric
features of this estuary are influenced by several factors, with the most important being climatic variations, the impact of human activities, and sea
hydrodynamics. The annual discharging capacity of a river into an estuarine
environment is related not only to the rainfall density in the river catchment
area, but also to natural and artificial barriers to river flow encountered between the river source and its point of discharge. In the following text, the
factors determining the historical development and modern characteristics
of the Nile estuary environment are reviewed and extended to include certain features of the River Nile itself. In this regard, it is appropriate to begin
this chapter with a brief introduction to the Nile, making special reference
to those parameters that have the greatest influence upon the Nile estuarine
environment.

1.1
The River Nile

Winding more than 6000 km from source to outfall, the Nile is the longest
river in the world. However, it is not only in its length that the Nile is dis-
tinguished amongst its great rivals. No other river traverses such a variety
of landscapes, such a medley of cultures, and spectrum of peoples, as does
the Nile. None has had such a profound historical and material effect upon
those who dwell along its banks, prescribing plenty or famine – the difference between life and death for multitudes since the beginning of man's
history.

The Nile basin extends from latitudes 4°S to 31°N and encompasses
parts of Burundi, Egypt, Eritrea, Ethiopia, Kenya, Rwanda, Sudan, Tanzania,
Uganda, and the Democratic Republic of Congo (Fig. 1). The Nile River is
sourced in Lake Victoria in east Central Africa. It flows generally northwards

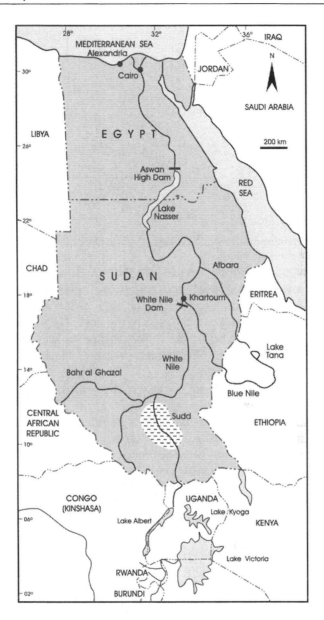

Fig. 1 Nile River trajectory from source to outfall

through Uganda, Sudan, and Egypt to reach the Mediterranean Sea. From its remotest head stream, the Luvironza River in Burundi, the river is 6671 km long, and its basin has an area of more than 2 590 000 km^2 [1].

The Nile flows from highland regions, with abundant moisture, to low-land plains with semiarid to arid conditions. Not only does the Nile provide

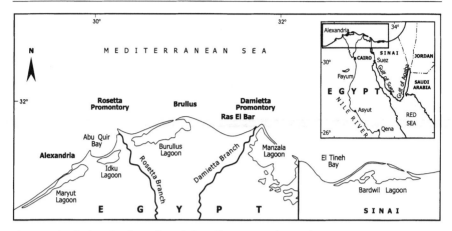

Fig. 2 Main discharging branches of the Nile River to the Mediterranean Sea

freshwater to millions, but within its basin there are five major lakes (Victoria, Edward, Albert, Kyoga, and Tana), vast areas of permanent wetland and seasonal flooding (The Sudd, Bahr al-Ghazal, and Macharmorches), and five major reservoir dams (from north to south: the Aswan High Dam, Roseires, Khashm El-Gibra, Sennar, and Jabel Aulia). Before the construction of the Aswan High Dam (AHD), the Nile annually delivered black mud to the Nile delta, making it fertile.

Egypt is the most downstream country of the Nile, with the last 1530 km of river length lying within Egyptian territory. At the city of Cairo (200 km from the Mediterranean coast), the River Nile bifurcates into two branches enclosing the delta region between them. These are the Rosetta (the western) branch and the Damietta (the eastern) branch that discharge Nile water into the Mediterranean through the Nile estuary (Fig. 2).

1.2
Nile Branches from Ancient to Modern Times

The Rosetta and Damietta branches of the Nile are similar in some respects but distinct in others. They differ in their discharging capacities of both water and sediments (throughout their history, both before and after the construction of the AHD) and in their geomorphology – a consequence of the variability of coastal and beach processes.

1.2.1
Rosetta Promontory

The Rosetta promontory began to develop sometime between 500–1000 CE when river water from earlier branches was naturally diverted and/or artificially redirected into an existing canal known afterwards as Rosetta [2].

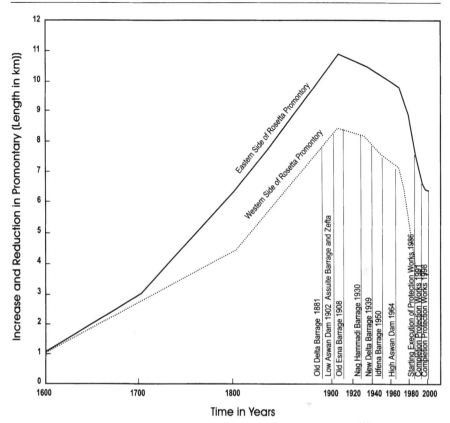

Fig. 3 Historical advance and retreat of the Rosetta promontory (modified from Fanos et al. [3])

The configuration of the shoreline has changed markedly during the past five centuries (1500–1998). The eastern and western shores of the promontory prograded seawards at an average rate of about 25 m year^{-1} during the period 1500–1900, though they retreated at variable rates during the period 1900–1998 [3]. The detailed history of advance and retreat of the Rosetta promontory is represented in Fig. 3. The gradual reduction of the promontory length was halted after protective measures were taken on both sides. However, wave erosion of the coastal areas on the western and eastern sides of the protective works has ensued (Fig. 4), and the rate of this erosion has reached 80–100 m year^{-1} [3, 4]. At present the Rosetta promontory extends for about 220 km (from Cairo to its discharging point) with an average width of 180 m and with a water depth of 2–4 m depending on the discharging strength. It covers an area of about 40 km^2, giving an estimated volume of 45×10^6 m^3 of sediment [5].

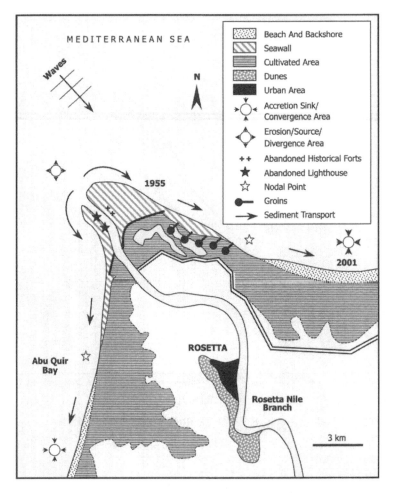

Fig. 4 Erosion and accretion features of the Rosetta promontory and the protective measures (modified from Frihy [4])

1.2.2
Damietta Promontory

The Damietta promontory was formed by the accumulation of sediments transported along the Damietta branch during the Holocene transgression [6]. It continues for 60 km west of Port Said at the entrance to the Suez Canal. During the period 1800–1998 the shoreline changes for this promontory were similar to those of the Rosetta. The promontory shoreline gradually advanced until 1895, and since then it has been retreating. The western side of the promontory advanced at a rate of 10 m year^{-1} between 1800 and 1895. Between 1895 and 1940 it retreated at an average rate of 35 m year^{-1}. On its eastern side the rate of advance of the shoreline between 1800 and 1912

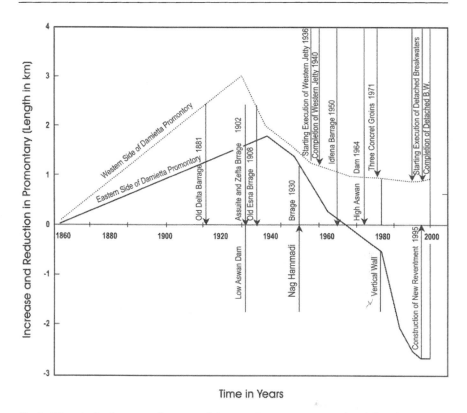

Fig. 5 Historical advance and retreat of the Damietta promontory (modified from Fanos et al. [3])

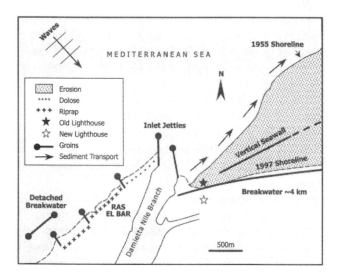

Fig. 6 Erosion features of the Damietta promontory and the protective measures (modified from Frihy et al. [6])

was about $20 \, \mathrm{m \, year^{-1}}$ (Fig. 5). Between 1912 and 1973 the rate of retreat for the Damietta promontory shoreline was about $40 \, \mathrm{m \, year^{-1}}$, increasing to $100 \, \mathrm{m \, year^{-1}}$ in the period between 1973 and 1995 [3].

Surveys of the progressive shoreline changes of the promontory between 1922 and 1995 have shown how the beaches have been affected by shoreline erosion (Fig. 6). It is estimated that $9.7 \, \mathrm{km^2 \, year^{-1}}$ of coastal area has been lost, as the shoreline has retreated at a rate of $0.044 \, \mathrm{km \, year^{-1}}$. This erosion is compensated along the flank of the promontory by the shoreline advancing at an average $0.008 \, \mathrm{km \, year^{-1}}$ and coastal area increasing by $13.3 \, \mathrm{km^2 \, year^{-1}}$ [6].

1.3
Climate, Geography, and Morphometry of the Nile Estuary

Egypt is the most downstream country traversed by the Nile River, and is well known for its arid climate. In Egypt the precipitation along the Mediterranean coastal strip (Nile estuary) is $200 \, \mathrm{mm \, year^{-1}}$, but declines dramatically inland, e.g., to $20 \, \mathrm{mm \, year^{-1}}$ near Cairo, 200 km from the northern coast. Farther inland, in Middle and Upper Egypt, rainfall is effectively zero. The semiarid climatic conditions of the northern African strip inevitably lead to heavy reliance on surface water resources. The Nile is thus the main source of freshwater in Egypt, and Egypt's agriculture is dependent on irrigation using Nile water released annually from the AHD [7].

It is not easy to quote precisely geographic coordinates for the Nile estuary. This is mainly due to the temporal displacements of its two branches and the annual variations in discharge into the Mediterranean since time immemorial. However, approximate eastern and western boundaries of the present day Nile estuary may be placed at longitudes 30°E and 33°E, with the northern and southern extremities at latitudes 31°N and 32°N, respectively. The discharging outlets of the Nile delta coastal lagoons and the sediment-laden freshwater of the Nile debouching into the Mediterranean Sea also lie within these boundaries (Fig. 7).

In their study of the Nile delta sediments in the Mediterranean, Bellaiche et al. [8] indicated that the leading tip of the Nile deep-sea sediment fan is located near 32°23′N/28°22′E. The authors did not report any recent sediments at that distal point, however, they demonstrated that deep-sea turbidities of mixed origin (Egyptian and Levantine), fill the sedimentary basin located south of Cyprus.

The Nile estuary, also known as the Nile delta coastal area, occupies the central part of the Egyptian northern coastal zone bordering the Mediterranean Sea. The Nile delta coast from Abu Quir bay to Port Said is arcuate (Fig. 3), and has a beach and contiguous coastal flat backed by coastal dunes or wide lagoons. The two main Nile promontories at Rosetta and Damietta interrupt the sandy shore line of the delta. The nearshore area is a hydrologically active zone characterized by a gentle slope varying from 1 : 50 to 1 : 100,

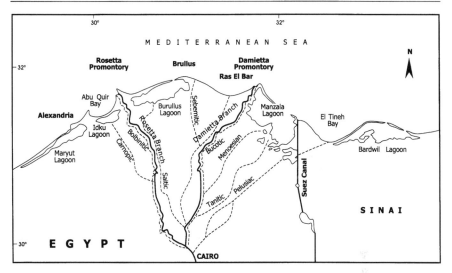

Fig. 7 Ancient and recent geographical boundaries of both the direct and indirect discharging outlets of the Nile delta

and a dissipative wide beach [4, 9]. On account of the high economic, ecological, aesthetic, and recreational importance of this zone, there are increasing levels of environmental stress from both natural (erosion, dune quarrying, and subsidence and rising water levels) and anthropogenic influences (population growth and increasing development) [10].

The coastal zone of the Nile delta is undergoing major contemporary changes due to the natural and anthropogenic activities noted above. Along the Nile delta coast natural influences include tectonic activity, climatic and sea level fluctuations, and fluvial and marine processes. The anthropogenic factors include the construction of Nile barrages, the AHD, networks of irrigation and drainage canals, and protective works.

Erosion has impacted on the agricultural and urban lands along the delta promontories of the Nile delta coast. Sediments accumulate within embayments and saddles between the Rosetta and Damietta promontories. A number of coastal protection structures such as jetties, groins, seawalls, and wave breaks have been built to combat beach erosion and to reduce shoaling [4].

Despite the high energy of the hydrologic and hydrodynamic processes of the Nile delta coast, it remains the shallowest part of the Egyptian Mediterranean shelf area. It has been mentioned that the hydrological processes along the Egyptian coastal area are mainly controlled by climatic factors (mainly wind and air temperature) and by the ambient currents in the southern Mediterranean [11]. The bathymetric map of the Egyptian shelf (Fig. 8) indicates a maximum depth of 300 m at latitude 32°N in the distal end of Nile delta [12].

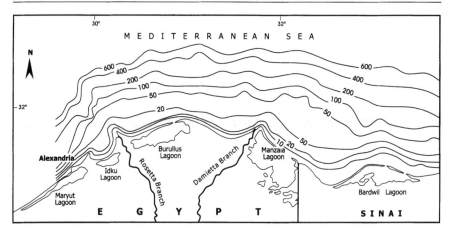

Fig. 8 Bathymetric configuration of the Egyptian Mediterranean shelf facing the Nile estuary (after Hamza [12])

1.4
Demographic Development and Nile Discharging Measures

The combined population of Nile basin countries is close to 300 million, with about half of this population being dependent on the Nile water [13]. Egypt has a total population of more than 67 million, representing about 22% of all Nile basin inhabitants, though this population is unequally distributed throughout the country. Egypt is divided into four geographic regions: the Nile valley and delta, the Western Desert, the Eastern Desert, and Sinai. The physiography and aridity of the deserts bordering the Nile valley and delta constitute a barrier obstructing the full utilization of Egyptian land. About 99% of the Egyptian population is concentrated within 5.5% of the area of the Nile valley and delta region [4]. About 50% of the Egyptian population is concentrated in the delta and coastal governorates, excluding the capital, Cairo, which accounts for more than 20% of the national population, and supports up to $25\,000$ person km^{-2}.

The point of entry of the Nile into Egypt is the southern part of Lake Nasser, at Wadi Halfa, south of Aswan (Fig. 1). From Aswan, the river is a meandering channel as far as 20 km north of Cairo. At that location the river bifurcates into two main branches, each of which meanders separately over the delta to the sea. On the Nile flood plain, extensive artificial drainage systems exist, especially in the traditionally cultivated land. These drainage systems discharge into one or another of the Nile branches or into the Northern delta lakes and the Mediterranean Sea.

The Nile provides Egypt with about 95% of its annual water requirements. According to historical records the average annual discharge of the Nile between 1899 and 1959 was estimated as 84 km^3 year^{-1} (84×10^9 m^3 year^{-1}). Record discharges during 1916 (120 km^3 year^{-1}) and 1984 (420 km^3 year^{-1})

demonstrate the dramatic fluctuations of the Nile flow. The agreement signed between Egypt and Sudan in 1959 endows Egypt with exclusive access to $55.5 \, km^3 \, year^{-1}$ of Nile flood water, to be withdrawn from Lake Nasser (The Aswan High Dam Reservoir).

With increasing population in Egypt, the per capita share of Nile water has decreased from $2561 \, m^3 \, year^{-1}$ in 1955 to $1123 \, m^3 \, year^{-1}$ in 1990, and then to $680 \, m^3 \, year^{-1}$ in 2000. As the population continues to grow it is expected that this per capita share will decline further to $500 \, m^3 \, year^{-1}$ in 2025 [15].

Approximately 85% of Egypt's water resources are committed to the irrigation of the 3.4×10^6 ha of cultivated land. Egypt is the only country in the Nile basin that has significant industrialization. Since Egypt is the last country that the Nile passes through en route to the Mediterranean, this industrialization has no effect on the quality of the river water in the other Nile basin countries [16]. The main industries in Egypt are food processing, textile and other manufacturing, pulp and paper, cement production, fertilizer production, and heavy industries such as steel, machinery and chemicals. Most of the industrial activity is concentrated along the River Nile and its main branches in the areas surrounding Cairo and Alexandria. The majority of these industries discharge any wastewater directly, without treatment, into the Nile river and therefore into the waterways that feed the Nile estuary, the coastal lakes, and finally the Mediterranean sea [17].

Due to the limited surface water resources in Egypt and the fast pace of growth in both agriculture and industrialization, and rapid population growth, the volume of Nile water received by the Mediterranean has diminished drastically. The summer of 1964 saw the last normal discharge of Nile flood water into the Mediterranean. The average total annual Nile flow for the 5 years prior to this event (i.e., 1959–1963), amounted to $42.9 \, km^3$ of freshwater delivered to the Nile estuary [18]. After 1964 the discharge de-

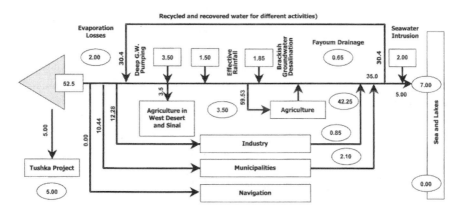

Fig. 9 Strategic balance model of the Egyptian surface freshwater for the year 2017 (modified from El-Arabawy [7])

creased to about 18 and 21 km^3 in the years 1982 and 1984, respectively. This quantity was discharged exclusively from the Rosetta Nile branch, as the Damietta branch remained closed at that time. The surplus of Nile fresh-water reaching the Mediterranean annually amounts to 2.5–4 km^3 [19]. This represents a considerable fraction (in these cases 15–25%) of the total land runoffs discharging annually into the Mediterranean through coastal lakes and other land effluents connected to the sea [18]. El-Arabawy [7] recently developed a water balance strategic model for the year 2017, implementing the Egyptian government plan to recycle as much drainage water as possible. El-Arabawy [7] estimated that some 6–7 km^3 of return flow into the Mediter-ranean and northern lakes is required to mitigate salt-water intrusion and preserve the Nile delta salt balance (Fig. 9).

2
Hydrology and Hydrochemical Parameters of the Nile Estuary

The basin area of the River Nile discharging to the Mediterranean sea is about 3×10^6 km^2. The flow rate is as high as 601 m^3 s^{-1} [18]. Nile water arrives at the Mediterranean not only through the Nile branches, Damietta in the west and Rosetta to the east, but also through coastal lakes outlets and various drainage effluents. These effluents continuously discharge water with a complex mixture of varied waste materials into the sea. The quantity and characteristics of these wastes mainly reflect the diversity of human ac-tivities of the Egyptian population in the Nile delta region (1000–1200 km from Aswan). As mentioned above, the main consumers of the Nile water in Egypt are (in decreasing order of demand) agriculture, municipalities, and industries. The principal effects of agricultural activities on water quality include changes in salinity, and deterioration of water quality due to fertil-izer and pesticide use. This leads to eutrophication of water bodies (coastal lakes) via an increase in nutrient loading. Thus agriculture may be considered a widespread source of pollution in the Nile estuary. Although they are dis-persed, the runoff from these areas is collected in agricultural drains which become point sources of pollutants for the coastal lakes. The main pollutants coming from these sources are salt, nutrients (nitrogen and phosphorus), and pesticides. Nile water salinity is measured at the AHD in order to mon-itor salinity increases before the water discharges into the Mediterranean. The average salt concentration in waters ahead of the dam is in the order of 150 mg L^{-1}. This concentration increases to 250 mg L^{-1} near Cairo, and fur-ther to 2000–3000 mg L^{-1} at the northern lakes and estuary mouth at the point of discharge into the Mediterranean [17]. The deterioration of Nile wa-ter quality is most pronounced in the Rosetta and Damietta branches due to the disposal of municipal and industrial effluents, in combination with agri-cultural drainage and decreasing flow as water arrives at the Nile estuary.

The quality of the Nile river waters, from Aswan to the Mediterranean sea (0–1200 km), is shown in Table 1.

In addition to nutrient-enriched waters, other pollutants such as trace metals and hydrocarbons of industrial origin are reaching the Nile estuarine environment. All of these pollutants have severely affected the Egyptian northern coastal ecosystem, especially seaward of the delta estuaries (Rosetta and Damietta). There are numerous reports of high concentrations of contaminants such as aluminium, iron, copper, zinc, cadmium, and lead, dissolved and in particulate forms, in waters contributing to the estuarine environment of the Nile. The particulate form is mostly associated with suspended matter (both organic and inorganic), which afterwards is deposited as sediments [20–23]. Similar results are found for nutrient salts in both of the Nile branches and their estuaries. The latter studies show that concentrations of nitrogen and phosphorus nutrients are present in high concentrations in the Nile branches upstream, reflecting the

Table 1 River Nile water quality (probes taken July 1991 to April 1992 [16])

	Concentration (mg/L)	Distribution along river course (Aswan to Mediterranean Sea)
Eutrophication		
NH_3, ammonia	< 0.1 – 0.6	Even distribution
NO_3, nitrate	1 – 4	Even distribution
NO_2, nitrite	< 0.05	Even distribution
P, total phosphorus	< 0.25	0–1000 km from Aswan
	0.1 – 1.6	1000–1200 km from Aswan
PO_4, ortho-phosphorous	< 0.1	0–1000 km from Aswan
	0.1 – 1	1000–1200 km from Aswan
Organic matter content		
Dissolved oxygen	2 – 10	Less than 8 from 1000 km onwards
BOD	< 4	0–1000 km from Aswan
	< 8	1000–1200 km from Aswan
COD	< 25	0–1000 km from Aswan
	< 45	1000–1200 km from Aswan
Coliform(Thousands/100 mL)		
–Total	2.5 for 30 of 53 probes	Even distribution
	> 18 for 6 of 53 probes	
–Faecal	< 2 for 42 of 55 probes	Even distribution
	2 – 6 for 13 of 55 probes	

BOD Biological oxygen demand, amount of oxygen consumed in 5 days under optimal conditions in biodegradation process, *COD* Chemical oxygen demand, amount of oxygen consumed by water sample release there of municipal and agricultural wastes [24, 25]. Nutrient concentration values then decrease gradually towards the estuary mouth and seaward.

2.1
Pollution Sources and its Influence on the Nile Estuary

Contamination of Nile estuary water by hydrocarbons is a consequence of the expanding petroleum and petrochemical industries in Egypt. The Nile delta area is now considered as one of the major oil- and gas-producing fields in Egypt. The release of oil wastes into the Nile estuary is inevitable, and oil products are harmful pollutants adversely affecting the biota of the Nile estuary ecosystem. In their studies on the levels of chlorinated hydrocarbons in living organisms from the Egyptian Mediterranean coast and Nile estuary, Abd-Allah et al. [25] have analyzed the tissue of fish (*Mugil cephalus*) and a bivalve (*Donax sp.*) for residues of 22 organochlorine pollutants. The results obtained have indicated that 2,2-bis(*p*-chlorophenyl)-1,1-dichlorethylene (*p,p*-DDE) is dominant in fish with concentrations of $2-4\,ng\,g^{-1}$. 1-Chloro-2,2-bis(*p*-chlorophenyl)ethylene (DDMM) dominated in the bivalves, which yielded concentrations ranging from $9-15\,ng\,g^{-1}$. Toxaphene was also detected in these fauna, with the maximum concentration of $9.7\,ng\,g^{-1}$ being found in bivalves. This compound may also be derived from pesticides washed into agricultural drains feeding the estuaries. Other investigations of hydrocarbon and oil contamination of the Nile delta coast environments and lakes have indicated highly toxic compounds in water, sediments, and living organisms [26–31].

The variable levels of pollutant concentrations in the Nile estuary environment are related to the river discharging capacity, the distribution of land-sourced effluents along the Nile delta region, and temporal variations in these factors. The discharging capacities of the Nile branches reach peak values during the winter season [12, 32]. The Rosetta estuary is the main discharging branch as the Damietta branch was dammed 20 km inland of the river mouth by an artificial dam (Farskur Dam), since the erection of AHD. The flow from the Damietta branch is limited to drainage coming from municipal, agriculture, and industrial polluted water emanating from the final 20 km of channel before the Mediterranean coast. This is one of the two main reasons for the Damietta estuarine environment being more polluted than the Rosetta. The other reason is related to the Mediterranean circulation in general, and more specifically to Egyptian coastal hydrology. The eastward flowing Mediterranean currents along the Egyptian coast carry pollutants from the western effluents (Rosetta branch and coastal lakes) to the eastern side of the delta, to mix with the concentrated pollutants from the low-discharging Damietta branch. In addition, the largest and the most polluted coastal lake discharges its water into the sea adjacent to the Damietta estuary.

2.2
Nile Delta Lakes as Part of the Nile Estuary

Also referred to as delta coastal lagoons, the delta lakes originally developed during the intense flooding period of the 19th century. The periodic advance and retreat of the shoreline has resulted in some of these lakes becoming directly connected to the Mediterranean Sea via narrow outlets (Fig. 10). The Nile delta lakes occupy a significant area ($> 1100\,km^2$) of the Egyptian Mediterranean coastal zone. From west to east, the lakes are Lake Mariut, Lake Edku, Lake Burallus, and Lake Manzalah. The largest in surface area is Lake Manzalah, while the smallest is Lake Mariut. All four lakes are shallow with an average depth of 1.10 m. Their salinity is known to vary from fresh to brackish at the southern lake shores, and they are saline to hypersaline in the northern shoreline areas bordering the wet lands [18].

Although the Nile is the main influx into the lake environment, the lakes also serve as collection basins for agricultural, sewage, and industrial drainage water. Consequently, severe degradation in lake water quality and ecosystem have occurred since the erection of the AHD. The surface area of the lakes has shrunk in response to silting caused by large quantities of sus-pended matter carried into the lakes along with untreated to partially treated sewage and agriculture drainage water. The lake basins have also been af-fected by urbanization, agriculture, and highway construction. In fact, the modern northern delta lakes cover < 50% of the area they occupied 35 years ago. The individual geographic position and hydrographic features of each lake are shown in Table 2.

The Nile delta lakes are important in that they have inherited the role of several pre-existing Nile tributaries at this location that supplied freshwater and sediments to the Mediterranean. Despite the degradation of their water quality, the lakes still supply the Egyptian estuarine coastal area with many nutrients. The Nile delta lakes are also regarded as optimal fishery grounds, where, until the end of 1985, fish production amounted to 50% of the annual

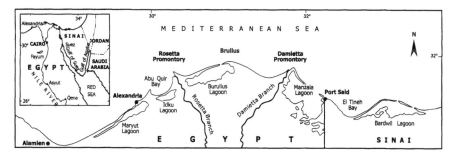

Fig. 10 Nile delta coastal lakes (lagoons) and their connections with the Mediterranean Sea

Table 2 Main geographic and hydrographic features of the Nile delta lakes

Parameter	Mariut	Edku	Burullus	Manzalah
Long.(E)	29.28(E)	30.20(E)	31.00(E)	31.48(E)
Lat.(N)	31.20(N)	31.33(N)	31.62(N)	31.46(N)
Surface area (km^2)	62.00	109.00	350.00	650.00
Depth range (cm)	50–150	40–220	50–200	50–140
Av. Water salinity (.‰)	6.3	3.8	2.5	5.9
Annual discharging volume ($\times 10^9$ m^3)	2.37	2.06	3.2	6.7
Discharging rate (m^3 s^{-1})	74.0	60.0	80.0	165.0
Water residence time (days)	10	21	42	32
Trophic status	Hypertrophic	Eutrophic	Mesotrophic	Hypertrophic
Water sources	A, I, S, G	A, S	A	A, I, S

A agriculture, *I* industrial, *S* sewage, *G* groundwater. Modified after Hamza [18]

Egyptian fish yield. As a result of declining water quality and shrinking lake surface area the fish yields have decreased markedly [18].

3
Impact of Aswan High Dam Construction on the Nile Estuary

3.1
Coastal and Fisheries Reduction

Since the construction in 1964 of the AHD, the continuing debate on the relative merits and disadvantages of this project have progressed from hydropolitical concerns to the socio-economic strategies amongst the Nile basin countries and other interested neighbors. A principal purpose for the damming of the river Nile by the Egyptian Government was to address the need to control flooding and to manage irrigation systems for national agricultural developments. The scheme has provided the additional benefits of hydroelectricity generation and the creation of a strategic freshwater reservoir to moderate water supply during low flood periods. Dam construction may also have unintended negative impacts on the surrounded environment, and these must be taken into account in any assessment of the value of the project. A full discussion of this issue is beyond the scope of this chapter. However, in the light of case studies of environmental impacts of dam construction in other locations from the USA to Africa it is clear that fluvial, sedimentary, estuarine, and ecological processes are complexly interlinked and it is no simple matter to predict the results of interfering with them. Planned changes to one part of the system leads to unexpected and often indeterminate effects on another.

Scientific research indicates that the reduction in freshwater discharge and fertile suspended matter are the main factors determining the impact of the AHD on the Nile estuary. These impacts include the erosion of the Nile delta coastal area and disturbance to the ecological equilibrium of the Levantine basin [3, 32–36]. The maximum extent of the Nile delta shoreline was a result of sediment build-up during the period of high floods in the 19th century. At that time, the shoreline advanced seawards due to the domination of sediment supply to beaches over the erosive activities of waves and currents. This period of advance was halted and the inshore line retreated in the year 1900, due to climatic changes in eastern Africa and extensive use of Nile sediments and water in perennial irrigation. Erosion of the delta shoreline accelerated after 1964 due to the construction of the AHD and the consequent reduction in sediment supply to the delta coast to only 5% of earlier average rates (Fig. 11).

According to Inman and Scott [37] the total sediment load (sand + silt + clay) carried by the Egyptian Nile waters, prior to the AHD construction, ranged between 160 and $178 \times 10^6 \, \text{t year}^{-1}$. The suspended fraction (silt + clay) accounted for $112 \times 10^6 \, \text{t year}^{-1}$, much of which was deposited on agricultural fields. The remaining sand load of 50–$66 \times 10^6 \, \text{t year}^{-1}$, represents

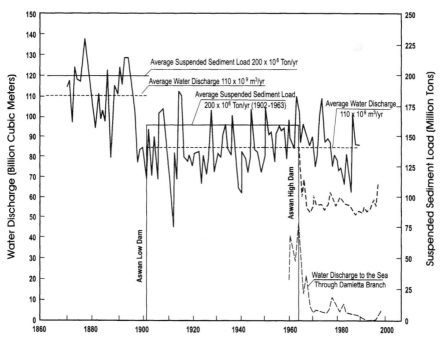

Fig. 11 Discharge of Nile water and suspended sediments to the Mediterranean before and after construction of the AHD (modified from Fanos et al. [3])

a sediment volume supply rate of $30-40 \times 10^6$ m^3 year^{-1}. This sediment volume found its way to the sea and compensated wholly or partially for the sediment losses resulting from coastal erosion. The inevitability of erosion of the Nile delta becomes obvious when we compare the original sediment supply rate of $30-40 \times 10^6$ m^3 year^{-1} with the delta coast erosion rate of 32×10^6 m^3 year^{-1} averaged for the period 1919/1922 to 1984 [3].

Stanley [38] observed that although sediment is being transported as far as Cairo, virtually no sediment is being supplied to replenish the coastline via the channels flowing into the Mediterranean. This situation results from the diversion of Nile water into more than 10 000 km of irrigation and drainage canals, north of Cairo. The water in these canals is either still or very slow moving. Consequently the suspended sediment either settles on the canal floor whence farmers recover it for addition to their fields, or is pumped along with the canal water into the four large freshwater coastal lakes near the outer edge of the delta (Fig. 10). The loss of sediment supply to replenish the Nile delta coast is a significant problem related to the construction of the AHD. Nevertheless, the AHD has provided undeniable benefits in the tremendous boon to Egyptian agriculture and to industry via the pollution-free provision of cheap hydroelectric power. It has also protected Egypt from flooding, and water from a year of plenty can be saved for a drought year.

In a detailed study of the subsidence of the northeastern Nile delta, Stanley [34] showed that delta areal loss is also due to continued land surface subsidence at rates of up to 40-50 cm per century. He has also warned that eustatic sea level rise, conservatively estimated at 4-8 cm during the next 40 years, and at least 50 cm by the year 2100, may compound the effects of delta subsidence and coastal erosion to submerge the delta region as far inland as 30 km from the present day coast. In this scenario, the Port Said–Northern Suez Canal–Lake Manzalah region, with a population of one million, would become particularly susceptible to flooding because it is located in one of the more rapidly subsiding parts of the delta [34].

The disturbance of the ecological equilibrium in the Levantine basin, due to the AHD construction, has also been demonstrated in scientific investigations. Before the AHD was built 50% of the Nile flow emptied into the Mediterranean. During an average pre-AHD flood the total discharge of nutrient salts was estimated to be approximately 5500 t of phosphate and 280×10^3 t of silicate. The nutrient-rich floodwater, or Nile stream, was approximately 15 km wide, had sharply defined boundaries, extended along the Egyptian coast, and was sometimes detected off the coast of southern Turkey [39]. The fertility of the southeastern Mediterranean has decreased markedly since the AHD construction. In fact the estimated post-AHD phosphate quantity discharged into the Mediterranean derived from the entire land runoff (not only through the Nile estuaries) amounts to 84.9 t year^{-1} [11]. A commensurate decrease in the average fish catch from nearly 35×10^3 t in 1962 and 1963 to less than one fourth of this in 1969,

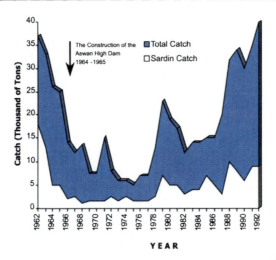

Fig. 12 Annual average fish yield and sardine catch from the Egyptian Mediterranean coast before and after construction of the AHD (after El-Sayed and Van Dijken [39])

reported by Egyptian Mediterranean marine fisheries, parallels the decrease in the discharged freshwater and fertile sediments. Hardest hit was sardine fishing, primarily *Sardinella aurita*, which is heavily reliant on increased phytoplankton growth during the flood season. Whereas a total of 18×10^3 t of sardines were caught in 1962, a mere 460 and 600 t of sardine were landed in 1968 and 1969, respectively [32, 39, 40]. In recent years there has been a noticeable increase in the sardine catch along the Egyptian coast (8590 t in 1992) with most of the landings coinciding with the period of maximum discharge from coastal lakes during winter. Since the 1980s the total fish catch (pelagic and bottom) for the Egyptian coast has been restored to pre-AHD levels (Fig. 12). El-Sayed and Van Dijken [39] question whether this is due to intensified fishing efforts or recovery of the fish stocks. Recent scientific investigations leave little doubt about the changes which have occurred in the pelagic ecosystem. However, the recovery of total fish landings of late (Fig. 12), particularly sardines, is puzzling and in stark contrast with the low levels of primary productivity. Sophisticated numerical simulations of both hydrodynamic and ecosystem functions along the Egyptian coast have already filled gaps in our knowledge regarding certain phenomena, such as the winter algal blooms. These may help to resolve the above conflicting results and expectations, as described in the next section.

3.2
Simulation of the Nile Delta Coastal Ecosystem

There has been much interest in recent years in the use of numerical models to simulate the ecosystem of the southeastern Mediterranean. This activity was realized in the Land-3 Project, financed by the World-Laboratory Agency in 1995, and later annexed by the MFSPP project (Mediterranean Forecasting System Pilot Project), financed by the EU commission during the period 1998–2001. The use of the FinEst (Finnish–Estonian) ecosystem numerical model within the Land-3 project succeeded in simulating the ecosystem parameters of the Egyptian coast between longitudes 29°5'E and 33°45'E. In addition, the model was able to simulate the influence of land-runoff on the productivity of the Nile delta coastal ecosystem. In the latter simulation, climatic conditions for the Egyptian coast were used as an external forcing factor that influenced coastal hydrodynamics. After setting up the model, it was used to simulate 60 Julian days (January and February) representing winter conditions. Using this model, Hamza et al. [41] showed that in the winter season (December–February) meteorological conditions play an important role in keeping high nutrient concentrations in the Nile delta area for long periods, in addition to the role played by the nutrient load (40% of annual discharge, coming mainly from the Rosetta Nile Branch and the delta drainages). The conditions of this season promote phytoplankton growth; this may explain the existence of winter algal blooms that have affected the Egyptian coastal area since the AHD construction [40].

The model simulation results have shown that high concentrations of both nutrient salts (e.g. phosphorus) and algal biomass (chlorophyll a) are expected to be found along the Nile delta coastal area during the winter season (Fig. 13a,b).

The simulation predicted relatively warm air temperatures with variations of 2–6 °C between the day and the night, based on available meteorological data. Interestingly, during the same winter period the winds blow both from the NE and NW quadrants, creating a central zone of calm on the coast off the Nile delta zone. The winds are mainly onshore (Fig. 14). Based on these results Hamza et al. [41] explained the winter algal bloom along the Nile delta coast as follows:

1. The Rosetta Nile branch flow and delta drainages constitute the main winter source of nutrients supplied to the coastal area off the Nile delta.
2. Meteorological conditions during winter consistently favor the development of phytoplankton blooms due to the quasi-stable conditions in the delta offshore area.
3. East-flowing counter currents characterize the southern part of the Mediterranean. The algal bloom may be dispersed by these eastward flowing currents.
4. The nutrient-enriched surface seawater layer (due to winter convections and eastern drainages), could maintain algal species-specific growth rates during their transportation, forming patches with different phytoplankton size-classes.

a-

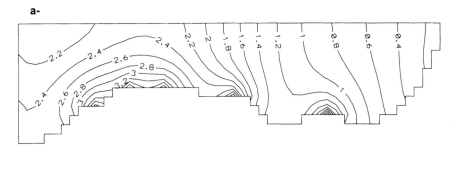

b-

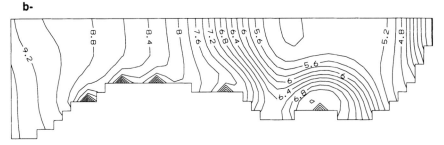

Fig. 13 Simulation of October–December distribution of **a** surface layer $PO_4 - P$ $(mg\,m^{-3})$ and **b** average 10 m layer chlorophyll a $(g\,m^{-2})$

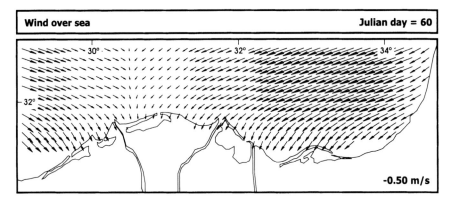

Fig. 14 Simulation of wind field average during the 60 Julian days along the Egyptian Mediterranean coastal waters

5. Due to the low grazing impact of both zooplankton and fish during the winter season, algal blooms may conserve their bulk for longer periods compared to the regular autumn blooms known in this area [32, 40].

Another application of the ecosystem model for the Nile delta coastal area involves the coupling between the hydrodynamic model (HYDRA), and the ecosystem FinEst model. This dual model gives real simulations and offers the possibility of using the model as an operational forecasting tool [42]. The results obtained using the dual model to simulate the Nile delta coastal area during winter show the extent to which the freshwater discharging from the Rosetta estuary mouth, and other shoreline drainages, may influence water salinity variations with depth in that area (Fig. 15). Moreover, for a randomly selected time period during winter, the model calculations have demonstrated the fertility of the Nile delta coastal ecosystem. This was the result obtained for the available physical, chemical, and biological parameters, as shown in Fig. 16. The concluding remarks of Hamza et al. [42] regarding the model run results are that in the Nile delta coastal area, especially close to the shore, phytoplankton communities are dominated by small size-classes of phytoplankton (< 20 μm), such as phytoflagellates and picophytoplankton. The reason for this may be that their reaction rates are much faster than those of net phytoplankton, so the former are able to utilize nutrients more efficiently. Alternatively, the seawater turbidity due to the continued land-source discharge and the consequent low PAR (Photosynthetic Available Radiation) levels, could also explain the small size class dominance.

Based on the simulations, it was also concluded that the decision of the Egyptian government to reduce the discharge of Nile water through the Nile

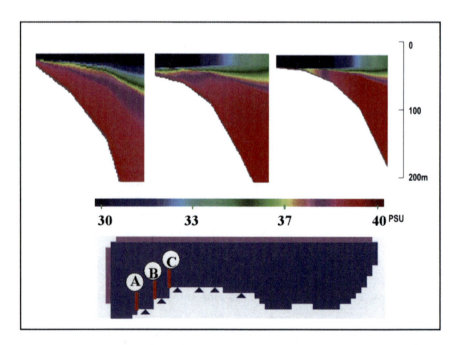

Fig. 15 Simulation of vertical distribution of salinity in three selected grids surrounding the Rosetta Promontory (after Hamza et al. [42])

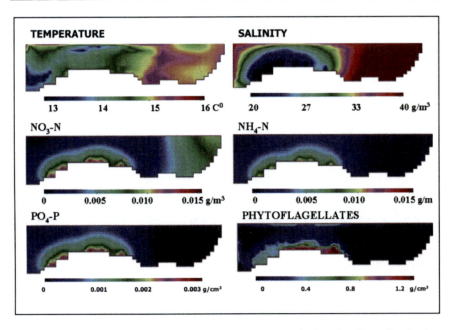

Fig. 16 Simulation of temperature, salinity, nutrients and phytoflagellate distribution along the Egyptian Mediterranean coastal area, showing the influence of the Nile water on the ecosystem parameters during winter (after Hamza et al. [42])

estuary may not significantly affect the nutrient concentration in this area. This is especially so if large volumes of primary treated sewage and agriculture drainage water are still being discharged. However, a slight increase in the average water salinity may occur, with the possible effect of modifying the community structure of the living biota. This may have a striking effect on fisheries in the area, particularly in the Levantine basin.

The above concluding points could explain the observed restoration of the ecosystem equilibrium after more than 30 years of disturbances associated with the AHD. It is well known that in restored unbalanced aquatic environments, chemical equilibrium may be reached after more than 20 years, while longer periods are necessary to reach biological equilibrium [41]. That may be the case for the estuarine environment of the Nile, the Nile delta coastal area, and by extension the southeastern Mediterranean environment.

From the preceding discussions it is obvious that the estuary of the Nile is not a simple environment, but one where both human interference and natural events play negative and positive roles in the final outcome. The future imperative is to balance human needs with nature conservation – a policy commonly referred to as sustainable development – which requires concerted effort to build a well-planned environmental strategy to conserve and manage the national (Egyptian) resources. To be sustainable that strategy requires the availability of necessary budgets, the collaboration of Nile basin countries, and finally, the development of infrastructures and coastal protection works that will preserve the Nile estuary environment.

References

1. Roskar J (2000) Geografskizbornik 53:32
2. CORI/UNESCO/UNDP (1978) Coastal protection studies. Final technical report, vol 1
3. Fanos AM, Khafagy AA, El-Kady MM (1999) Proceedings of the 7th Nile 2002 conference, Cairo, Egypt
4. Frihy OE (2001) Ocean Coast Manag 44:489
5. Abdel-Sattar AM, Elewa AA (2001) 2nd international conference and exhibition for life and the environment, Alexandria, Egypt
6. Frihy OE, Dewidar KhM, Nasr SM, El-Raey MM (1998) Int J Rem Sens 19:1901
7. El-Arabawy MM (2002) Proceedings of the 9th Nile conference, Kenya
8. Bellaiche G, Lonke L, Gaullier V, Mascle J, Courp T, Moreau A, Radan S, Sardou O (2001) Earth Planet Sci 333:399
9. Nafaa ME, Frihy OE (1993) Egypt J Coast Res 9:423
10. Stanley DJ, Warne AG (1993) Science 260:628–634
11. Hamza W, Tamsalu R, Ennet P, Kullas T (1996) Proceedings of the Canadian coastal zone management conference (CZC'96), Rimouski, Quebec, Canada
12. Hamza W (2001) Mediterranean Forecasting System Pilot Project UALEX-FS-DES, final scientific report. MFSPP-ISAO-CRN-EU Project
13. Ashok S (1998) Arab Studies Quart 20:1
14. Attia FA (1999) Proceedings of the 7th 2002 Nile conference, Cairo, Egypt
15. Abdel Ghany MB, Abdel Dayem MS (1996) Misr J Ag Eng Cairo Univ Irr Conf Cairo, Egypt
16. Mason S (2000) Water usage in the Nile basin. Project on environmental cooperation in the Nile basin (ECONILE), working paper no 1. EAWAG, ETH, SPF, Switzerland, p 53
17. El-Kady M, Millette J (2002) Proceedings of the 9th Nile conference, Kenya
18. Hamza W (2000) Economic development and compensatory measures related to the management of the Egyptian water resources. In: Water security in the third millennium. Mediterranean countries as a case. UNESCO Science for Peace Series 9:453
19. Halim Y (1991) The impact of human alternation of the hydrological cycle on ocean margin. In: Mantura RFC, Martin JM, Wollast R (eds) Ocean margin process in global changes. Wiley, New York, p 302
20. El-sayed MA, Aboul Naga WM, Halim Y (1993) Estuar Coast Shelf Sci 36:463
21. El-Nady FE, Dowidar NM (1997) Estuar Coast Shelf Sci 45:345
22. Abdel-Moati MA, El-Sammak AA (1997) Water Air Soil Pollut 3–4:413
23. Saad MH, Hassan EM (2002) Hydrobiologia 1–3:131
24. Dessouki SA, Soliman AI, Deyab MA (1993) J Environ Sci Mansoura Univ 6:159
25. Abd-Allah AMA, Hassan AA, El-Sebae A (1998) Lebensm Unters Forsch 20:25
26. Abbass MM, Abd-Allah AM, El-Gendy K, Ali HA, Tantawy G, El-Sebae AH (1991) Bull Inst Oceanogr and Fish 17:71
27. El-Gendy KS, Abd-Allah AM, Tantawy G, El-Sebae AE (1991) J Environ Sci Health 26:15
28. Abd-Allah AMA (1992) Toxico Environ Chem 3:89
29. Abd-Allah AMA, El-Sebae AE (1995) Toxico Environ Chem 47:15
30. Abu-El-Ela N, Abd-Allah AM (1997) J Egyptian Public Health Assoc LXXII:215

31. Abd-Allah AMA (1999) JAOAC 82:391
32. Dowidar NM (1984) Deep-Sea Res 31:983
33. Caputo R (1985) National Geographic 17:577
34. Stanley DJ (1988) Science 240:497
35. Frihy OE (1988) J Coast Res 4:597
36. Blodget HW, Taylor PT, Roark JH (1991) Mar Geol 99:67
37. Inman DL, Scott AJ (1984) The Nile littoral cell and man's impact on the coastal zone of the southeastern Mediterranean. SIO Reference Series 48–31, University of California
38. Stanley DJ (1996) Mar Geol 129:189
39. El-Sayed SZ, Van Dijken GL (1995) Ocean Dept Texas A & M Univ Quarterdeck 3.1
40. Dowidar NM (1988) Productivity of the south-eastern Mediterranean. In: El-Sabh MI, Murtty TS (eds) Natural and man-made hazards. p 477
41. Hamza W, Ennet P, Tamsalu R (1998) The ecosystem calculation for the Egyptian part of the Mediterranean. In: Tamsalu R (ed) The coupled 3D hydrodynamic and ecosystem model FinEst. MERI 35:143
42. Hamza W, Ennet P, Tamsalu R, Zalesny V (2003) J Aquat Ecol 37(3):307

Hdb Env Chem Vol. 5, Part (2006): 175–195
DOI 10.1007/698_5_028
© Springer-Verlag Berlin Heidelberg 2005
Published online: 6 October 2005

Environmental Quality of the Po River Delta

Alfredo Provini (✉) · Andrea Binelli

University of Milan, Biology Department, Via Celoria 26, 20133 Milan, Italy
alfredo.provini@unimi.it

Abstract The Po River collects the discharges of the most populated and industrialized area of Northern Italy and enters the Adriatic Sea with a mean flow of $1470 \, m^3 \, s^{-1}$ spreading in nine branches along the final stretch and forming a delta originating about 50 km from the sea. According to the last systematic survey carried out in 1989–1990, the Po delta waters suffer a low–moderate pollution from heavy metals and organic micropollutants. However, the Po River carries a nutrient load high enough to cause a severe marine eutrophication problem south of its delta. Provisional models have shown that even a substantial reduction on civil and industrial waste water discharges coupled with an optimal use of fertilizers in agriculture would not be sufficient to solve the problem. The Po River represents an important source also, for heavy metals and marine sediments collected at the river mouth are more polluted than are those offshore, especially for Cu, Zn and Hg. On the basis of the European guidelines the delta sediments have a medium to high contamination as far as the concentration of Ni and DDT are concerned, while data on persistent organic compounds indicate a moderate ecological risk for the biota living in the delta. These same conclusions can be obtained using the "sediment quality benchmark" procedure proposed by the NOAA in the United States.

Keywords Heavy metals · Nutrient loads · Po River delta · Persistent organic compunds · Water quality

Abbreviations

NOAA	National Oceanic and Atmospheric Administration
TDP	Total dissolved phosphorus
DOP	Dissolved organic phosphorus
PP	Particulate phosphorus
DRP	Dissolved reactive phosphorus
TP	Total phosphorus
TDN	Total dissolved nitrogen
TDIN	Total dissolved inorganic nitrogen
DON	Dissolved organic nitrogen
PN	Particulate nitrogen
TN	Total nitrogen
ISTAT	Italian National Statistical Institute
EU	European Union
POP	Persistent organic pollutants
FDA	Food and Drug Administration
d.w.	Dry weight
EBI	Extended biotic index
BMWPS	Biological Monitoring Working Party score
GEMS	Global environmental monitoring system

1
The Po Basin

The Po basin lies in Northern Italy beneath the Alps (Fig. 1). Its surface area (70 000 km^2) corresponds to about 25% of the entire Italian land area. The resident human population (about 16×10^6 peoples) is concentrated in 3500 administrative districts of three States (France, Switzerland, and Italy). Agricultural land occupies about 20% of the total and this supports about 50% of the Italian farm animal rearing (48% of cattle, 43% of pigs). About 34% of Italian industry operates in this territory and these consume 48% of energy demand of the country.

In the Po River Plain there is a volume of freshwater of more than $117\,930 \times 10^6$ m^3 due to the presence of the five largest Italian lakes (Garda $49\,030 \times 10^6$ m^3, Maggiore $37\,500 \times 10^6$ m^3, Como $22\,500 \times 10^6$ m^3, Iseo 7600×10^6 m^3, and Orta 1300×10^6 m^3), which account for 81% of the volume of all Italian lakes and reservoirs (145×10^9 m^3). Their outlets join the Po River.

The mean annual meteoric contribution measured at over 500 pluviometric stations is 1200 mm, which corresponds to 80×10^9 m^3.

The Po hydrographic system is inserted in this context and the main axis, represented by the Po, has a length of 677 km. The river course is conditioned by alluvial materials derived from a land surface of about 41 000 km^2 and transported by about 20 tributaries. Those coming from the Alps have a smaller turbid transport due to the presence of the lakes and have a more regular flow, whereas those from the south have more irregular flow and

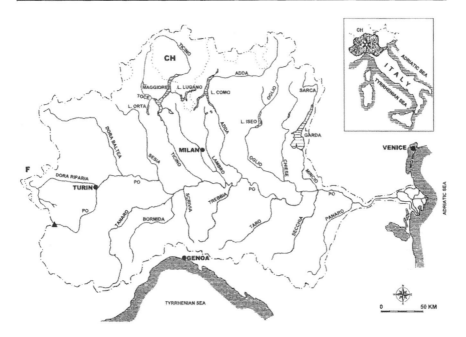

Fig. 1 River Po catchment basin

transport a large quantity of suspended material, which causes a migration of the Po river bed toward the north.

The watershed closes at Pontelagoscuro, 91 km distant from the Po mouth, and downstream from this point the river bed is higher than the surrounding plain.

Water abstraction for drinking purposes varies from a minimum of 100 L capita^{-1} day^{-1} to a maximum of 400 L capita^{-1} day^{-1}, making a total water uptake of 3×10^9 m^{3}year^{-1}, 70% of which is from underground sources [1]. The same total quantity of water is used by industry. Agriculture uses a water volume of 20×10^9 m^{3}year^{-1} (including 2×10^9 m^{3}year^{-1} abstracted from underground sources) most of which is used for irrigation of 70% of the rural land in the plain, equal to about 13 000 km^2.

The Po basin contains 273 hydroelectric and 11 power plants (two other nuclear power plants are presently closed) that produce a total of about 17 000 MW of nominal power; ten of these plants are located directly on the Po River.

2
The Po Delta

A peculiar element of the Po River hydrography is represented by its delta, which originates about 50 km from the sea. The present shape of the delta is due to natural events and anthropogenic interventions, especially during the seventeenth century. The delta has a surface area of about $400 \, km^2$, and its length increases by about $70 \, m \, year^{-1}$ corresponding to an area increase of about $50-60 \, ha \, year^{-1}$.

The official hydrological data published by the Po Hydrographic Office refers to measurements taken at Pontelagoscuro, mentioned above. This location is at 8.35 m above sea level (hydrometric zero) and characteristic daily flow data $(m^3 \, s^{-1})$ are as follows:

- Mean 1470
- Maximum 8940
- Minimum 275
- Absolute maximum 11 580

The annual hydrological regime is usually characterized by a period of low flow in the winter months (January–February) due to most precipitation over the watershed being unavailable in the form of snow and ice; a period of maximum flow (May–June) due to the subsequent thawing of snowfall; a second period of low flow (August) more accentuated than in the winter months, and finally a second period of high flow in October–November due to the heavy rainfalls occurring during these months.

The Po River enters the Adriatic Sea spreading out in nine branches along the final stretch (Fig. 2), the major branch (Po della Pila) carrying about 50% of the flow. The hydrodynamics of the Po River waters into the North Adriatic Sea varies according to the prevailing winds and are illustrated in Fig. 3 [2].

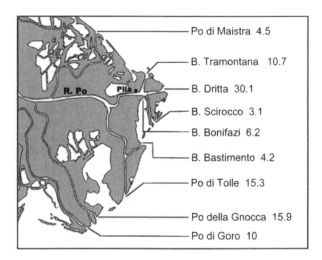

Fig. 2 River Po delta. Numbers refer to the percentage of the mean flow carried by each branch

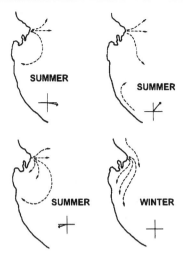

Fig. 3 Simplified scheme of the distribution of the Po River waters into the North Adriatic Sea

Thus:

- In summer with no winds or with winds having a velocity less than 5 m s^{-1}, the Po waters disperse in the open sea with an anticyclonic vortex; probably in this condition the southern Po branches carry the major part of the flow.
- In summer with low wind from SW or NW the situation previously described is modified slightly with openings appearing in the anticyclonic vortex. Only strong winds of more than 10 m s^{-1} from the NE can disperse the weak circulation condition in this zone.
- A drastic change occurs during winter time, when the Po River waters spread almost exclusively through the southern zone.

3
Water Quality of the Delta

Water quality of the Po River from its source to its mouth was assessed on two separate dates by the Italian National Council of Researches, taking into consideration the pollution entering the river. After that no other systematic researches were performed on water pollution. For this reason, data from the last study carried out in 1989–1990 are presented here relating to the closing station of Pontelagoscuro, which is representative of the water conditions of the whole delta.

Po waters were characterized by a fairly moderate undersaturation of O_2 (about 10%). Large variations were recorded depending on flow rate that can resuspend sediments and the overall respiration of the biotic community settled on stable substrates (macrophytes, algae and epiphytic or epilytic fauna, bacteria). No significant correlation was found between oxygen concentration and BOD or organic carbon [3], as already reported for other rivers (GEMS, 1989) [4].

The ionic spectrum (Fig. 4) determined during 1990 showed the presence of limestone, gypsums, and silicates of Ca, Fe, and Mg [5]. The spectrum was similar to the one presented by Allegrini et al. [6] in relation to the geomorphological aspects of the high–medium reach of the Po River, with minor variations that do not modify the relationships between variables.

Regarding nutrients, in four sampling points of the delta (Fig. 5) the mean total dissolved phosphorus (TDP) was made up mainly by inorganic forms (90%) and had a value similar to that of particulate phosphorus (PP) [7]. The orthophosphate phosphorus (DRP) represented 50% of the total phosphorus (TP). The P level reached a maximum in 1981, when the average DRP was as

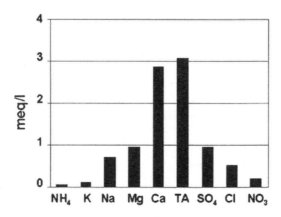

Fig. 4 Ionic spectrum of the Po River waters at the closing section of the basin

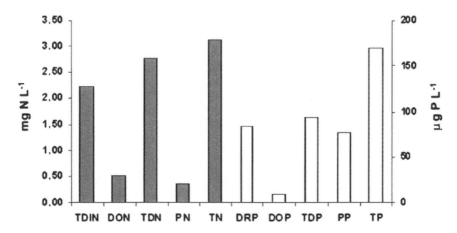

Fig. 5 Nutrient concentration in the Po River waters at the closing section of the basin

high as $160 \,\mu g\,L^{-1}$ [8] but, after the introduction of phosphate-free detergents and improvement in the quality of effluents and sewage treatment plants, the P levels decreased constantly, levelling off at about $60 \,\mu g\,L^{-1}$ as from 1994 [9].

Almost all of the total dissolved nitrogen (TDN) was inorganic (Table 1) and the nitrogen in particulate form (PN) was only about 10% of the total (TN) [7]. A seasonal trend was observed for ammonia, with a minimum during May–August (about $0.1-0.2 \,mg\,L^{-1}$) and a maximum in winter because of the influences both of flow and agricultural practices in the basin. No downward trend similar to that with phosphorus can be observed in the nitrogen concentration, as values around $3 \,mg\,L^{-1}$ were measured in 1997 [10]. Today values still seem to be rising, after a small reduction in 1990–1991, because of the intense agricultural exploitation of the Po Valley [9].

Data on the concentration of heavy metals both in water and in suspended sediments are reported in Table 1 [11]. To have a picture of the changes due to anthropogenic impacts requires the calculation of the mean enrichment factor defined as the ratio between the experimental metal concentrations and the corresponding reference values [12, 13]. In this respect Cd, Hg, and Pb were the elements that showed a higher degree of enrichment compared to the expected natural levels, followed by B, Cr, Cu, Zn, Ni, and As. The concentration of dissolved metals in Po River waters depends on various factors in addition to such input variations in the basin from production cycles, domestic discharges, changes of flow, and erosion processes. For example, an interesting positive correlation between metal concentrations and tem-

Table 1 Mean concentration and standard deviation of metals at the closing section of the Po River basin

Metal	n of data	Water ($\mu g/L$)	Suspended solids ($\mu g/g$)
Al	35	1789.3 ± 4245	21206 ± 9523
Fe	35	1577.8 ± 2964.9	22420 ± 818
Mn	35	78.7 ± 106.0	1353 ± 281
B	10	15.2 ± 24.9	102.5 ± 50
Ba	35	7.5 ± 12.8	109.0 ± 32
As	34	0.41 ± 0.44	7.45 ± 2.3
Cd	35	0.075 ± 0.05	1.74 ± 0.72
Co	35	0.72 ± 0.93	11.9 ± 2.8
Cr	35	8.21 ± 11.8	124.0 ± 48
Cu	35	4.52 ± 5.95	73.0 ± 14
Ni	35	7.50 ± 11.52	112.0 ± 41
Pb	35	4.49 ± 4.01	75.0 ± 102
Zn	35	16.30 ± 16.01	342.0 ± 217
Hg	24	0.066 ± 0.059	0.002 ± 0.002

perature or orthophosphate phosphorus was found, indicating that biota can contribute to the control of the dissolved metal concentrations in a significant way. Such biotic effects have also been observed in lakes and oceans [14, 15].

Data on the concentration of organic micropollutants have a high heterogeneity regarding the presence and type of active principles. Among herbicides, molinate, atrazine, simazine, and alachlor were the most often present compounds in the delta waters [16]. These occur especially from May to June, according with their use in agricultural practices, with maximum values of $3.94 \, \mu g \, L^{-1}$, $0.88 \, \mu g \, L^{-1}$, $0.26 \, \mu g \, L^{-1}$, and $0.21 \, \mu g \, L^{-1}$, respectively. Use of atrazine was forbidden in 1988.

The only two organophosphate insecticides found were diazinon and phorate. These compounds are the most used in agriculture, according to the sales data given by the Italian National Statistical Institute (ISTAT, various years). The levels of phorate were around $4 \, ng \, L^{-1}$ with a maximum of $19 \, ng \, L^{-1}$ while those of diazinon, present only in 22% of samples, were between 1 and $4 \, ng \, L^{-1}$. These concentrations are below the maximum values set by the EU Directive ($1 \, \mu g \, L^{-1}$) for superficial waters usable for drinking purposes.

With regard to organochlorine pesticides, only one value of lindane ($0.11 \, \mu g \, L^{-1}$) exceeded the criteria set by the EU Directive for aquatic life [16]. Other values were low with a mean of $2.25 \, ng \, L^{-1}$. The time trend in concentration values of this pesticide is downward since the use of this compound has been limited by law since the mid 1970s. Pollution by HCB was detected only in 11% of analyzed samples, with the concentration being very low (between 1 and $2 \, ng \, L^{-1}$).

4
Heavy Metals
and Persistent Organic Pollutants in Living Organisms and Sediments

4.1
Living Organisms

Few researches have been carried out on organisms living along the Po River and especially in the delta. Available data refer mainly to fish and these are difficult to compare because different species of organisms have different accumulation capacities, depending on their position in the trophic chain and their lipid content.

Leonzio et al. [17] and Focardi et al. [18] measured Hg, Cd, and Pb levels in muscular tissue and in the liver of ten species of fish from the Po delta (Table 2). Data showed that Hg concentrations are very roughly similar in the two examined tissues, with a range between 0.06 and $1.06 \, mg \, kg^{-1}$ d.w. On the other hand, Cd and Pb tend to accumulate mainly in the liver with maximum values considered across all species of 0.78 and $2.3 \, mg \, kg^{-1}$ d.w., respectively.

Table 2 Heavy metal concentration (mg kg^{-1} d.w.) in fish of the Po delta

Species	Tissue	Hg	Cd	Pb
Eel	M	0.12 (0.08)	< 0.02	< 0.10
(*Anguilla anguilla*)	L	0.40 (0.21)	0.52 (0.45)	1.89 (0.59)
Carp	M	0.25 (0.06)	< 0.02	< 0.10
(*Cyprinus carpio*)	L	0.19 (0.05)	0.70 (0.29)	1.29 (0.06)
Pike	M	0.27 (0.07)	< 0.02	< 0.10
(*Esox lucius*)	L	0.25 (0.04)	0.08 (0.02)	0.74 (0.46)
Black bullhead	M	0.14 (0.09)	< 0.02	< 0.10
(*Ictalurus melas*)	L	0.06 (0.06)	0.29 (0.19)	0.41 (0.22)
Golden gray mullet	M	0.18 (0.15)	< 0.02	< 0.10
(*Liza aurata*)	L	0.20 (0.06)	0.64 (0.11)	2.32 (1.56)
Pumpkinseed	M	0.49 (0.16)	< 0.02	< 0.10
(*Lepomis gibbosus*)	L	0.08 (0.06)	0.78 (0.44)	0.66 (0.49)
Mullet	M	0.32 (0.09)	< 0.02	< 0.10
(*Mugil cephalus*)	L	0.58 (0.21)	0.63 (0.63)	2.20 (1.65)
Bass	M	0.59 (0.38)	< 0.02	< 0.10
(*Micropterus salmoides*)	L	0.25 (0.03)	0.51 (0.13)	0.42 (0.25)
Shi drum	M	0.77 (0.35)	< 0.02	< 0.10
(*Umbrina cirrosa*)	L	0.84 (0.68)	0.03 (0.01)	0.83 (0.25)
Grass goby	M	1.06 (0.68)	< 0.02	< 0.10
(*Zosterisessor ophiocephalus*)	L	0.89 (0.42)	0.08 (0.05)	1.57 (1.09)

standard deviations are in parenthesis; M muscle, L liver

Trace metal levels showed a moderate contamination of Hg, Cd, and Pb. This was interpreted to be due to the fact that the industrial inputs are located only in limited areas far from the delta. Moreover, salinity, particulate matter, and algal biomass are able to bind metallic ions, also diminishing their bioavailability.

In detail, the results in Table 2 show that the highest Hg levels (0.59–1.06 mg/kg^{-1} d.w.) were measured in muscle of bass, shi drum, and grass goby, whilst the lowest concentrations were detected in those of eel and bullhead (0.12–0.14 mg kg^{-1} d.w.). The value of Hg in liver ranges from one to six times that measured in the muscle. This is probably because of the different percentage of the metal in organic form present in animal diets. A high percentage of organic Hg causes a low Hg level in the liver and a high accumulation in the muscle. In contrast, the inorganic Hg form is more easily metabolized and thus is present at high concentrations in the liver but not in the muscle.

Levels of Cd and Pb in fish muscle were lower than those of Hg, due to the metabolic capability of controlling these metals. Concentrations in muscle were always below the detection limits (Table 2). In comparison, Cd and Pb values in the liver were much higher than in muscle and ranged from 0.03 to 0.78 mg kg^{-1} d.w. and 0.41–2.32 mg kg^{-1} d.w., respectively.

Some trace metals were analyzed by Camusso et al. to assess bioaccumulation and interspecies differences along the different trophic levels of the deltaic food web [19]. The highest values for essential (Cu = 61.3 µg g^{-1}, Zn = 369 µg g^{-1} d.w.) and more toxic (Hg = 0.44 µg g^{-1}, Cd = 0.75 µg g^{-1}, Pb = 13.03 µg g^{-1} d.w.) trace metals were found in the scavenger *Cyclope neritea*, which is closely tied to the sediment. The highest Ni and Cr concentrations were mainly measured in the bivalve *Spisula subtruncata* (10.7–20.4 µg g^{-1} d.w.) and in the macroalga *Ulva rigida* (9.8–11.6 µg g^{-1} d.w.), indicating a higher presence of these metals in water and suspended matter rather than in sediments.

Table 3 shows the average concentrations of some trace metals found in different omnivorous fish species and filter-feeding organisms collected at the closing section and in the Po River delta (Pila) [20]. No clear differences were noticed among metal concentrations measured in the filter-feeder soft tissues at the two sampling stations, showing the lack of point and diffuse sources in the deltaic area. Concentrations of Cd, Pb, Cr, Ni, and Zn were always higher in soft tissues of bivalves than in fish muscle sampled from Pontelagoscuro. Hg shows opposite behaviour, as has also been found in France in the River Seine [21] and Loire estuary [22]. Contamination levels of all metals in the aquatic biota seem to reflect the environmental concentrations of the habitat occupied by the organisms (sediments, suspended matter, or water) more closely than their place in the food chain [20].

Regarding organic micropollutants, most data refers to organochlorine pesticides and PCBs. Such attention to this category of pollutants has occurred not only in the Po basin but also in all aquatic ecosystems, mainly because the environmental risk of these compounds is due to their high bioaccumulation potential.

Table 3 Heavy metal concentration (µg g^{-1} d.w.) in some organisms of the Po delta

	Sampling site	Cd	Hg	Pb	Cr	Ni	Zn
Omnivorous fish	Pontelagoscuro	0.02 (0.01–0.05)	0.69 (0.34–1.31)	1.62 (0.51–4.27)	0.34 (0.17–0.53)	0.65 (0.37–1.40)	23 (10–37)
Filter feeders	Pontelagoscuro	1.82 (0.27)	0.18 (0.02)	3.79 (0.55)	3.96 (0.56)	2.7 (0.8)	443 (16)
	Pila	0.77 (0.57–0.97)	0.11 (0.08–0.13)	3.09 (3.01–3.18)	2.67 (1.66–3.68)	3.6 (2.9–4.2)	332 (278–385)

Since end of the 1980s most organochlorine pesticides have been forbidden for agricultural uses and PCBs were limited to non-dispersive uses. Hence, it would be expected that their environmental contamination should have diminished after this time. However, from the examination of data published by Galassi and Gandolfi [23] and Galassi et al. [24] it is possible to observe that DDE and HCB concentrations were stable ten years after their prohibition; lindane was reduced but PCBs have increased (Table 4).

Focardi et al. (1991) [18] analyzed some persistent organic pollutants (POP) in several fish species sampled in the Po delta. Their results showed high levels of PCBs in all the species with concentrations ranging from 0.1 to 6.7 mg kg^{-1} d.w., while levels of HCB, pp'DDT and its main metabolite pp'DDE were very low (Table 5).

The use of bivalve *Dreissena polymorpha* by Binelli et al. [25] confirmed the presence of high levels of PCBs and DDTs still in the middle of the 1990s (Table 6), suggesting the lack of environmental recovery after the limitation by law of these compounds. A more recent study showed a high PCB contam-

Table 4 Concentration of organochlorine pesticides and PCBs (mg kg^{-1} lipids) in fish in the Po delta

Species	Year	Lindane	pp'DDT	pp'DDE	pp'DDD	HCB	PCB
Chub	1980	0.078	n.d.	2	0.4	0.16	21.2
	1990	0.02	n.d.	2.2	n.d.	0.2	119.3
Bleak	1980	0.102	n.d.	1.9	0.8	0.15	18.6
	1990	0.02	n.d.	1.3	n.d.	0.05	63.6
Pike	1989	–	0.4	2.7	–	0.2	13

Table 5 Concentration of organochlorine pesticides and PCBs (mg kg^{-1} d.w.) in the Po delta in 1989

Species	Specimens	HCB	pp'DDE	pp'DDT	PCBs
Carp (*Cyprinus carpio*)	6	0.008 (0.007)	0.1 (0.05)	0.018 (0.01)	3.089 (2.437)
Pike (*Esox lucius*)	4	0.002 (0.001)	0.024 (0.014)	0.004 (0.002)	0.114 (0.069)
Bass (*Micropterus salmoides*)	8	0.004 (0.002)	0.094 (0.039)	0.028 (0.014)	6.691 (2.762)
Pumpkinseed (*Lepomis gibbosus*)	8	0.004 (0.003)	0.033 (0.026)	0.017 (0.016)	1.569 (1.299)
Black bullhead (*Ictalurus melas*)	8	0.003 (0.002)	0.025 (0.015)	0.006 (0.005)	0.261 (0.21)

Table 6 Concentration of organochlorine pesticides and PCBs (mg kg^{-1} d.w.) in *Dreissena* exposed in the Po delta during 1994

Sampling site	PCBs	HCB	Lindane	pp'DDE	pp'DDD	pp'DDT
Pila	2.24	0.0075	0.002	0.11	0.011	0.019

ination in the muscle of eel (*Anguilla anguilla*) caught in the Po delta, with mean levels of about 250 μg kg^{-1} w.w. [26]. These values were similar to those found in other European and North American areas that are not considered to be heavily contaminated. A real hazard for human health is not expected because these values are lower than the PCB level for edible fish (2 mg kg^{-1} w.w.) proposed by the FDA [27]. Also pp'DDT was detected in the eel soft tissues at rather elevated concentrations (0.004 mg kg^{-1} w.w.), although this insecticide has been banned in Italy since 1978. The most abundant DDT metabolite was pp'DDE, which reached a concentration of 0.027 mg kg^{-1} w.w., seven times higher than that of the parental compound, followed by the pp'DDD at 0.020 mg kg^{-1} w.w. Taking into account that the lipidic percentage in this fish was less than 5% of the body dry weight, the sum of the pp' isomers (0.051 mg kg^{-1} w.w.) is higher than the limits of Σ DDT (0.05 mg kg^{-1} w.w.) set by the European Union for edible fish (Directive 99/71/CE), keeping also in mind the lack of the op'DDT, DDD, and DDE values.

4.2
Sediments

Data on the contamination of sediments of the Po delta are more frequent and up-to-date in comparison with those on organisms. Camusso et al. [20] pointed out that the metal concentration in sediments at the closing section of the river is similar to that found in Po River sediment before its confluence with the Lambro River, close to Milan, which carries the waste waters of one of the most industrialized and urbanized areas of Italy (Table 7). This result would suggest a fairly good environmental condition of the last section of the Po River.

The picture of contamination changes in the delta area because Hg, Cd, Pb, Cu, and Zn levels measured at Pila, located along the main branch of the Po River (Fig. 1), are higher. This increase probably is a result of the intrusion of the tidal salt wedge, which can transport heavy metals from other coastal zones or can induce a higher deposition than at Pontelagoscuro. Metal values measured at Pila are absolutely comparable to those observed after the confluence with the Lambro River, this latter being an area of low environmental quality.

Recent research by Camusso et al. [28] showed a not-negligible DDT contamination in delta sediments (Table 8), with values that were even higher than those measured in a previous study carried out in 1990 in the same area.

The parental compound and its isomer op'DDT, which in 1990 were lower than the detectable limit, were readily detectable in 2001 indicating a new input of this insecticide. There is some evidence that this contamination derived from the transport along the Po River of DDT from Lake Maggiore, which in 1996 suffered a heavy DDT contamination by accidental industrial discharges (DDT production is not forbidden in Italy). Additional inputs may result from other uses of this compound in floriculture around the lake.

In contrast to the above findings with DDT, PCB concentrations were lower in 2001 than those detected in the study of 1990. This result was also the case at other sampling points along the river.

In recent years Galassi et al. [29] investigated the sediments at the closing section and found a PAH contamination of petrogenic origin (Table 9). The compounds that were mainly detected were those with three condensed rings, while the heaviest ones were poorly present. The main sources of this contamination are chiefly due to accidental leakages from pipelines or other petrochemical activities, but also from vehicular traffic and water runoff from local road surfaces.

Table 7 Heavy metal concentration ($\mu g\,g^{-1}$ d.w.) in fine-grained sediment ($< 63\ \mu m$)

Heavy metal	Pontelagoscuro	Pila
Hg	0.172 (0.002)	0.344 (0.004)
Cd	0.256 (0.041)	0.928 (0.076)
Pb	27.9 (0.5)	53.0 (3.4)
Cu	56.3 (4.1)	98.3 (6.2)
Cr	227 (19)	185 (16)
Ni	126 (4)	100 (8)
Zn	138 (6)	206 (14)
Total C (%)	2.0 (0.10)	3.10 (0.01)

Table 8 DDT and PCB concentrations ($ng\,g^{-1}$ d.w.) in sediment of the delta

Total PCBs	pp'DDT	pp'DDE	pp'DDD	op'DDT	op'DDE	op'DDD
23.1	2.3	6.7	0.8	0.1	1.5	11.4

Table 9 Level of PAHs ($ng\,g^{-1}$ d.w.) in the delta

Total PAHs		4–6 Rings		2–3 Rings		Fluoranthene	
Sep 1996	Mar 1997	Sep 1996	Mar 1997	Sep 1996	Mar 1997	Sep 1996	Mar 1997
609	489	307	182	302	307	27	64

Meador et al. [30] state that PAH concentrations equal to 120 ng g^{-1} d.w. are low and that values higher than 17 µg g^{-1} d.w. are an indication of heavy pollution. In this respect, values of these contaminants in the delta area are therefore of moderate importance, a conclusion confirmed by the findings of Galassi [31] in sediments of some Italian lakes regarded as not very contaminated.

5
Measured Loads to the Adriatic Sea

5.1
Nutrients

Due to its geomorphologic and hydrodynamic characteristics, the northwestern Adriatic Sea is the largest area of the Mediterranean Sea that suffers most from eutrophication. Its coastal waters have frequently experienced algal blooms leading at times to living organism kill and unpleasant conditions for tourism [32]. Even if this phenomenon can be partially attributed to the natural characteristics of the North Adriatic Sea [33], several studies have indicated that eutrophication problems derive from the increasing nutrient inputs, mainly phosphorous, of the tributary rivers [2–34]. Among these tributaries, the Po River is the most important, accounting for 50% of the total freshwater discharge into the Adriatic Sea. In 1978 it was estimated to carry 27% of P and 22% of N of the whole Italian production [35]. Moreover, it was shown that harmful eutrophication effects appear in coastal waters south of the Po delta when salinity is low [2], demonstrating that Po discharge and counterclockwise sea currents are involved in the phenomenon. Since 1968 many researches have been performed to evaluate the nutrient load of the Po River [36–39]. Functional relationships between nutrient concentration and flow rates showed a good correlation between these two variables. For reasons of uniformity between the various sets of data, the daily mean flow was used in the calculations of the annual load, which was achieved through integration of the product of the flow by the concentration over the period of time considered. Unfortunately, a lack of data prevented the quantification of the solid transport on the bottom as well as the possibility of taking into account the heterogeneity of concentrations in the transverse section. The load data evaluated from the four most recent investigations, covering a wide range of flow conditions, are presented in Table 10.

The eutrophication problems in the North Adriatic Sea led to the setting of directives, aimed at a reduction of the nutrient load of the Po River, in accordance with European Union directives that force the EU members to reduce their nutrient discharges into the aquatic environment. De Wit and Bendoricchio [44] tried to estimate the effect of emission reduction at the outlet of the Po River up to the year 2020. Four scenarios were defined:

Table 10 Nutrient loads carried by the Po River into the Adriatic Sea in different periods

Period	NumberQ (m³s⁻¹)		TDP t(year⁻¹)	TP t(year⁻¹)	TDN t(year⁻¹)	TN t(year⁻¹)
1982–1987 [a]	73	1394	5319 [e]	12 221	95 000 [f]	106 562 [h]
1988–1990 [b]	37	1362	3978	7300	118 900	144 175
1990–1993 [c]	92	1495	3650 [e]	9855	117 165 [g]	163 885
1995–1996 [d]	12	1596	3310	5770	136 100	147 200

[a] [40], [b] [41], [c] [42], [d] [43], [e] $P - PO_4$, [f] $N - NO_3$, [g] TDIN, [h] TIN

Table 11 Predicted loads for the Po River in 2015–2020

Nutrient load (t year⁻¹)	Scenario 1	Scenario 2	Scenario 3	Scenario 4
Total P	7700	6800	5400	4500
Total N	125 000	95 000	89 000	59 000

1. *No change* assumes that the load in 2020 will be reached by a linear change from 1990–1995 to 2015–2020
2. *Balance farming* assumes that nutrient surplus has been limited, so that it does not exceed 25 kg N and 5 kg P year⁻¹ ha⁻¹
3. *Optimal waste water treatment* assumes that the waste water treatment technology can remove 90% of the nutrient load
4. *Balance farming and optimal waste water treatment*) is simply a combination of scenarios 2 and 3

The results of this evaluation are presented in Table 11 and show that only scenario four, a substantial reduction in waste water discharges, coupled with an optimal use of fertilizers in agriculture (the major source of nutrients), can lower the N load of the Po River. However, the reduction of P concentration, which is the limiting factor for algal blooms in the Adriatic Sea, does not seem so effective. This last point needs to be considered along with the theoretical calculations of Chiaudani et al. [45], who demonstrated that even a P load of about 3000 t year⁻¹ does not guarantee complete oligotrophic conditions for the coastal marine waters.

5.2
Heavy Metals

Heavy metal transport to the Adriatic Sea has been studied to a lesser extent than has nutrient loading because of the complexity of the exchange phenomena between water and suspended solids, which means that metals are

Table 12 Mean particulate and dissolved loads of the Po River at the closing section of the basin (Pontelagoscuro)

Element	Particulate load 1988–90 (tyear^{-1})	Particulate load 1995–96 (tyear^{-1})	Dissolved load 1995–96 (tyear^{-1})
Fe	91 000	118 000	99.8
Mn	4332	5906	109.6
As	70	127	70.3
Cd	6.9	3.9	0.94
Co	40	134	1.31
Cr	536	971	51.6
Cu	314	421	69.6
Ni	515	576	66.7
Pb	171	147	4.36
Zn	1026	3179	168.7

not always present in the dissolved phase. The results for heavy metal loads measured in two separate surveys are listed in Table 12 [43].

From Table 12 it is evident that heavy metals are transported mainly in the solid phase, especially Fe, Mn, Co, Cr, and Zn. The loads seem to be steady or slowly increasing over time, but the changing hydrological conditions and levels of solid transport may have influenced these findings. As an example, it was found that the heavy metal load was 16–50% of the yearly load during the flood of November 1994 [46]. Camusso et al. [43] state that trace metals have short residence times in the waters of the Po delta. This implies that riverine loads probably represent the net fluxes to the North Adriatic and that dilution appears to be the main factor controlling the transport of most trace metals in low to medium flow conditions. A more recent study showed that the Po River represents an important source of heavy metal contamination to the North Adriatic Sea and marine sediments collected at the river mouth are more polluted than those of the open sea (Table 13) [47].

Coastal sediment contamination is most marked with regard to Cu, Hg, and Zn, whose concentrations are heavily affected by anthropogenic sources. These metals have an enrichment factor in the coastal sampling stations of seven times higher than those located in the open sea; similarly, Cd, Cr, and Ni are enriched by a factor of three. Heavy metal contamination measured off-shore has levels equal to or even lower than those found in sediment cores collected by Guerzoni et al. [48] in the same sampling area and considered since that time to be the background values for the Adriatic Sea.

Organic carbon (OC) plays an important role in the transport and distri-bution of metals in the delta area. Also the quantity and the composition of transported suspended solids are not of negligible importance in influenc-ing the contamination level [49]. Camusso et al. [47] have found a significant

Table 13 Heavy metal concentrations ($\mu g\,g^{-1}$ d.w.) in sediments of the Po River delta and in the coastal area facing it

Heavy metal	Riverine	Estuarine	Coastal	Off-shore
As	3.0 (0.1)	3.8 (0.4)	10.6 (0.8)	5.5 (0.9)
Cd	0.12 (0.04)	0.07 (0.03)	0.35 (0.15)	0.13 (0.02)
Co	14 (1)	14 (1)	17 (2)	11 (1)
Cr	115 (1)	105 (18)	120 (31)	54 (5)
Cu	6.1 (0.2)	14 (9)	52 (5)	8.0 (2.4)
Hg (ng/g)	11 (3)	30 (16)	461 (223)	72 (11)
Ni	71 (1)	70 (5)	75 (24)	27 (3)
Pb	17 (1)	25 (7)	62 (4)	34 (7)
Zn	19 (1)	22 (17)	109 (15)	14 (9)
V	38 (9)	32 (6)	94 (12)	41 (3)
Mn	588 (82)	465 (32)	640 (80)	452 (82)
Fe (mg/g)	20.2 (3.1)	15.4 (3.0)	27.8 (3.1)	15.7 ($\pm$ 1.0)
OC (%)	0.1 (0.01)	0.3 (0.4)	1.0 (0.3)	0.4 (0.1)

correlation ($p < 0.001$) between metal concentrations and organic carbon, suggesting that the remarkable differences between levels measured in the riverine environment and the two marine areas (Table 13) could be really due to the different OC percentage as well as to the size of the sedimented material. Several studies carried out in the North Adriatic Sea [48] have pointed out that trace elements in marine sediments depend on the mineralogy and the size of suspended solids and that they are associated with the silt and clay fraction. Effectively, sediments transported by different rivers into the Po are quite different because of their provenance (Alps or Apennines) and the variability of the hydrologic conditions. Additionally, the morphology of the Po delta is generally quite dynamic and variable. Hence, the sediment average composition at the closing section of Pontelagoscuro is 40–50% silt and the remaining part sand with fine or medium granulometry.

6
Sediment Risk Assessment

The environmental quality of the Po River delta can be inferred mainly from the sediment risk assessment since only data on this compartment have been frequently collected in recent years.

The evaluation of the sediment quality, performed by the use of the European guidelines, showed that the deltaic zone suffers medium to high contamination as far as the concentrations of Ni and DDT are concerned, and medium to low contamination for Cu and PCB. In contrast, there is no contamination for As, Cd, Cr, Hg, Pb, Zn [28]. In detail, the levels of several heavy metals are lower than Dutch reference standards [47], even though the concentrations of Cd, Cr, Cu, and Pb in the deltaic zone were about twice, and Hg about seven times, those at Pontelagoscuro. On the other hand, data on persistent organic compounds indicate a moderate ecological risk, especially in some periods of the year, being higher than Dutch standards [32].

Galassi et al. [50] showed that the value of the EC_{50} for *Daphnia magna* required about a 50-fold concentration of Po water. However, lower concentration factors could lead to 100% immobility of *Daphnia* at certain times of the year.

Two recent biomonitoring campaigns measured the levels of some POPs and trace metals in the Po Delta. These studies used a variety of contemporary acute and chronic ecotoxicological tests with several organisms, both with whole sediment samples and with sediment extracts [51]. No acute toxicity was detected in sediments of the delta as indicated by the results obtained with trout larvae and the crustacean *Ceriodaphnia dubia*. The EC_{50} values for these species were even higher than those measured at the Po headwaters, considered as the control site (Table 14). The results obtained with bioluminescent bacteria (*Vibrio fischeri*) were controversial. The values obtained with whole sediment were about 50% lower than those revealed near the Po

Table 14 Ecotoxicological tests and biotic indices for the sediment risk assessment

| Parameter | Control site (Po springs) | | Delta | |
	Summer	Winter	Summer	Winter
Whole sediment tests				
O. mykiss growth (% TU, 7 days)	100	100	101.9	100.7
C. dubia biomass production (μg, 7 days)	21.4	21	18.1	19.6
V. fischeri (30 min EC_{50}, g/L)	30.3	8.6	12.6	3.4
Sediment extract tests				
C. dubia (48 h EC_{50}, g/L)	6.2	24.7	22.8	15.2
V. fischeri (30 min EC_{50}, g/L)	1.25	4.14	6.65	36.8
Biotic indexes				
EBI	9	8	9	7
BMWPS	86.5	68	97	65.5
Taxonomic units	19	13	21.5	13

EBI extended biotic index, *BMWPS* biological monitoring working party score, *TU* toxicity units headwaters, while toxicity decreased dramatically in the sediment extracts (Table 14). Such findings suggest that the more toxic compounds are probably tied to the organic fraction. As an additional point, the use of macroinvertebrates also showed the good quality of the deltaic benthic community,, which was shown to be comparable to that found in the upstream control station [10].0

Viganò et al. [51] showed that at sub-toxic doses, sediment samples collected in the final stretch of the Po River were not mutagenic in tests with *Salmonella typhimurium his⁻* strains TA98, TA100 and TA102, irrespective of the presence of activator agent S9 extract.

In conclusion, on the basis of the "sediment quality benchmark" procedure proposed by the NOAA (National Oceanic and Atmospheric Administration) in the 1990s and subsequently developed [52, 53], levels of the examined pollutants are moderate to low and they are associated to a moderate "possible" risk for the biota living in the Po River delta [54]. Only Cr and Ni should be associated with a "high probable" risk, but their geological origin needs further studies.

References

1. Marchi E (1983) Genio Rurale 46:27
2. Marchetti R (1984) Quadro di sintesi delle indagini svolte dal 1978 sul problema dell'eutrofizzazione delle acque costiere dell'Emilia-Romagna. In: Regione Emilia-Romagna (ed) Eutrofizzazione dell'Adriatico – Ricerche e linee di intervento. Regione Emilia-Romagna, Bologna, Italy, p 471
3. Provini A, Crosa G (1991) Ossigeno e fattori di deossigenazione delle acque. In: CNR-IRSA (ed) La qualità delle acque del Fiume Po negli anni '90. Quaderni 92:5.1
4. GEMS (Global environmental monitoring system) (1989) Biodegradable organics and oxygen balance. In: Meybeck M, Chapman D, Helmer R (eds) Global freshwater quality. A first assessment. WHO and UNEP, Alden Press, Oxford,p 306
5. Milan C, Tartari G, Previtali L (1991) Caratteristiche idrochimiche. In: CNR-IRSA (ed) La qualità delle acque del Fiume Po negli anni '90. Quaderni 92:2.1
6. Allegrini M, Lanzola E, Marchetti R, Piantino P, Vanini GC, Verde L (1977) La qualità delle acque nell'alto Po e nel tratto medio superiore. In: Indagini sulla qualità delle acque del Fiume Po. Quaderni 32:145
7. Tartari G, Milan C, Elli M (1991) Idrochimica dei nutrienti. In: CNR-IRSA (ed) La qualità delle acque del Fiume Po negli anni '90. Quaderni 92:6.1
8. Marchetti R, Provini A, Gaggino GF (1988) Eutrophication of inland and coastal waters in Italy. In: Maltoni C, Selikoff IJ (eds) Living in a chemical word. Ann NY Acad Sci 534:950
9. Camusso M, Vignati D, van de Guchte C (2000) Aquat Ecosys Health Mgmt 3:335
10. Buffagni A, Bordin F, Pieri A, Occhipinti A (2000) Comunità macrobentoniche (Parte II): applicazione di indici biotici e qualità biologica. In: CNR-IRSA (ed) Caratterizzazione dei sedimenti e qualità ecologica nel Fiume Po. Quaderni 113:226

11. Pettine M, Camuso M, Cogliati N, Ferrara R, Martinetti W, Macerati E, Mastroianni D (1991) Fattori di variazione della concentrazione dei metalli. In: CNR-IRSA (ed) La qualità delle acque del Fiume Po negli anni '90. Quaderni 92:3.1

12. Salomons W, Forstner U (1984) Limnol Oceanogr 32:112

13. Meybeck M, Chapman DV, Helmer H (1989) Global freshwater quality. A first assessment. WHO and UNEP, Alden Press, Oxford

14. Sigg L, Sturm M, Kistler D (1987) Metal transfer mechanisms in lakes. Thalassia Yugoslav 18:293

15. Bruland KW (1980) Earth Planet Sci Lett 47:176

16. Baraldi O, Baldi M, Davì ML, Gazzella L (1991) Pesticidi e cloro-derivati. In: CNR-IRSA (ed) La qualità delle acque del Fiume Po negli anni '90. Quaderni 92:8.1

17. Leonzio C, Mattei N, Franchi E, Morelli P, Fossi C, Focardi S (1990) Ambiente Risorse Salute 104:123

18. Focardi S, Fossi C, Leonzio C, Marsili L, Mattei N (1991) Idrocarburi clorurati e metalli pesanti in varie specie ittiche del delta del Po. In: CNR-IRSA (ed) La qualità delle acque del Fiume Po negli anni '90. Quaderni 92:SP/21.1

19. Camusso M, Martinotti W, Balestrini R, Guzzi L (1998) Chemosphere 37:2911

20. Camusso M, Balestrini R, Martinotti W, Arpini M (1999) Aquat Ecosys Health Mgmt 2:39

21. Teil MJ, Blanchard M, Carru AM, Chesterikoff A, Chevreuil M (1996) Sci Total Environ 181:111

22. Amiard JC, Amiard-Triquet C, Metayer C, Marchand J, Ferrè L (1980) Water Res 14:665

23. Galassi S, Gandolfi G (1981) Nuovi Annali di Igiene e Microbiologia XXXII:393

24. Galassi S, Mingazzini M, Battegazzore M (1992) Sci Total Environ 132:399

25. Binelli A, Galassi S, Mariani M (1996) Acqua & Aria 7–8:689

26. Bressa G, Sisti E, Cima F (1997) Mar Chem 58:261

27. FDA (2001) Fish and fisheries products hazards and controls guidance, 3rd edn. Center for food safety and applied nutrition, Washington DC

28. Camusso M, Galassi S, Vignati D (2002) Water Res 36:2491

29. Galassi S, Guzzella L, Lopez A (2000) Microinquinanti organici nei sedimenti. In: CNR-IRSA (ed) Caratterizzazione dei sedimenti e qualità ecologica nel Fiume Po. Quaderni 113:50

30. Meador JP, Stain JE, Reichert WL, Vanasi U (1995) Residue Rev 143:79

31. Galassi S (1992) Mem Ist ital Idrobiol 50:481

32. Vollenweider RA, Rinaldi A, Montanari G (1992) Sci Total Environ, Supplement 1992:63

33. Tomasino MG (1996) Ecol Model 84:189

34. Barmawidjaja DM, Van der Zwaan GJ, Jorissen FJ, Puskaric S (1995) Mar Ecol 122:367

35. Provini A, Mosello R, Pettine M, Puddu A, Rolle E, Spaziani FM (1979) In: Marchetti R, Castoldi F, Cappelletti E (eds) Metodi e problemi per la valutazione dei carichi di nutrienti. CNR, Ambiente, Rome, p 472

36. Fossato VU (1971) Ricerche idrologiche e chimico-fisiche sul Fiume Po a Polesella, giugno 1968-giugno 1970. Arch Oceanogr Limnol 17:125

37. Provini A, Pacchetti G (1982) Ingegneria Ambientale 11:173

38. Provini A (1984) Il carico di fosforo convogliato a mare dal Fiume Po. In: Regione Emilia-Romagna (ed) Eutrofizzazione dell'Adriatico – Ricerche e linee di intervento. Regione Emilia-Romagna, Bologna, Italy, p 471

39. Marchetti R, Provini A, Crosa G (1989) Mar Poll Bull 20:168

40. Provini A, Crosa G, Marchetti R (1992) Sci Total Environ, Supplement 1992:291

41. Marchetti R (1991) Quadro complessivo delle condizioni delle acque del Fiume Po negli anni '90 e tendenze evolutive. In: CNR-IRSA (ed) La qualità delle acque del Fiume Po negli anni '90. Quaderni 92:16.1
42. Pagnotta R, Caggiati G, Piazza D, Ferrari F (1995) Inquinamento 4:8
43. Camusso M, Bonacina M, Pettine M (1998) Fresenius Envir Bull 7:51
44. de Wit M, Bendoricchio G (2001) Sci Total Environ 273:147
45. Chiaudani G, Gaggino GF, Vighi M (1982) 5th AIOL (Associazione Italiana di Oceanologia e Limnologia) Congress, Stresa, Italy
46. Camusso M, Vignati D, Pagnotta R (2001) Inquinamento 24:50
47. Camusso M, Bonacina M, Martinotti W, Pettine M (2000) Verh Internat Verein Limnol 27:1260
48. Guerzoni S, Frignani M, Giordani P, Frascari F (1984) Environ Geol Water Sci 6:11
49. Mingazzini M, Camusso M, Trimarchi E, Onorato L (2001) Fres Environ Bull 10:291
50. Galassi S, Guzzella L, Battegazzore M, Carrieri A (1992) Ecotoxicol Environ Saf 29:174
51. Viganò L, Arillo A, Buffagni A, Camusso M, Ciannarella R, Crosa G, Falugi C, Galassi S, Guzzella L, Lopez A, Mingazzini M, Pagnotta R, Patrolecco L, Tartari G, Valsecchi S (2003) Water Res 37:501–518
52. Long ER (1992) Mar Poll Bull 24:38
53. Jones DS, Suter II GW, Hull RN (1997) Toxicological benchmarks for screening potential contaminants of concern for effects on sediment-associated biota, 1997 Revision. Technical Report ES/ER/TM-95/R4. Oak Ridge National Laboratory, TN, USA
54. Viganò L (2000) Analisi combinata degli indicatori studiati nei sedimenti e valutazione del rischio per la comunità acquatica. In: CNR-IRSA (ed) Caratterizzazione dei sedimenti e qualità ecologica nel Fiume Po. Quaderni 113:269

Hdb Env Chem Vol. 5, Part H (2006): 197–232
DOI 10.1007/698_5_030
© Springer-Verlag Berlin Heidelberg 2005
Published online: 18 November 2005

The Sacca di Goro Lagoon and an Arm of the Po River

Pierluigi Viaroli[1] (✉) · Gianmarco Giordani[1] · Marco Bartoli[1] ·
Mariachiara Naldi[1] · Roberta Azzoni[1] · Daniele Nizzoli[1] · Ireneo Ferrari[1] ·
José M. Zaldívar Comenges[2] · Silvano Bencivelli[3] · Giuseppe Castaldelli[4] ·
Elisa A. Fano[4]

[1]Department of Environmental Sciences, University of Parma,
Parco Area delle Scienze 33A, 43100 Parma, Italy
pierluigi.viaroli@unipr.it

[2]Institute for Environment and Sustainability, JRC-CCE, Ispra, Italy

[3]Environment Office, Province of Ferrara, Italy

[4]Department of Biology, University of Ferrara, Italy

Abstract The Po di Volano canal–Sacca di Goro lagoon is a small hydrographic system partially located in the southern part of the Po River Delta. The total surface area is $830\,km^2$ for the watershed and $26\,km^2$ for the lagoon, respectively. The watershed is exploited for agriculture, whilst the coastal lagoon is one of the most important European sites for clam (*Tapes philippinarum*) farming. The lagoon and small inland zones are also included in the Po River Delta Regional Park. Since the mid 1980s, in the Sacca di Goro lagoon abnormal macroalgal blooms occurred, mainly due to the proliferation of the green seaweed *Ulva rigida*. The enormous macroalgal biomass production was often followed by summer anoxia and dystrophic crises.

In this paper, a review of the main studies concerning altered nutrient cycling and water and sediment pollution is presented. Special attention is paid to the discussion of the different aspects of the watershed–coastal lagoon interactions: main features of the watershed and its evolution, anthropogenic pressures, river runoff influence, nutrient and other contaminant cycles, shellfish farming, and macroalgal blooms. Finally, a brief presentation of possible scenarios is given in an ecological economics perspective.

Keywords Aquaculture · Coastal lagoon · Eutrophication · Pollution · Watershed

Abbreviations

PV	Po di Volano canal
CB	Canal Bianco canal
GI	Giralda canal
PG	Po di Goro River
SG	Sacca di Goro
N	Nitrogen
DIN	Dissolved inorganic nitrogen
DON	Dissolved organic nitrogen
PON	Particulate organic nitrogen
TN	Total nitrogen (DIN+DON+PON)
P	Phosphorus
SRP	Soluble reactive phosphorus
DIP	Dissolved inorganic phosphorus
$Ca{\sim}PO_4$	Calcium-bound phosphate
$Fe{\sim}PO_4$	Iron-bound phosphate
Exch. PO_4	Exchangeable phosphorus
pw PO_4	Pore-water phosphate
COD	Chemical oxygen demand
BOD_5	Biological oxygen demand at five days
SOD	Sediment oxygen demand
AR	Ammonium nitrogen recycling
B	Biomass
LOICZ	Land Ocean Interactions in Coastal Zone, is a core project of the International Biosphere Geosphere Programme
NEM	Net ecosystem metabolism
C	Carbon
POPs	Persistent organic pollutants
DDT	Dichloro-Diphenyl-Trichloroethane, common name of the 1,1,1-trichloro-2,2-bis(4-chlorophenyl)ethane (name adopted by the International Union of Pure and Applied Chemistry)

TBT	Tributiltin
Hg	Mercury
Pb	Lead
Ni	Nickel
Cd	Cadmium
Cr	Chromium
Cu	Copper
Fe	Iron
^{134}Cs	Radioactive isotope of cesium
^{137}Cs	Radioactive isotope of cesium
Bq	Becquerel, measure unit of radioactivity

1
Foreword

The Po River drains a large part of northern Italy with a surface of $67\,000$ km^2 and about 17 million inhabitants. In 1991, a total load of 59.5 million of equivalent inhabitants was estimated, of which 69% was from the industrial compartment (mainly chemical and food industries), 26% from humans and 5% from agriculture [1]. In the 1980s, annual loads discharged by the Po River were in the range of $82\,000$–$188\,000$ t of total nitrogen and 9000–$23\,000$ t of total phosphorus, respectively [2]. The Po River Delta has a surface area of 730 km^2 and comprises five main arms and a number of coastal lagoons. The Po di Volano canal–Sacca di Goro lagoon is a small hydrographic system partially located in the southern part of the Po River Delta, which accounts for approximately 1% of the total area of the Po River catchment. Only the eastern part of this watershed belongs to the delta, whilst the western part is for the most part a human-regulated system resulting from a centuries-old reclamation (Figs. 1 and 2).

The scientific relevance of this small system depends upon the number of research programs that have been realized since 1987. The Sacca di Goro lagoon has been considered as an experimental field site for analyzing interactions among river basins and related coastal zones. Several international and national projects have been carried out dealing with water quality and eutrophication phenomena, ecosystem structure and processes, aquatic resource exploitation, relations between anthropogenic activities, impacts and preventive and remedial measures. In this paper, the main features of the watershed–lagoon system and its evolution are first presented. A review of the main studies concerning altered nutrient cycling and water and sediment pollution is also presented. Special emphasis will be devoted to the discussion of the different aspects of the watershed–coastal lagoon interaction: river runoff influence, nutrient and other contaminant cycles, shellfish farming, macroalgal blooms, as well as the main economical implications.

(a)

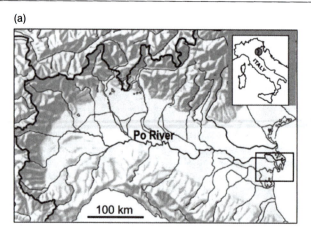

(b)

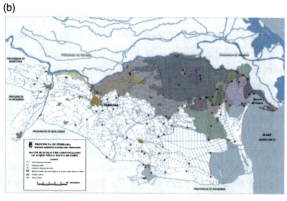

(c)

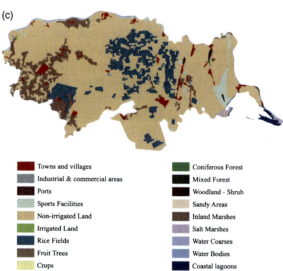

▮ Towns and villages	▮ Coniferous Forest
▮ Industrial & commercial areas	▮ Mixed Forest
▮ Ports	▮ Woodland - Shrub
▮ Sports Facilities	▮ Sandy Areas
▮ Non-irrigated Land	▮ Inland Marshes
▮ Irrigated Land	▮ Salt Marshes
▮ Rice Fields	▮ Water Coarses
▮ Fruit Trees	▮ Water Bodies
▮ Crops	▮ Coastal lagoons
▮ Complex Cultivation	▮ Sea
▮ Broad-leaved Forest	

◀ **Fig. 1** **A** Map of the Po River plain. The *square* comprises the watershed of the Po di Volano–Sacca di Goro and the Po River Delta. **B** Map of the hydrographic network of the Po di Volano–Sacca di Goro watershed. The main rivers and canals are evidenced: Po di Goro (PG), Po di Volano (PV), Canal Bianco (CB), Giralda (GI); Acque Alte (A), Acque Basse (B). SG: Sacca di Goro Lagoon, PH: Porto Garibaldi Harbor. **C** Map of soil and land uses of the Po di Volano–Sacca di Goro watershed

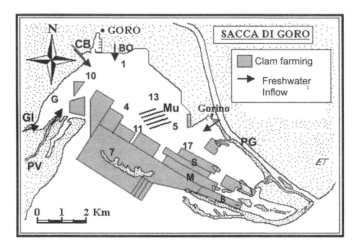

Fig. 2 Map of the Sacca di Goro indicating the sampling stations. Legend: PV = Po di Volano river, GI = Giralda canal; CB = Canal Bianco, BO = Bonello canal, GO = Po di Goro river, Mu = mussel farm

2
Landscape and Features of a Typical River Basin – Coastal Lagoon System in the Po River Delta

The Po di Volano-Sacca di Goro watershed has developed mostly in the last three centuries, when the modern Po River Delta has grown approximately 30 km seawards [3 and references therein]. The Sacca di Goro originated as a bay in the second half of the 18th century, when the Po di Goro arm was prograding southeastward. The current shape was attained in the first half of the 20th century, whilst in the last 20 years the lagoon has been reshaped several times in order to avoid dystrophic outbreaks and allow internal navigation.

All the deltaic area is influenced by sea eustatism (5–12 cm in the last century) and is subjected to a natural subsidence that was accelerated in the last 50 years by marshland reclamation and, especially, by groundwater and natural gas extraction [3]. Areas below the sea level can be found up to 40 km inland. The whole Sacca di Goro watershed lies in a depressed area; it can be estimated that in the last century the relative sea level has increased 107 cm at Codigoro town, which is approximately 15 km inland from the shoreline, and 139 cm at Volano village that is on the coast.

2.1
Hydrography of the Po di Volano Watershed

The Po di Volano–Sacca di Goro watershed is 860 km^2 and composes of three sub-basins: the Po di Volano (PV), the Canal Bianco (CB) and the Giralda (GI), which account for 670, 117, and 67 km^2, respectively (Fig. 1b and c).

Until the 12th century, the PV was one of the southernmost arms of the Po River Delta. Afterwards, water discharge decreased and, as a consequence of land reclamation and flow regulation, the PV became an artificial canal [4]. In 1987, the PV was further regulated by diverting the water flow into the Canale Navigabile, which outflows into the Adriatic Sea near the Porto Garibaldi harbor. At present, the PV watershed is restricted to the northeast part of the Province of Ferrara and lies in a depressed area. Here the drainage system is composed of a network of small canals that flow into two major canals (Acque Basse and Acque Alte). Water is then discharged into the PV by a system of pumping stations (scooping plants). The CB and GI canals are also man-regulated and partially connected with the PV. The overall hydrographic system is man regulated and, as a consequence, freshwater flows are partially independent of rain events. During the rainy period, approximately from late summer through late spring, the canals are kept empty in order to avoid the flooding risk. In summer, the drainage system is filled with freshwater that is released from the Po di Goro (PG) river for irrigation purposes. The freshwater which drains from the agricultural system is then delivered to the PV, CB and GI leading to summer peaks of freshwater discharge in the lagoon. In the last decade, the annual water discharge from the above-mentioned canals ranged from 336×10^6 (1998) to 594×10^6 (1996) m^{3}y^{-1}.

2.2
The Sacca di Goro Lagoon: A Human-Regulated Aquatic Ecosystem

The Sacca di Goro lagoon (SG) is a shallow-water embayment of the Po River Delta (44°47′–44°50′ N and 12°15′–12°20′ E); the surface area is 26 km^2, the total water volume is approximately 39×10^6 m^3 and the average depth is approximately 1.5 m (Fig. 2). The lagoon is surrounded by embankments. The main freshwater inputs are the PV, CB and GI canals. Freshwater inlets are also located along the Po di Goro arm (PG), but there are no direct estimates of the freshwater input from it [5]. Until 1994, input from PG was usually assumed to be equivalent to 10% of the discharge of the PG. After 1994 the PG connections to the SG were regulated by sluices and the total PG discharge into the SG was considered negligible compared to the other freshwater sources. On average, the water retention time ranges between 1 and 4 days, although in the sheltered areas water stagnation can occur [6]. At present, water exchanges between the Sacca di Goro and the adjacent Adriatic Sea depend on two openings in the southern sand barrier, which at present are approximately equivalent in width (about 900 m). The recent evolution of the Sacca di Goro was characterized by a continuous accretion of the sand barrier, which led to the formation of the westernmost sea mouth [7]. In the last 20 years, this principal mouth has suffered a progressive narrowing from 2580 m in 1980 to 1350 m in 1992. To relieve water stagnation, a channel was cut in 1993, which evolved into the second mouth, whereas the main mouth has narrowed even further. Numerical models demonstrated that in the early 1990s, the eastern area of the lagoon was sheltered and was separated from the western and central zones influenced by freshwater inflow from the Po di Volano and by the sea, respectively [8]. Moreover, the eastern sub-basin was very shallow (maximum depth 1 m) and accounted for one half of the total surface area and for one-fourth of the water volume. Here, due to the silting of the principal mouth, water stagnation was one of the major causes of the summer dystrophyc crises, which frequently occurred from 1987 to 1992 and in 1997 [9, 10].

Since 1998, the hydrodynamics has been further and radically modified with the dredging of internal canals [11]. During this intervention, the dredged material was disposed of in shallow areas in the eastern part of the lagoon to create permanent islands for preventing macroalgal growth.

The bottom of the lagoon is flat and the sediment is composed of typical alluvial mud with high clay and silt contents in the westernmost and northern zones. Sand is more abundant near the southern shoreline and along the main canals, whilst sandy mud occurs in the central and easternmost areas [12]. However, the upper sediment horizon is highly dynamic and is undergoing rapid changes also as a consequence of human exploitation.

2.3
Natural Heritages

The eastern part of the Po di Volano watershed and the Sacca di Goro lagoon belong to the Regional Park of the Po River Delta. The natural park was established in 1989 by the Emilia Romagna Region and comprises the historical delta and the southern part of the modern Po River Delta as well as aquatic and terrestrial environments along the coast (Parco Regionale del Delta del Po, http://www.parcodeltapo.it/). The natural park is composed of a wide variety of habitats and biotopes which are sites of national and international interest (Ramsar convention, UNESCO heritage sites, Sites of Interest for the European Community). Among these, there are the Sacca di Goro lagoon and an ancient Mediterranean forest (Gran Bosco della Mesola). The Sacca di Goro, with minor aquatic ecosystems, is important for waterfowl protection and conservation. The Gran Bosco della Mesola lies between the Po di Goro and the Po di Volano and has a surface area of about 10 km^2. The vegetation is formed mainly by Mediterranean species, namely holm and English oaks, ash trees, elms, white poplars, etc., with a relict population of pine at the edges.

Residual traces of Roman and Etruscan remains are still visible; whilst several monuments are well preserved, namely the monastic site of Pomposa with the abbey and a series of Byzantine mosaics, the Castle of Mesola and the scooping plants that were designed in the 16th century as the former hydraulic system for managing the water drainage.

2.4
Main Climate Features

The climate of the region is Mediterranean with some continental influence (wet temperate Mediterranean). The minimum monthly mean air temperature is 1.3 °C in January and the maximum monthly mean is 24.7 °C in July. Wet deposition is approximately 600 mm per year, with late spring and autumn peaks. In the last decade, significant changes with an increase of short-term intense events have occurred. An increased variability of rainfall, up to 15% within 2020, has been also predicted by the long-term scenarios of the Intergovernemental Panel on Climate Change (IPCC), that has foreseen an increase of air temperature in summer, no significant changes in mean precipitation values, but an increase of its variability (http://www.ipcc.ch/).

3
Anthropogenic Activities and Inherent Pressures

3.1
Population, Main Activities and Related Potential Nitrogen and Phosphorus Loadings in the Po di Volano Watershed

At present, the Po di Volano watershed comprises ten municipalities with approximately 67 000 inhabitants. The population density is relatively low, with a negative trend in the 1991–2001 decade that is typical of rural areas in Italy (Table 1). The main economical activities are agriculture in the terrestrial part of the basin and aquaculture in the lagoon.

The population decrease, along with aging, is leading towards a deep transformation of zootechnical and agricultural practices. Approximately 80% of the Po di Volano watershed (about 650 km^2) is exploited for agriculture (Fig. 1c). At present, there are few main crop types: maize and wheat, rice, sugar beet, soybean, fruit trees and vegetables. Among vegetables, some relevant products are also exported: salad, asparagus, pumpkin, strawberry, carrot, melon and watermelon. Maize, wheat, sugar beet and soybean are cultivated with conventional techniques and need pesticides and fertilizers, therefore they can have a potential impact on water quality. Rice is cultivated in paddy soils with permanent submersion and needs pesticides and fertilizers. In the last decade, rice-field surface has increased and attained up to

Table 1 Human population and land use changes from 1991 to 2001 in the Po di Volano watershed [13]. Surface data considered here account for approximately 80% of the total surface area of the watershed

	Year	1991	2001	Difference %
Human population (No individuals)		72 114	67 086	−7.0
Livestock (No. animals)	Cattles	10 158	2502	−75.4
	Pigs	8546	14 023	+64.1
	Poultry	930 588	326 451	−64.9
agriculture main crops (surface, km²)	Cereals	268.7	278.6	+3.7
	Grasses	23.3	27.0	+15.9
	Vegetables	53.5	35.7	−33.3
	Fruit trees	44.8	34.7	−22.5
	Industrial	107.8	122.3	+13.5
	other	45.4	35.4	−22.0

Table 2 Potential loadings ($t\ y^{-1}$) that are generated by different activities in the Po di Volano watershed. See text for explanation

	Nitrogen		Phosphorus	
Year	1991	2001	1991	2001
human population	316	294	63	59
livestock	1100	452	367	152
agriculture	747	747	26	26
atmosphere	594	594	0	0
total loading	2757	2088	456	237

65 km² in 2001. The permanent flooding keeps pesticides and fertilizers in solution and increases the risk of pollution of the lagoon. Vegetables are grown either with conventional techniques or with organic practices, therefore their impact varies from high (green houses) to negligible (organic). Livestock have a minor relevance, with a decrease in poultry and cattle rearing.

The potential loadings of BOD_5, N and P (Table 2) generated by the above-mentioned activities were calculated using the census data reported in Table 1 and the conversion coefficient applicable for Italy [1], in agreement with the approach that was pursued for assessing the nitrogen loading from the coastal watershed to the receiving coastal waters [14]. Inputs from atmosphere were also considered [6, 15]. In the decade 1991–2001, the potential loadings from agriculture and atmosphere remained constant, whilst the livestock source decreased sharply. The contribution by the human population diminished slightly, due both to population decrease and improvement of the performances of waste water treatment plants. The overall potential loading decrease was 24% for nitrogen and 48% for phosphorus, respectively.

3.2
Aquatic Resource Exploitation in the Sacca di Goro Lagoon

The shallow lagoons of the Northern Adriatic Sea are the most important sites for the Manila clam (*Tapes philippinarum*) within the European Community with an estimated crop ranging between 50 000 and 60 000 t y^{-1} [16]. The Manila clam was first introduced in the Venice lagoon in 1983 [17, 18] and in the Sacca di Goro 3 years later in 1986 [19]. At present, the Sacca di Goro is one of the top European sites for clam rearing. About one-third of the lagoon surface (8 km^2) is exploited for clam farming (see Fig. 2), with an annual production that reached a maximum of approximately 15 000 t y^{-1} in the early 1990s (Fig. 3). The corresponding economic revenue has been oscillating between 50 and 100 million Euros each year. At present, clam farming is managed by cooperatives of fishermen that exploit licensed areas, under the control of regional and local authorities. Up to 1500 fishermen are associated in cooperatives and are primarily employed in clam farming in the Sacca di Goro and in mussel farming in the adjacent sea.

Young clams (5–10 mm in shell length) are continuously collected along the southern sand barrier and then sown in the licensed areas. Due to the high food availability, the commercial size is attained in a few months. Adult clams usually attain shell length of approximately 4 cm and a total weight of about 10 g. Within the licensed areas clam densities are generally maintained at approximately 500 adult individuals m^{-2} with peaks of 1000 individuals m^{-2}, even if densities up to 2000 individuals m^{-2} are not infrequent [20]. A good revenue was and is currently obtained by controlling the product

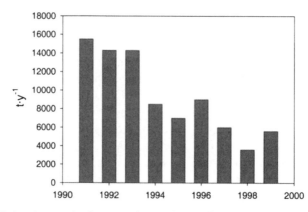

Fig. 3 Annual clam harvest in the Sacca di Goro lagoon from 1991 to 1999. Modified from Zaldivar et al. [73]

delivered to the market. For this reason, each fisherman and/or fisherman co-operative can harvest a fixed quota that is established day by day, based on the market demand. Harvesting is performed with manual dredging. This causes local alterations in the natural sediment stratification and increases sediment resuspension, nutrient release and mobility of reduced compounds [21, 22]. Some attempts at mechanical harvesting were performed and research was started in order to optimize the manual harvesting technique, which at present is indeed very primitive. This could potentially increase the harvesting capacity with unpredictable effects on the marketing performances and revenue as well as on sediment and water quality.

In the late 1980s and early 1990s, but also in recent years, clam farming has suffered serious setbacks due to massive clam mortality in summer. The blooming of the seaweed *Ulva rigida* and the occurrence of dystrophic events triggered by the decomposition of macroalgal biomass were proposed as one of the main factors determining the decline of clam farming in the lagoon [10]. Recent studies demonstrated that clams themselves can contribute to the oxygen depletion, increase inorganic nutrient recycling and cause an accumulation of organic matter which stimulates sulphate reduction and sulphide production in the surface sediments [20, 23–25]. Moreover, the dominance of clams within the benthic community resulted in the displacement of other organisms and the loss of their associated functions [26, 27].

4
Eutrophication, Macroalgal Blooms and Dystrophic Crises in the Sacca di Goro Lagoon

4.1
Water Quality, Nitrogen and Phosphorus Loadings Discharged from the Po di Volano Watershed

Since 1991, the freshwater chemistry in the watershed was monitored fortnightly with standard techniques by the Environmental Services of Province of Ferrara [28–34] and the Regional Agency for Environmental Protection [35].

In the Po di Volano, a sharp decrease of dissolved inorganic nitrogen (DIN) concentrations was observed from early to late 1990s. DIN maximum was usually attained in December–January, whilst total phosphorus peaked in Autumn. For example, in 1992 and 1997, the nitrate peaks were 18.0 mg N L^{-1} (annual mean = 2.8 mg N L^{-1}) and 5.4 mg N L^{-1} (annual mean = 1.4 mg N L^{-1}), respectively. Ammonia nitrogen was almost constant throughout the decade, with a peak of 6.5 mg N L^{-1} (annual mean = 2.0 mg N L^{-1}) in 1992 and 4.5 mg N L^{-1} (annual mean = 1.7 mg N L^{-1}) in 1997. An evident

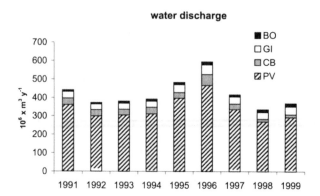

Fig. 4 Freshwater discharge in the Sacca di Goro lagoon from the Po di Volano canal and minor sources. Data on water discharge were kindly provided by the Consorzio di Bonifica, 1° Circondario Polesine Ferrarese. PV: Po di Volano, CB: canal Bianco, GI: Giralda canal, BO: Bonello canal

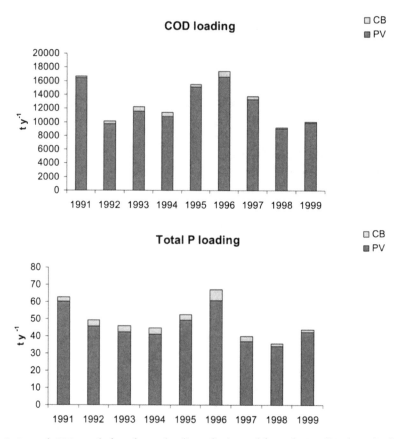

Fig. 5 Annual COD, total phosphorus loadings discharged from the Po di Volano (PV) and Canal Bianco (CB) in the Sacca di Goro lagoon from 1991 to 1999

decrease was also observed for total phosphorus from an annual mean of 180 µg P L^{-1} (peak = 560 µg P L^{-1}) in 1992 to 110 µg P L^{-1} (peak = 220 µg P L^{-1}) in 1997. On average, the COD underwent an increase from 26 mg O$_2$ L^{-1} (peak = 45 mg O$_2$ L^{-1}) up to 41 mg O$_2$ L^{-1} (peak = 55 mg O$_2$ L^{-1}), whilst BOD$_5$ remained almost constant.

The COD, total phosphorus and dissolved inorganic nitrogen loadings have been estimated from the chemistry data and the freshwater discharge values of the PV and CB canals that account for approximately 95% of the hydraulic loadings delivered by the Po di Volano watershed to the Sacca di Goro (Fig. 4). From 1991 till 1999, the annual COD loading of PV + CB decreased from 16 663 to 10 025 t (– 40%), total phosphorus loading from 63 to 44 t (– 29%) and DIN loading from 1761 to 411 t (– 77%) (Figs. 5 and 6). The DIN diminution was mainly due to the abatement of the oxidized nitrate, which fell from 1312 to 165 t y^{-1} (– 87%); whilst the ammonium reduction was from 449 to 247 t y^{-1} (– 45%). This is coherent with the potential loading reduction that also accounts for total phosphorus and COD removal (Table 2). However, this enormous nitrate diminution also depended upon the implementation of wastewater works that were ameliorated by coupling oxidation with denitrification processes.

In 1997, a monthly sampling was performed in order to assess the contribution of organic dissolved and particulate nitrogen to the total nitrogen load and compare it with the inorganic N species [36]. The resulting total nitrogen loading was 1324 t y^{-1}, of which 49% was dissolved organic-N, 13% was particulate organic-N, 19% ammonium and only 18% nitrate and nitrite. Although the uptake of the inorganic oxidized species could have a relatively small effect on the overall nitrogen loading, the removal of the most reactive species should result in a limitation for primary producers, mostly for the nitrophilous macroalgal species [37].

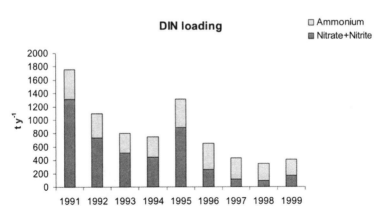

Fig. 6 Annual DIN loading discharged from the Po di Volano and Canal Bianco in the Sacca di Goro lagoon from 1991 to 1999

4.2
Successional Stages of Macroalgal Growth and Decay
and their Effects on Trophic Status, Water Quality and Dystrophic Events

In the coastal lagoons of the northern Adriatic coast, the pristine aquatic vegetation was mainly composed of submerged aquatic phanerogams, namely *Zostera noltii*, *Cymodocea nodosa* and *Ruppia cirrhosa* [38, 39]. The increased eutrophication of the last 30 years [40] was followed by the disappearance of the submerged aquatic vegetation, the occurrence of phytoplankton blooms, and in the last stage, by the growth of ephemeral macroalgae [9, 41, 42]. This is indeed a common feature of the most threatened coastal areas in the world [43, 44].

In the Sacca di Goro lagoon, the nitrogen loading was considered responsible for the abnormal growth of the green nitrophilous seaweed *Ulva rigida* and, to a lesser extent, of the red macroalgae *Gracilaria verrucosa* (Fig. 7). *Gracilaria* blooms were important mainly in late 80s–early 90s [39, 45]. Af-

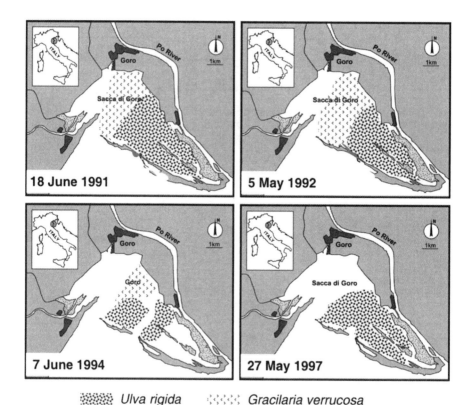

Fig. 7 Maximum spreading of macroalgal mats in the Sacca di Goro lagoon. Data from [10, 37, 46, 47, 49, 57]

terwards, *Gracilaria* almost disappeared and monospecific *Ulva* blooms were dominating. The maximum spreading was observed from 1987 to 1992 [9, 45–47] and in 1997 [10]. From 1996 to 1998, secondary peaks of *Cladophora* spp. were observed in late summer and early autumn [48]. In the eastern sheltered zone of the lagoon, which is approximately $10\,km^2$, macroalgal blooms attained the maximum development in late spring, with biomass peaks towards the end of May. Macroalgal beds were usually patchy in late winter, when they grew in the shallower and most confined areas. When meteorological conditions were favorable with high temperatures and calm wind, *Ulva* attained net growth rates up to 15% d^{-1}, with peaks up to 20% d^{-1} [10, 49]. Under these circumstances, in 1991–1992 and 1997 a standing stock comprised between 50 000 to 70 000 t as wet weight was estimated. The biomass peak was then followed by a sudden collapse of macroalgal mats, which started to decompose causing anoxia and sulphide release to the water column.

The correlation between *Ulva* biomass and dissolved oxygen in the water column is shown for station 17 (Figs. 2 and 8). In spring, when the macroalgal productivity was at its maximum, the total metabolism resulted in oxygen concentrations from around 100% saturation at night to over 300% saturation in the afternoon; therefore oxygen tended to escape the water mass. During this phase of the macroalgal life cycle, the net community oxygen production frequently rose up to $20\text{–}30\,mmol\,m^{-2}\,h^{-1}$, which was 2–3 times higher than the coupled respiration rates [50, 51]. Oxygen production was also coupled with the accumulation of biomass within the aquatic ecosystem that led to a subsequent increase of decomposition processes and the inherent oxygen demand [49, 52, 53]. Under these circumstances, macroalgal photosynthesis became limited, due to self shading and nutrient depletion, and oxygen production was not sufficient to compensate for the oxygen demand. The decomposition of the accumulated biomass rapidly switched the metabolism from hypereutrophic to dystrophic causing the onset of persistent anaerobic processes. The sudden crash of macroalgal productivity with oxygen deficiency in the water column and dystrophy usually occurred in June–July, but early spring and autumn episodes were also recorded. During neap tides and under calm wind conditions, anoxia persisted for several days, mostly in the eastern sheltered area. Here water stagnation was one of the most important causes of the dystrophic outbreaks [9, 47]. The anoxia was more persistent when macroalgal biomass was higher, since oxygen deficiency depended primarily on the biomass decomposition. For example, during the dystrophic outbreaks in July 1992 a decay rate of about 4% d^{-1} was estimated, which corresponded to a potential oxygen uptake of approximately $400\,mmol\,m^{-2}\,d^{-1}$ [50]. The more intense or productive became the growth phase, the more intense was the subsequent heterotrophic activity and consequently the anoxic event, as demonstrated for 1991–1992 and 1997, respectively.

During the macroalgal collapse, the persistent anoxia was accompanied by a significant release of sulphide to the water column. The sulphide was pro-

Ulva biomass

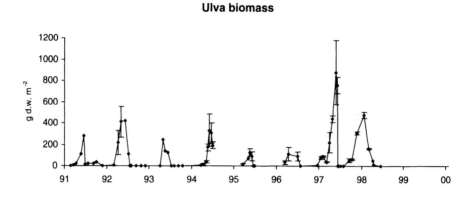

Dissolved oxygen

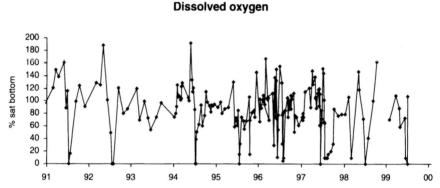

Fig. 8 Annual variations in *Ulva* biomass and dissolved oxygen in the bottom water at station 17 in the Sacca di Goro lagoon from 1 January 1991 until 31 December 1999 (modified from Viaroli et al. [10] with the addition of unpublished data for 1998 and 1999)

duced during the decomposition of *Ulva* biomass by both the mineralization of organic sulphur compounds and the bacterial sulphate reduction activity [54]. Usually, in the anoxic sediments of coastal aquatic ecosystems dissimilatory bacterial sulphate reduction is the dominant decomposition pathway that accounts for more than 50% of the organic matter oxidation [55]. The sulphides produced by sulphate reduction can be removed from the pore-water by precipitation with metallic ions, e.g., ferrous iron. In highly eutrophic lagoons, such as the Sacca di Goro, a considerable production of free sulphide may occur associated with the mineralization of macroalgal biomass in anoxic bottom waters, where its concentration is independent of the potential iron buffering capacity of the sediment [52]. In the first decay phase, an oxic water mass can coexist with a strongly reduced layer, located beneath the decomposing macroalgal bed [56]. Within the anoxic organic slime, a large amount of sulphides accumulated and then migrated upwards (Fig. 9). During the dystrophic outbreak sulphides can diffuse also in the superficial

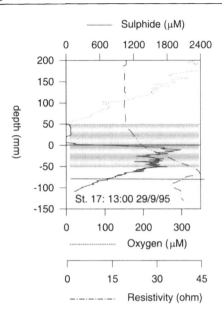

Fig. 9 Profiles of oxygen (*dotted line*) and sulphides (*continuous line*) through a decomposing *Ulva* layer at the water-sediment interface at station 17 in the Sacca di Goro lagoon at 1.00 pm on 29 September 1995. The macroalgal layer and organic slime comprised approximately between + 50 and – 50 mm (*shaded area*), as shown by the resistivity profile (modified from Bartoli et al. [56])

water mass, as demonstrated in 1997 when a concentration of 85 μM was attained [57].

Macroalgal growth and outbreak were likely due to multiple causes that included both bottom-up controls and internal biochemical and physiological regulators [58]. A general precept is that the relative abundance of the main nutrients, especially nitrogen, controls the macroalgal productivity and the onset of biomass collapse [59, 60]. In systems with a high nitrate input, as in the Sacca di Goro lagoon, nitrate was considered as one of the key factors in controlling macroalgal blooms [37]. On the other hand, *Ulva* growth can regulate nutrient retention and recycling within the lagoon ecosystem. The strict relationship between *Ulva* biomass and nitrate and DIN in the water column can be inferred by comparing Figs. 8 and 10.

Usually, nitrate and DIN attained the highest concentrations in autumn and winter, when macroalgal uptake was minimum. Afterwards they decreased rapidly leading to prolonged nitrogen deficit conditions in which only *Ulva* was able to grow, out-competing other primary producers as well as denitrification processes [51]. A slight increase of ammonium concentrations was usually observed in summer, when the ammonium release from sediment ranged from 10 to 20 mmol m^{-2} d^{-1} [51, 61]. During the spring bloom of *Ulva*, most of the available nitrogen flowed through the *Ulva* biomass (Table 3). In

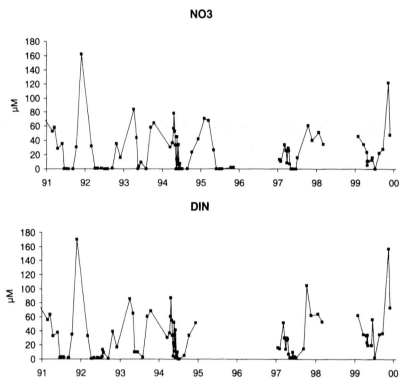

Fig. 10 Annual variations of dissolved nitrate (NO3) and dissolved inorganic nitrogen (DIN) in the water column at station 17 in the Sacca di Goro lagoon from 1 January 1991 until 31 December 1999 (modified from Viaroli et al. [10] with the addition of unpublished data for 1998 and 1999)

Table 3 Main nitrogen fluxes at station 17 in the Sacca di Goro lagoon in 1997 [10, 51, 57]. Units in mmol m^{-2} d^{-1}. TN: total nitrogen (DIN + DON + PON)

	Ulva uptake	Total Denitrification	Ammonium efflux	External TN loading
January	6.8	0.9	1.5	38.2
April	44.2	1.5	5.7	5.1
August	0.0	2.6	9.7	5.6

April 1997, when the net growth rate of *Ulva* was 0.14 ± 0.04 d^{-1} a net nitrogen uptake of 44.2 mmol m^{-2} d^{-1} was detected, whilst total denitrification was only 1.5 mmol m^{-2} d^{-1} [51, 57]. Under these circumstances, macroalgal uptake largely exceeded denitrification rates and nitrogen was retained within the lagoon. Before the macroalgal collapse, the overall *Ulva* standing biomass accounted for an average of 174 t of nitrogen (range 133–221 t). This nitrogen bulk corresponded to approximately 50% of the annual DIN loading from the

Po di Volano watershed. The nitrogen accumulation within *Ulva* was shown to be only a temporary sink. Usually, after the macroalgal crash, biomass decomposition caused a rapid nutrient loss and recycling [49, 52].

SRP concentrations followed an opposite pattern, with very low concentrations throughout the year and peaks that coincided with the sudden collapse and decomposition of the macroalgal biomasses (Fig. 11). This pattern was considered to be a result of phosphorus speciation that was mainly controlled by sedimentary processes [62, 63]. The sedimentary refractory organic-P and calcium-bound phosphate were the main sedimentary pools, whilst the iron bound phosphate was almost negligible (Table 4). The calcium-bound pool can be considered as a sink for phosphorus due to its low reactivity over a wide pH range, whilst the reactivity of the organic-P pool depended on the organic matter recalcitrance. Although the iron-bound phosphate was quantitatively less important, iron played an important role in controlling the

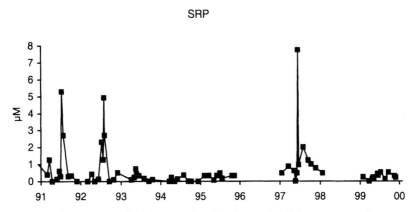

Fig. 11 Annual variations of soluble reactive phosphorus (SRP) in the water column at station 17 in the Sacca di Goro lagoon from 1 January 1991 until 31 December 1999 (modified from Viaroli et al. [10] with the addition of unpublished data for 1998 and 1999)

Table 4 Concentrations of the main phosphorus species in the surficial sediment horizon (0-2 cm) of Sacca di Goro from April 1997 to March 1998. Total and organic-P and calcium-bound phosphate (Ca~PO_4) are indicated as annual mean with the corresponding standard deviation. Iron-bound (Fe~PO_4), exchangeable (Exch. PO_4) and porewater (pw PO_4) phosphate are reported as seasonal ranges (min–max.). Units are in $\mu mol\,cm^{-3}$ of sediment except pw PO_4 ($nmol\,cm^{-3}$)

Station	Total P	Organic P	Ca ~ PO4	Fe ~ PO4	Exch. PO4	pw PO4
G	17.8 ± 1.7	3.1 ± 2.0	13.7 ± 1.5	0.11–0.45	0.5–1.9	1.2–118.7
4	10.7 ± 2.4	3.4 ± 2.2	7.0 ± 1.6	0.02–0.09	0.2–0.9	1.5–30.2
17	16.3 ± 4.0	5.4 ± 4.8	9.9 ± 3.4	0.01–0.12	0.3–0.8	1.9–42.0

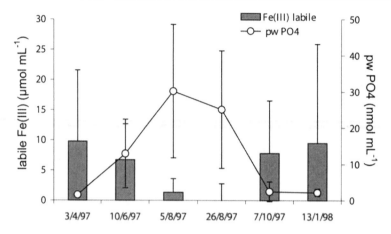

Fig. 12 Relationship between sedimentary oxidized iron species [labile-Fe(III)] and SRP in the pore-water (pw PO4) at station 4 in the Sacca di Goro lagoon from April 1997 to January 1998. Giordani and Azzoni, unpublished data. For analytical details see [62, 63]

pore-water phosphate concentrations through the iron hydroxide-sulphide system [64]. Usually, during summer dystrophic events, the reduction of Fe(III) to Fe(II) was followed by an increase of the pore-water SRP concentration (Fig. 12). When SRP concentration exceeded certain thresholds, SRP diffused to the water column. The SRP release was often enhanced by sediment bulking due to the gas release during macroalgal decomposition. From early autumn through the winter period, oxidation processes took place with the reoxidation of Fe(II) to Fe (III). In turn, the Fe(III) increase was accompanied by a decrease of pore-water SRP (Fig. 12).

5
Assessment of the Impact of Aquaculture on Water Quality and Benthic Fluxes

The trophic status and water quality in the Sacca di Goro lagoon was considered to depend not only on external loadings but also on aquaculture and biological resources exploitation. In the last decade, the large development of the Manila clam farming induced a deep transformation of the benthic communities with an overall biodiversity loss, which was dramatically enhanced by the frequent dystrophic crises [26, 27, 65]. Moreover, the zones surrounding the exploited sites were colonized by opportunistic species, viz. the polychaete *Capitella capitata*, which was recognized as an indicator of organic matter enrichment of the sediment [66].

The high clam density in the farmed areas led to the filtration of large amounts of water that acted as a control of the seston concentration and delivered an easily degradable material on the sediment surface as faeces and pseudofeces. The faecal material represented in turn a labile substratum for bacteria, thus causing an increase of the sediment oxygen demand [20, 23, 67]. In the farming area of the Sacca di Goro lagoon a significant correlation between sediment oxygen demand (SOD, mmol $m^{-2} h^{-1}$) and clam biomass (B, g dry flesh m^{-2}) was found in late June with the following regression equations

23 June 1999, SOD = - 2.97-0.030*B, R^2 = 0.978, P < 0.01 [20]

19 June 2000, SOD = - 1.95-0.031*B, R^2 = 0.944, P < 0.01 [22]

The background SOD resulting from the above equations comprised between - 1.95 and - 2.97 mmol $m^{-2} h^{-1}$. The average clam density in the licensed areas was approximately 500 individuals m^{-2}, which corresponded to 200 g dry flesh m^{-2}. Under these circumstances, a mean increase of - 6 mmol $m^{-2} h^{-1}$ of SOD can be inferred. The high clam density also stimulated ammonium and phosphate regeneration across the sediment interface as a result of the direct clam excretion, changes in the redox conditions, and the mineralization of the labile faecal materials. A significant correlation between ammonium regeneration (AR, μmol $m^{-2} h^{-1}$) and clam biomass (B, g dry flesh m^{-2}) was found in late June with the following equations

23 June 1999, AR = 318 + 6.2*B, R^2 = 0.945, P < 0.01 [20]

19 June 2000, AR = 22 + 9.8*B, R^2 = 0.956, P < 0.01 [22]

The background AR was usually low with a high variability. At 500 individuals m^{-2}, AR increased to 1.2-1.9 mmol $m^{-2} h^{-1}$. In June 1999, in the control area, soluble reactive phosphorus (SRP) fluxes were negligible (10 μmol $m^{-2} h^{-1}$), whilst in the sediment with the highest clam densities it was 150 $\pm$ 20 μmol $m^{-2} h^{-1}$ [20]. In the farming areas, the sediment was enriched by biogenic carbonates that are recognized to control phosphorus recycling through adsorption and co-precipitation processes [64]. As a consequence, ammonium efflux was greater than SRP release and the atomic N to P ratio in the benthic fluxes tended to be close to 30 : 1, which seemed beneficial for macroalgae [58, 68]. Under summer conditions, with high temperatures and photon fluxes, high regeneration rates of nutrients can stimulate phytoplankton growth as well as sustain macroalgal blooms. Preliminary experiments demonstrated that the photosynthetic activity of *Ulva rigida* was stimulated at relatively low clam densities [24].

In the Sacca di Goro lagoon, clams are normally harvested by sediment dredging with a manual rake. In order to control product quality and prices, each fisherman and/or fisherman cooperative can harvest a fixed quota that is established day by day, based on the market demand. Approximately 1000 workers per day are simultaneously involved in the harvesting activity. On average, each fisherman is expected to harvest approximately 30 kg of clams per day, which correspond to the exploitation of about 50 m^2 of sediment. Therefore, in an average working day the harvested surface area is approximately 50 000 m^2; this represents less than 1% of the whole farming area (8 km^2). Results of mesocosm experiments in an area where the clam culture

Table 5 Benthic fluxes measured before harvesting (Pre-H), during harvesting (H) and 8 h after harvesting (post-H) in cores sampled in a clam farming site in the Sacca di Goro lagoon on 19 June 2000. Mean values of five replicates (± standard deviation) are reported. Negative values indicate an uptake, positive values indicate a release. Units in mmol m^{-2} h^{-1}. Modified from Viaroli et al. [22]

	Pre-H	H	Post-H
Oxygen	−3.05±2.39	−60.90±8.91	−5.78±1.16
Ammonium	0.22±0.30	11.40±2.31	0.10±0.23
SRP	−0.017±0.068	1.04±0.48	−0.019±0.056

was left undisturbed for approximately 6 months demonstrated that, under summer conditions, sediment dredging increased SOD 20 times, and ammonium and SRP regeneration approximately 50 and 100 times, respectively (Table 5) [22]. At the whole lagoon scale, in summer harvesting activities accounted for an additional oxygen demand which was less than 3% of the background value, a phosphorus excess which was approximately 15% of the background loading and an ammonium excess that was always less than 2%. The sediment resuspension and the inherent increase of benthic fluxes were very rapid and transient events that had no large-scale persistent effects on the water column [21]. Moreover, harvesting effects were appreciable only in summer [67]. Manual harvesting had also minor effects on the macrofaunal benthic community. Only rare species disappeared or underwent a density reduction [26]. The macrofaunal community was already compromised and simplified with a small number of highly tolerant and opportunistic species. This seemed to depend more on the frequent dystrophic outbreaks and clam metabolism, rather than upon harvesting disturbance [26, 27, 65, 69].

6
A General Model of the Sacca di Goro Lagoon and its Watershed

Since nitrogen was identified as the most important factor in determining the lagoon eutrophication, an ecological network analysis was performed in order to assess the influence of trophic status and primary producers forms in determining nitrogen cycling [70, 71]. Nitrogen cycling and transformations within, and transport through the lagoon were found to be closely related to the form of the dominant primary producers. Phytoplankton components, due to their short life cycle and rapid turnover time, fostered fast recycling within the water column and transfer to sediment or the adjacent sea. Ephemeral opportunistic macroalgae, namely *Ulva* and to a lesser extent *Gracilaria*, induced lags and pulses within the ecosystem by sequestering nitrogen during the growth phase and releasing it during bloom collapse and decomposition.

A 0D biogeochemical model of the Sacca di Goro lagoon was then implemented and tested [72]. The model considered the nutrient cycles in the water column and in the sediment, phytoplankton, zooplankton, and *Ulva* dynamics, as well as shellfish farming. Nutrient loadings from the watershed, wet and dry deposition, temperature, light intensity and wind speed were considered as forcing functions. The exchange with the Northern Adriatic Sea was also considered. The simulated results were compared with the available data showing that the model was able to represent the essential dynamics and the magnitude of the different processes that occurred in the lagoon. A preliminary analysis of the influence of the forcing functions and the exchange with the boundaries, in particular the influence of the watershed, was also considered.

Since clam rearing was demonstrated to play a relevant role in the lagoon functioning, a discrete stage-based model for the growth of *Tapes philippinarum* was coupled with the continuous 0D biogeochemical model of the lagoon [73]. As forcing functions for the clam growth, temperature and nutrient concentrations in the water column derived from the biogeochemical model were introduced with a variable stage duration for each stage in the population matrix. Furthermore, the effects of harvesting as well as the mortality due to anoxic crises were taken into account. The simulated clam productivity conformed to the measured data. Moreover, the model allowed assessing the impact of clams on nutrients and oxygen dynamics. This model has been coupled with a watershed model able to simulate point and diffuse pollution that is able to allocate pollutant loads according to different types of activities and land use in the catchment [?]. This will allow the analysis of the impact of management practices on quality of the water arriving to the lagoon and link anthropogenic activities with the nutrient cycles in Sacca di Goro. Furthermore, it will be used for studying several scenarios that will be developed below.

The vulnerability of the lagoon was also assessed with an integrated simulation model that considered the main physical forcings, macroalgae and sediment at different clam densities [25]. Basically, oxygen production by macroalgae, oxygen respiration by sediment, macroalgae and clams, and physical reaeration were considered and validated with experimental data. Each clam density was then associated to a probability of occurrence of hypoxia and its duration. Therefore, the model provided a rapid and valuable tool for assessing farming policies, i.e. the maximum clam yield that minimizes the hypoxia risk. The model results showed that clams had a considerable impact on the oxygen budget, relatively low commercial densities (500 adult individuals per square meter) being associated with intermediate risk of hypoxia in June and September. Under these circumstances, with water temperatures above 25 °C and 25 psu salinity, an oxygen demand of $20\text{--}30\,\mathrm{mmol\,m^{-2}\,h^{-1}}$ can turn a shallow water column almost anoxic instantaneously [20, 22].

The overall metabolism of the Sacca di Goro lagoon was analyzed with the LOICZ biogeochemical model [6, 74]. This model is based on mass balance of water, salt and nutrients and its outputs are estimations of water and nutrient fluxes and the net ecosystem metabolism (NEM) [75]. The water turnover time of the lagoon ranged from 1 to 4 days, depending on freshwater inputs and the water exchanges between the lagoon and the adjacent sea. Budgets of dissolved inorganic phosphorus (DIP) and nitrogen (DIN) were calculated on a seasonal basis for the 1989–1997 period [74]. Positive and negative internal fluxes of DIP were both observed in relation with macroalgal biomass and riverine discharges. Data for 1997 indicated that DIP and DIN internal fluxes conformed to the seasonal trend of macroalgae with a pronounced uptake during the growth phase and a sudden release during the subsequent senescence phase [6]. The NEM followed a similar trend with an autotrophic phase during the macroalgal growth; however on an annual basis, the lagoon can be considered as heterotrophic with a NEM of about $- 50 \, \text{mmol} \, \text{C} \, \text{m}^{-2} \, \text{d}^{-1}$.

7
Xenobiotics and Micropollutants

Coastal lagoons are recognized to be sensitive to the accumulation of persistent contaminants due to their hydrogeomorphic conditions. In the late 1980s, due to the growing concern of environmental pollution in the Sacca di Goro lagoon, a research and monitoring programme was launched by the Environmental Office of the Province of Ferrara [76, 77]. Along with ecosystem studies, a monitoring of the distribution of the main pollutants within different ecosystem components was also performed. Further monitoring programs were carried out in the subsequent years, but they were specially designed for controlling clam and mussel farming, which is constrained by regional and national rules dealing with contamination thresholds for commercial seafoods. The monitoring effort focused on the most common pesticides and persistent organic pollutants (POP) and heavy metals that were commonly found in the water of the Po River [78, 79], as well as on the main radionuclides that were delivered by the fall-out of the Chernobyl accident.

7.1
Pesticides and Persistent Organic Pollutants (POPs)

The Po river watershed is heavily exploited for agriculture and industry, with several hazardous activities that were recognized to release pesticides and POPs in the aquatic systems [79]. As far as the Po di Volano is concerned, a study on the organic contaminant dispersion was carried out only in the easternmost zone of the watershed [80]. However, one may expect levels of contamination similar to those found in the main arms of the Po river. Among a large number of chemicals that can be potentially delivered in the environment, few herbicides (molinate, atrazine, simazine and other s-triazines) and insecticides (DDT, aldrin and γ-lindane) were considered in a monitoring campaign carried out from November 1987 until May 1990 [81, 82].

The atrazine and simazine concentrations in the water followed a seasonal trend, with peaks in May and, to a lesser extent, in late autumn (Table 6). The May peak coincided with the maize sowing, these herbicides being typically applied in pre-emergence during maize cultivation. The autumn peak can be explained by the use of these herbicides for the control of annual broad-leaved weeds and annual grasses. Although atrazine has been banned it is still included in the priority substances list under the Water Framework Directive of the European Union (http://europa.eu.int/comm/environment/water/waterframework). The few data dealing with molinate indicated a late spring peak of concentrations in the water column (Table 6). This is in agreement with the rice cultivation, the molinate being used as a herbicide in the early phase of rice cultivation, when rice-fields were inundated. Approximately two weeks after the sowing, the contaminated waters were discharged into the canals and then into the lagoon, to allow for a short drying phase. Aldrin, γ-lindane, DDT and its metabolites were found episodically at low levels, mainly in the stations that are close to the freshwater outlets. Most of these compounds were banned before the 1970s or were strongly regulated and constrained. However they are still present in the Po River due to the accumulation in the sediment and their slow decay rates [83, 84].

Both herbicides and POPs accumulated in the superficial sediment horizon of the Sacca di Goro lagoon. Here concentrations were approximately 50 times higher than in the water column. However, contamination levels were much lower than in areas impacted by heavy industrial activities [85, 86]. Few analyses were performed on the biota, with the exception of a seasonal survey on mussels (*Mytilus galloprovincialis*). Herbicides had a low impact on mussels, whilst an appreciable contamination by DDT and its metabolites was

Table 6 Maximum concentration and period of occurrence of six pesticides that were found in the Sacca di Goro lagoon from November 1987 to May 1990 [81, 82]. ND: not detectable

	Water ($\mu g\,L^{-1}$)		Sediment ($\mu g\,kg^{-1}$)		Mussel ($\mu g\,kg^{-1}$)	
Atrazine	0.16	May	10.20	November	0.24	November
Simazine	0.14	May	8.00	March	0.22	November
DDT+ metabolites	0.16	March	4.30	November	3.50	November
γ-lindane	0.03	May	1.50	November	ND	–
aldrin	0.14	November	ND	–	ND	–
molinate	0.98	May	ND	–	ND	–

detected, with concentrations in the range found in the Adriatic Sea [87]. Clams were not considered, whilst clam contamination by POPs was extensively studied in the Venice Lagoon [85, 86]. Further data concerning POPs, especially organochlorine pesticides and chlorobiphenyls, in sea-foods were measured in the Adriatic Sea, but no specific information on the Sacca di Goro were reported [83, 87, 88].

Most of the monitoring effort addressed land-based impacts, whilst little concern was devoted to other chemicals such as organotin compounds. These compounds have a short life in the environment when compared with POPs and some of them have a seasonal behavior depending on their use. Tributyltin (TBT), which is used as a paint additive to prevent biofouling (e.g. tube worms, algae and barnacles) on ship hulls, was banned since 1982 on ships under 25 meters in length. This has occurred because testing indicated that TBT, even at concentrations as low as $ng\,L^{-1}$, may be responsible for the weakening of oyster and mussel shells as well as for slowing the growth of various species of aquatic snails. Furthermore, no studies on highly soluble chemicals such as octyl- and nonylphenols (in the Priority list of the Water Framework Directive) or pharmaceuticals have been carried out in the Sacca di Goro, whilst they have been found in the Po River [78, 79, 84].

7.2
Trace Metals

From November 1987 to March 1991, mercury, lead, cadmium, total chromium, copper, zinc, nickel, cobalt, aluminium, iron, and manganese were analyzed in water, suspended particulate matter, surficial sediment and shellfish [89–92]. Geochemical studies on sedimentary distribution of both main and trace elements were further conducted in 1997 [93] and in 1998 [94]. Copper, zinc, iron, and manganese were also monitored in the *Ulva* biomass throughout an annual cycle in 1997, as well as analyses of different ecosystem components were performed in 2000 [95].

In the water column, mercury was always below the detection limit $(0.3\,\mu g\,L^{-1})$, whilst other trace metals ranged from undetectable up to peaks of $16.3\,\mu g\,L^{-1}$ (Pb), $11.7\,\mu g\,L^{-1}$ (Ni), $3.3\,\mu g\,L^{-1}$ (Cd), $1.7\,\mu g\,L^{-1}$ (Cr) that were several times higher than those measured in the open sea. Trace metals were highly concentrated in the suspended solid fraction and, especially, in the sediment surface, where they attained values 10–1000 greater than those measured in the water column. The sedimentary concentrations were in a range that was commonly determined in several lagoons of the Northern Adriatic coast [96]. Nevertheless, an enrichment of the superficial sediment horizon was found for Hg, Cd, Pb, and Cu, indicating a diffuse contamination of the Sacca di Goro [93]. As a result, high trace metal concentrations were detected in mussels and clams that were feeding on the particulate fraction. However, a significant fraction of the ingested metals was lost after the purification procedure that was usually adopted before the shellfish marketing [89].

The partitioning of copper and iron in different environmental components is presented as an example (Table 7). Cu and Fe concentrations in the dissolved and particulate fraction in the water column and in the superficial sediment showed a high variability, without significant seasonal trends. By contrast, metal contents in the *Ulva* biomass underwent a clear seasonality with winter peaks, when macroalgal growth and biomass were low, and minima in late spring, when the macroalgal biomass attained its maximum. As far as iron is concerned, seasonal patterns were also influenced by the external inputs from the PV, CB and GI canals, that delivered a total annual loading of 21 t, with spring minima and summer peaks. Metal behavior in the Sacca di Goro is presumably similar to that evidenced for the lagoon of Venice and the coastal system in the Northern Adriatic area [96, 97]. Here, metal distribution is strongly affected by sources, affinity for suspended particles and sediment granulometry. Due to the high eutrophication level, metal fluxes depend also on uptake and release by macrophytes [98]. Finally, metal speciation is influenced by a suite of redox reactions that depend upon organic matter decomposition and microbial activity. Under these circumstances, microbial metabolism is dominated by sulphate reduction processes with the production of sulphide; in turn, sulphides can react with metals also influencing their toxicity [55, 99, 100]. For further details on sedimentary reactions see also Sect. 4.2.

Table 7 An example of metal distribution within different ecosystem components in the eastern zone of the Sacca di Goro lagoon that was affected by macroalgal blooms. Data of stations 5, 7 and 8 [90, 91] and station 17 [?] are reported. LG: low *Ulva* growth (January–February), MG: maximum *Ulva* growth (May–June); DY: dystrophic outbreak (July). * Units as %. NS: not sampled

	Copper			Iron		
	LG	MG	DY	LG	MG	DY
Water-dissolved ($\mu g\, l^{-1}$)	<0.20–0.54	0.21–0.29	0.24–0.41	5.9–6.5	12.1–13.6	16.4–27.7
Water-seston ($\mu g\, g^{-1}$)	114–256	81–103	41–99	2262–3491	1131–7317	<470–2864
Sediment (0–5 cm) horizon ($\mu g\, g^{-1}$)	17–39	17–37	17–45	1.47–3.72*	0.46–3.01*	2.45–3.46*
Mussel ($\mu g\, g^{-1}$)	11.1–23.9	5.4–6.4	NS	317–2000	63–240	NS
Clam ($\mu g\, g^{-1}$)	5.8–7.8	4.3–7.1	NS	265–877	236–2226	NS
Ulva rigida ($\mu g\, g^{-1}$)	13.1–13.2	4.1–7.2	19.5–20.2	1480–1860	320–420	1240–1570

7.3
Radionuclides

In the Northern Adriatic Sea, two groups of anthropogenic radionuclides were detected since the mid 1980s: fission products and transuranic elements derived from nuclear atmospheric testing [101] and short- and long-lived radionuclides delivered from the Chernobyl accident [102]. The maximum deposition was attained in May 1986, with a peak of $3500\,\mathrm{Bq\,m^{-2}}$ for $^{137}\mathrm{Cs}$ and of $1740\,\mathrm{Bq\,m^{-2}}$ for $^{134}\mathrm{Cs}$ [102]. This caused an increase up to $60\,\mathrm{Bq\,m^{-3}}$ in the water column that was followed by a rapid sedimentation. In the Sacca di Goro lagoon monitoring campaigns on radiocontamination were carried out from 1989 to 1991 [103, 104]. Based on the $^{137}\mathrm{Cs}$ to $^{134}\mathrm{Cs}$ ratio and their different natural decay rates, it was estimated that the Chernobyl fall out caused an increase of $^{137}\mathrm{Cs}$ of 3 to 9 times. Most of the radioactivity due to both the fall out and riverine loadings was stored in the superficial sediment horizon (0–10 cm), as demonstrated by core profiles [105]. However, sediment resuspension and repeated dystrophic crises reversed the storage capacity of the sediment causing a radiocesium release to the water column and export to the open sea. Cesium losses were accelerated by the scavenging capacity of *Ulva*, that acted as a temporary sink during the active growth, concentrating Cs into the biomass, and as a source during the decomposition phase [105]. In the latter, microbial activity played a crucial role determining both detritus mineralization and changes of redox equilibria. As a result, the $^{134}\mathrm{Cs}$ activity was rapidly removed, whilst the $^{137}\mathrm{Cs}$ activity was kept lower than that expected from the natural decay kinetic (Table 8).

Table 8 Activity of $^{137}\mathrm{Cs}$ ($\mathrm{Bq\,kg^{-1}}$) in different environmental components in eastern zone of the Sacca di Goro lagoon that was affected by *Ulva* blooms. References are given in brackets. ND: not detectable

Year	1989 [103, 104]	1997 [106]
Sediment (0–5 cm horizon)	53.1 ± 4.5	16.0 ± 0.4
Ulva	3.9 ± 1.1	2.3 ± 0.8
Mussel	0.4 ± 0.1	ND

8
Concluding Remarks:
Scenarios and Tools for Managing Water Pollution and Resource Exploitation

The main forcing factors that resulted to affect ecosystem processes in the Sacca di Goro lagoon are summarized in Table 9. Based on a comparison of

Table 9 Maximum benthic fluxes of oxygen (SOD), ammonium (NH_4^+) and soluble reactive phosphorus (SRP) measured in dark incubation in late spring/summer and external loadings in the Sacca di Goro lagoon. External loading of COD, DIN and total P were considered. Negative values indicate an uptake, positive values indicate a release. Units are in $mmol\,m^{-2}\,h^{-1}$. References are given in brackets

	SOD	NH_4^+	SRP
Sediment colonised by *Ulva* at 500 gm^{-2}	-10.22 ± 1.09 [51]	-0.37 ± 0.05 [51]	0.00 [51]
Sediment after macroalgal crash	-6.60 ± 1.10 [46]	0.48 ± 0.28 [46]	0.13 ± 0.03 [46]
Bare sediment in clam farming	$-1.97 \div -2.97$ [22, 61]	$0.02 \div 0.14$ [23,62]	0.01 ± 0.01 [62]
Sediment colonised by 500 clams m^{-2}	$-6.95 \div -7.97$ [20, 22]	$1.56 \div 1.96$ [21,23]	0.15 ± 0.02 [21]
External Loading (Figs. 5 and 6)	-2.3	0.55	0.01

benthic fluxes and external loadings, one can infer that from late 1980s until 1998 the trophic status and dystrophic outbreaks were presumably driven by *Ulva*. From 1999 onwards, due to the decrease of macroalgal blooms, clam metabolism prevailed also due to the width of the farmed area that is approximately one third of the whole lagoon surface [20]. Likely, ecosystem health depends upon multiple stressors as well as internal buffers. Although in the last few years macroalgal bloom have almost disappeared, it should be considered that anthropogenic disturbances and natural forcings often cause unpredictable non-linear responses in coastal lagoons [107]. From 1990 to 1997, when macroalgal blooms had severe impacts on environmental quality and shellfish farming, the macroalgal biomass was harvested as an emergency intervention. Further, a first generation of stochastic and bioeconomic analyses were started in order to provide a scientific support to the management of macroalgal harvesting [108, 109]. A new generation of models is under development in order to design integrated watershed-lagoon management and policies [72, 73].

The Po di Volano–Sacca di Goro system is currently included as a testing site in an European project that deals with scientific tools for managing coastal lagoon in southern Europe (http://www.dittyproject.org). For this purpose, potential scenarios have been identified considering not only biogeochemical/ecological factors, but also socio-economical constraints. Basically, this approach allows the identification of the major drivers, their changes in time and space, and potential indicators for assessing environmental impacts and feedbacks. In terms of scenarios, one can consider changes in the intensity/quality of land uses and lagoon exploitation as well as changes in climatic factors and sea level.

At present, clam farming is performed in an $8 \, km^2$ area in the southern part of the lagoon along the sand barrier. Since 1999, after the new canals were dredged and the water circulation in the lagoon was ameliorated, the fisherman organizations requested an additional area to be exploited. This zone is approximately $6-8 \, km^2$ and extends in the eastern part of the lagoon and on the marine side of the newly formed sand barrier. An assessment of the potential risk/impact derived from the enlargement of the farmed area should be made considering the above-mentioned results dealing with oxygen consumption and internal nutrient recycling.

The eastern part of the lagoon belongs to the Regional park of the Po River and is under constraints for natural resources and ecosystem exploitation. In the long term, this can potentially cause a conflict for the lagoon exploitation. However, since the natural landscape of the Po Delta has a valuable potential for recreation and tourism, a touristic development can be considered as a way to compensate for losses due to clam farming restrictions as well as an interesting alternative to this primary activity. Tourism facilities are located along the coast in the same district and basically serve for summer holidays. However, due to the poor water quality and infrastructures, they are less competitive than the southernmost sites of the Adriatic Sea beaches. Moreover, summer holidays last for less than 2 months. The conventional tourism could be integrated with eco-tourism, that potentially covers also autumn and winter periods that are suitable for bird-watching and other naturalistic purposes.

The watershed of the Sacca di Goro is exploited for agriculture. At present there are few main crop types and more than 50% of the cultivated land is basically heavily impacted by pesticides and fertilizers. A further development of such activities (e.g., rice fields) can be foreseen as a potential impact for the lagoon ecosystem. Organic practices are less developed, although they are a promising alternative also in terms of revenue.

Since the Po di Volano-Sacca di Goro watershed has large depressed areas, it is expected to be very sensitive to climate changes and sea level rise [3]. General conclusions of the Intergovernmental Panel for Climate Control (IPCC) studies concerning Southern Europe identified an increase of atmospheric temperature, precipitation variability and superficial runoff (http://www.ipcc.ch/). The increase of water temperature and its duration could have effects on clam metabolism and recruitment, increase of pathogen diffusion, risk of harmful microalgal blooms, increase of oxygen demand and sulphide production, and biological invasion by exotic species. The increase of extreme events, namely flooding/drought alternance—as from 2000 to 2003—could affect both the watershed and the lagoon. Prolonged flooding could determine submersion in the depressed areas of the watershed and a decrease of salinity in the lagoon. The latter is dangerous for clams. Prolonged drought could cause a decrease in freshwater discharge. On the land side, it could cause an increase of costs (e.g., for irrigation) and decrease in vegetal production, as in 2003. On the lagoon side, less freshwater discharge could cause a decrease in phytoplankton production and biomass with effects on clam and mussel crops.

Overall, the behavior of the Po di Volano-Sacca di Goro system depends upon multiple stressors that are driven by both local and global processes. Likewise, an assessment of future management policies of water quality and resource exploitation has to consider an integrated watershed-lagoon-adjacent sea framework.

Acknowledgements This research was supported by the European Commission's Energy, Environment and Sustainable Development Programme (1998–2002) under contract No EVK3-CT-2002-00084, as a part of the project DITTY. This paper represents the contribution No 520/58 from the thematic Network ELOISE (European Land-Ocean Interaction Studies).

References

1. Barbiero G, Carone G, Cicioni GB, Puddu A, Spaziani FM (1991) Valutazione dei carichi inquinanti potenziali per i principali bacini idrografici italiani: Adige, Arno, Po, Tevere. Quaderni IRSA. IRSA-CNR, Rome, Italy
2. Provini A, Crosa G, Marchetti R (1992) Sci Total Environ, suppl 1992:291
3. Bondesan M, Castiglioni GB, Elmi C, Gabbianelli G, Marocco R, Pirazzoli PA, Tomasin A (1995) J Coast Res 11:1354
4. Gabbianelli G, Del Grande G, Simeoni U, Zamariolo A, Calderoni G (2000) Evoluzione dell'area di Goro negli ultimi cinque secoli (Delta del Po). In: Simeoni U (ed) La Sacca di Goro. Studi Costieri 2, Firenze, Italy, p 45
5. Viaroli P (1992) Eutrophication of the Po Delta lagoons: evolution and prospects for restoration. In: Finlayson M, Hollis T, Davis T (eds) Managing Mediterranean wetlands and their birds. IWRB Spec Publ 20:159
6. Viaroli P, Giordani G, Cattaneo E, Zaldívar JM, Murray CN (2001) Sacca di Goro Lagoon. In: Dupra V, Smith SV, Marshall Crossland JI, Crossland CJ (eds) Coastal and estuarine systems of the Mediterranean and Black Sea regions: carbon, nitrogen and phosphorus fluxes. LOICZ Reports & Studies No. 19. LOICZ, Texel, The Netherlands, p 36
7. Ciavola P, Gatti M, Tessari U, Zamariolo A, Del Grande C (2000) Caratterizzazione della morfologia di spiaggia lungo lo Scanno di Goro tramite tecniche GPS e rilievi batimetrici. In: Simeoni U (ed) La Sacca di Goro. Studi Costieri 2, Firenze, Italy, p 175
8. O'Kane JP, Suppo M, Todini E, Turner J (1992) Sci Total Environ, suppl 1992:489
9. Viaroli P, Pugnetti A, Ferrari I (1992) *Ulva rigida* growth and decomposition processes and related effects on nitrogen and phosphorus cycles in a coastal lagoon (Sacca di Goro, Po River Delta). In: Colombo G, Ferrari I, Ceccherelli VU, Rossi R (eds) Marine eutrophication and population dynamics. Olsen and Olsen, Fredensborg, Denmark, p 77
10. Viaroli P, Azzoni R, Bartoli M, Giordani G, Tajé L (2001) Evolution of the trophic conditions and dystrophic outbreaks in the Sacca di Goro lagoon (Northern Adriatic Sea). In: Faranda FM, Guglielmo L, Spezie G (eds) Structure and processes in the Mediterranean ecosystems. Springer, Berlin Heidelberg New York, p 443

11. Brath A, Gonella M, Polo P, Tondello M (2000) Analisi della circolazione idrica nella Sacca di Goro mediante modello matematico. In: Simeoni U (ed) La Sacca di Goro. Studi Costieri 2, Firenze, Italy, p 105
12. Simeoni U, Fontolan G, Dal Cin R, Calderoni G, Zamariolo A (2000) Dinamica sedimentaria dell'area di Goro (Delta del Po). In: Simeoni U (ed) La Sacca di Goro. Studi Costieri 2, Firenze, Italy, p 139
13. ISTAT (2001) V censimento generale dell'agricoltura. Istituto Nazionale di Statistica, Roma
14. Valiela I, Collins G, Kremer J, Lajtha K, Geist M, Seely B, Brawley J, Sham CH (1997) Ecol Appl 7:358
15. Marchetti R, Verna N (1992) Sci Total Environ, suppl 1992:315
16. Rossi R, Paesanti F (1992) Successful clam farming in Italy. In: Proceedings XXIII Annual Shellfish Conference. Fishmongers' Hall, London, United Kingdom, p 62
17. Breber P (1985) Oebalia 9:675
18. Cesari P, Pellizzato M (1985) Bollettino Malacologico 21:237
19. Carrieri A, Paesanti F, Rossi R (1992) Oebalia 17:97
20. Bartoli M, Nizzoli D, Viaroli P, Turolla E, Castaldelli G, Fano EA, Rossi R (2001) Hydrobiologia 455:203
21. Castaldelli G, Mantovani S, Welsh DT, Rossi R, Mistri M, Fano EA (2003) Chem Ecol 19:161
22. Viaroli P, Bartoli M, Giordani G, Azzoni R, Nizzoli D (2003) Chem Ecol 19:189
23. Sorokin YI, Giovanardi O, Pranovi F, Sorokin PI (1999) Hydrobiologia 400:141
24. Bartoli M, Naldi M, Nizzoli D, Roubaix V, Viaroli P (2003) Chem Ecol 19:147
25. Melià P, Nizzoli D, Bartoli M, Naldi M, Gatto M, Viaroli P (2003) Chem Ecol 19:129
26. Mistri M (2002) Estuaries 25:431
27. Sei S, Rossetti G, Villa F, Ferrari I (1996) Hydrobiologia 329:45
28. Alvisi A, Marzocchi D (1991) Indagine quali-quantitativa delle acque superficiali del bacino Burana-Volano. Dimensione Ambiente, Provincia di Ferrara, Ferrara, Italy
29. Alvisi A, Marzocchi D, Edelvais S (1992) Indagine quali-quantitativa delle acque superficiali del bacino Burana-Volano. Dimensione Ambiente, Provincia di Ferrara, Ferrara, Italy
30. Alvisi A, Edelvais S, Marzocchi D (1993) Indagine quali-quantitativa delle acque superficiali del bacino Burana-Volano. Dimensione Ambiente, Provincia di Ferrara, Ferrara, Italy
31. Alvisi A, Edelvais S, Marzocchi D (1994) Indagine quali-quantitativa delle acque superficiali del bacino Burana-Volano. Dimensione Ambiente, Provincia di Ferrara, Ferrara, Italy
32. Alvisi A, Edelvais S, Marzocchi D (1995) Indagine quali-quantitativa delle acque superficiali del bacino Burana-Volano. Dimensione Ambiente, Provincia di Ferrara, Ferrara, Italy
33. Alvisi A, Edelvais S, Marzocchi D (1996) Indagine quali-quantitativa delle acque superficiali del bacino Burana-Volano. Dimensione Ambiente, Provincia di Ferrara, Ferrara, Italy
34. Alvisi A, Edelvais S, Marzocchi D (1997) Indagine quali-quantitativa delle acque superficiali del bacino Burana-Volano. Dimensione Ambiente, Provincia di Ferrara, Ferrara, Italy
35. ARPA Emilia Romagna (2000) Sistema di gestione ed elaborazione dati acque superficiali interne. ARPA Sezione Provinciale di Reggio Emilia
36. Orlandi S (1998) Valutazione del carico dei nutrienti e prospettive di risanamento in un'area del Delta del Po Ferrarese. Graduate thesis, University of Parma, Italy

37. Naldi M, Viaroli P (2002) J Exp Mar Biol Ecol 269:65
38. Giaccone G (1977) Bollettino Museo Civico Storia Naturale Venezia 26:87
39. Piccoli F, Merloni N, Godini E (1991) Carta della vegetazione della Sacca di Goro. In: Bencivelli S, Castaldi N (eds) Studio integrato sull'ecologia della Sacca di Goro. FrancoAngeli, Milano, Italy, p 173
40. Vollenweider RA, Rinaldi A, Montanari G (1992) Sci Total Environ, suppl 1992:63
41. Castel J, Caumette P, Herbert R (1996) Hydrobiologia 329:ix
42. Sfriso A, Pavoni B, Marcomini A, Orio AA (1992) Estuaries 15:517
43. Morand P, Briand X (1996) Botanica Marina 39:491
44. Valiela I, McLelland J, Hauxwell J, Behr PJ, Hersh D, Foreman K (1997) Limnol Oceanogr 42:1105
45. Piccoli F, Godini E (1994) Ricerche qualitative e quantitative sulla vegetazione della Sacca di Goro, anni 1989–1990. In: Bencivelli S, Castaldi N, Finessi D (eds) Sacca di Goro: studio integrato sull'ecologia. FrancoAngeli, Milano, Italy, p 227
46. Viaroli P, Naldi M, Christian R, Fumagalli I (1993) Internationale Vereinigung für Theoretische und Angewandte Limnologie: Verhandlungen 25:1048
47. Viaroli P, Bartoli M, Bondavalli C, Naldi M (1995) Fresenius Environ Bull 4:381
48. Bondavalli C, Naldi M, Taje' L, Viaroli P (1996) Decomposizione di diverse alghe marine in relazione alla rigenerazione di azoto e fosforo ed al rilascio di solfuri. In: Virzo De Santo A, Alfani A, Carrada GC, Rutigliano FA (eds) Proceedings 7th congress Italian Society of Ecology, Napoli, Italy, p 367
49. Viaroli P, Naldi M, Bondavalli C, Bencivelli S (1996) Hydrobiologia 329:93
50. Viaroli P, Christian RR (2003) Ecol Ind 3:237
51. Bartoli M, Castaldelli G, Nizzoli D, Gatti LG, Viaroli P (2001) Benthic fluxes of oxygen, ammonium and nitrate and coupled and uncoupled denitrification rates in three eutrophic coastal lagoons with different primary producers. In: Faranda M, Guglielmo L, Spezie G (eds) Structures and processes in the Mediterranean ecosystems. Springer, Berlin Heidelberg New York, p 227
52. Viaroli P, Bartoli M, Bondavalli C, Christian RR, Giordani G, Naldi M (1996) Hydrobiologia 329:105
53. Castaldelli G, Welsh DT, Flachi G, Zucchini G, Colombo G, Rossi R, Fano EA (2003) Aquat Bot 75:111
54. Giordani G, Azzoni R, Bartoli M, Viaroli P (1997) Water Air Soil Pollut 99:363
55. Howart RW, Stewart JWB (1992) The interactions of sulphur with other element cycles in ecosystems. In: Howart RW, Stewart JWB, Ivanov MU (eds) Sulphur cycling on the continents: wetlands, terrestrial ecosystems and associated water bodies. SCOPE 33. Wiley, New York, p 67
56. Bartoli M, Barbanti A, Castaldelli G, Giordani G, Viaroli P (1996) Analisi di processi all'interfaccia acqua-sedimento con tecniche di microprofilazione. In: Virzo De Santo A, Alfani A, Carrada GC, Rutigliano FA (eds) Proceedings 7th congress, Italian Society of Ecology, Napoli, Italy, p 539
57. Viaroli P, Bartoli M, Giordani G, Azzoni R, Tajè L, Agazzi G, Nizzoli D (1999) Growth and decomposition of Ulva spp. in relation to dystrophy, nutrient retention and the sedimentary buffering capacity in the Sacca di Goro lagoon. In: De Wit R (ed) ROBUST: the role of buffering capacities in stabilising coastal lagoon ecosystems. Final Report. Commission of European Communities, Environment and Climate-ELOISE, Arcachon, France, p 75
58. Lapointe BE (1997) Limnol Oceanogr 42:1119
59. Lavery PS, McComb AJ (1991) Estuar Coast Shelf Sci 33:1
60. Pedersen MF, Borum J (1997) Mar Ecol Prog Ser 161:155

61. Bartoli M, Cattadori M, Giordani G, Viaroli P (1996) Hydrobiologia 329:143
62. Azzoni R, Giordani G, Bartoli M, Welsh DT, Viaroli P (2001) J Sea Res 45:15
63. Giordani G, Cattadori M, Bartoli M, Viaroli P (1996) Hydrobiologia 329:211
64. Roden EE, Edmonds JW (1997) Archiv für Hydrobiologie 139:347
65. Ceccherelli VU, Ferrari I, Viaroli P (1994) Bollettino di Zoologia 61:425
66. Weston DP (1990) Mar Ecol Prog Ser 61:233
67. Nizzoli D (2002) Regolazione dei processi biogeochimici in ambienti marini costieri con elevata densità della fauna bentonica. PhD Thesis, University of Parma, Parma, Italy
68. Atkinson MJ, Smith SV (1983) Limnol Oceanogr 28:568
69. Fiordelmondo C, Manini E, Gambi C, Pusceddu A (2003) Chem Ecol 19:173
70. Christian RR, Fores E, Comin F, Viaroli P, Naldi M, Ferrari I (1996) Ecolog Model 87:111
71. Christian RR, Naldi M, Viaroli P (1998) Construction and analysis of static, structured models of nitrogen cycling in coastal ecosystems. In: Koch AL, Robinson JA, Milliken JA (eds) Mathematical modeling in microbial ecology. Chapman & Hall, Microbiology Series, New York, p 162
72. Zaldívar JM, Cattaneo E, Plus M, Murray CN, Giordani G, Viaroli P (2003) Cont Shelf Res 23:1847
73. Zaldívar JM, Plus M, Giordani G, Viaroli P (2003) Modelling the impact of cultivated *Tapes philippinarum* population in the biogeochemical cycles of a Mediterranean lagoon (Sacca di Goro, Northern Adriatic, Italy). In: E Ozhan (ed) Sixth International Conference on the Mediterranean Coastal Environment MEDCOAST 03, 7–11 October 2003, Ravenna, Italy, p 1291
74. Cattaneo E, Zaldívar JM, Murray CN, Viaroli P, Giordani G (2001) Application of LOICZ Methodology to a Mediterranean Coastal Lagoon: Sacca di Goro (Italy). EUR Report n. 19921 EN. JRC.CEC, Ispra, Italy
75. Gordon DC Jr, Boudreau PR, Mann KH, Ong J-E, Silvert WL, Smith SV, Wattayakorn G, Wulff F, Yanagi T (1996) LOICZ Biogeochemical modelling guidelines. LOICZ Reports & Studies No 5. LOICZ, Texel, The Netherlands
76. Bencivelli S, Castaldi N (eds) Studio integrato sull'ecologia della Sacca di Goro. FrancoAngeli, Milano, Italy
77. Bencivelli S, Castaldi N, Finessi D (eds) Sacca di Goro: studio integrato sull'ecologia. FrancoAngeli, Milano, Italy
78. Crosa G, Marchetti R (1993) La qualità delle acque: asta principale e affluenti. In: Marchetti R (ed) Problematiche ecologiche del sistema idrologico padano. Acqua & Aria, Milano, p 609
79. Galassi S, Provini A (1993) Metalli pesanti e microinquinanti organici nei sedimenti e negli organismi. In: Marchetti R (ed) Problematiche ecologiche del sistema idrologico padano. Acqua & Aria, Milano, p 619
80. Piccapietra L, Bidoglio G, Elorza FJ, Rindone B (1998) Valutazione dell'influenza dell'uso di pesticidi sulla qualità di acque e di suoli in un bacino idrografico costiero. EUR Report n 18052 IT. JRC.CEC, Ispra, Italy
81. Baldi M, Coppi S, Davì ML, Benedetti S, Bovolenta A, Penazzi L, Previati M (1991) Valutazione del carico di erbicidi ed insetticidi organoclorurati nelle acque e nei sedimenti nella Sacca di Goro. In: Bencivelli S, Castaldi N (eds) Studio integrato sull'ecologia della Sacca di Goro. FrancoAngeli, Milano, Italy, p 91
82. Baldi M, Davì ML, Penazzi L (1994) Valutazione del carico di erbicidi ed insetticidi organoclorurati nelle acque e nei mitili nella Sacca di Goro (1989–1990). In: Bencivelli S, Castaldi N, Finessi D (eds) Sacca di Goro: Studio integrato sull'ecologia. FrancoAngeli, Milano, Italy, p 155

83. Binelli A, Provini A (2003) Mar Pollut Bull 46:879
84. ARPA (2003) Supporto tecnico alla Regione Emilia-Romagna, alle Province ed alle Autorità di Bacino per l'elaborazione del piano regionale di tutela delle acque e piano territoriale di coordinamento provinciale. Elaborato Relazione Generale. ARPA Ingegneria Ambientale, Bologna, Italy
85. Di Domenico A, Turrio Baldassarri L, Ziemacki G, De Felip E, Ferrari G, La Rocca C, Cardelli M, Cedolini G, Dalla Palma M, Grassi M, Roccabella M, Volpi F, Ferri F, Iacovella N, Rodriguez F, D'Agostino O, Sansoni R, Settimo G (1998) Organohalogen Compounds 39:199
86. Di Domenico A, Turrio Baldassarri L, Ziemacki G, De Felip E, Ferri F, Iacovella N, Iacovella N, Rodriguez F, Volpi F, D'Agostino O, Sansoni R (1997) Organohalogen Compounds 34:61
87. Bayarri S, Turrio Baldassarri L, Iacovella N, Ferrara F, Di Domenico A (2001) Chemosphere 43:601
88. Stefanelli P, Di Muccio A, Ferrara F, Attard Barbini D, Generali T, Pelosi P, Amendola G, Vanni F, Di Muccio S, Ausili A (2004) Food Control 15:27
89. Fagioli F, Landi S, Locatelli C, Vecchietti R (1991) Valutazione del carico dei metalli nelle acque, nei sedimenti e nei principali organismi accumulatori. In: Bencivelli S, Castaldi N (eds) Studio integrato sull'ecologia della Sacca di Goro. Franc Angeli, Milano, Italy, p 135
90. Fagioli F, Righini F, Landi S, Locatelli C (1994) Valutazione del carico dei metalli nelle acque, nel particolato sospeso, nei sedimenti e nei principali organismi accumulatori. (april 1990–marz 1991). In: Bencivelli S, Castaldi N, Finessi D (eds) Sacca di Goro: Studio integrato sull'ecologia. FrancoAngeli, Milano, Italy, p 177
91. Fagioli F, Locatelli C, Landi S (1994) Annali di Chimica 84:129
92. Locatelli C, Fagioli F, Bighi C (1996) Annali di Chimica 86:605
93. Dinelli E, Gabbianelli G, Tessari U (2000) Caratteri geochimici dei depositi attuali della Sacca di Goro (Delta del Po). In: Simeoni U (ed) La Sacca di Goro. Studi Costieri 2, Firenze, Italy, p 189
94. Covelli S, Fontolan G, Sartore L, Simeoni U, Tesolin V, Zamariolo A (2000) Aspetti geochimici dei sedimenti della Sacca di Goro (Delta del Po, Adriatico settentrionale). In: Simeoni U (ed) La Sacca di Goro. Studi Costieri 2, Firenze, Italy, p 81
95. Locatelli C, Torsi G (2002) Environ Monit Assess 75:281
96. Frignani M, Bellucci LG, Langone L, Muntau H (1997) Mar Chem 58:275
97. Sfriso A, Marcomini A, Zanette M (1998) Mar Pollut Bull 30:116
98. Jackson LJ (1998) Sci Total Environ 219:223
99. Berry WJ, Hansen DJ, Mahony JD, Robson DL, Di Toro DM, Shipley BP, Rogers B, Corbin JM, Boothman WS (1996) Enviro Toxicol Chem 15:2067
100. Rozan TM, Taillefert M, Trouwborst RE, Glazer BT, Ma S, Herzage J, Valdes LM, Price KS, Luther GW (2002) Limnol Oceanogr 47:1346
101. Nonnis Marzano F, Triulzi C (1994) Mar Pollut Bull 28:244
102. Nonnis Marzano F, Triulzi C (2000) Int J Environ Pollut 13:1
103. Parisi V, Mezzadri MG, Occhipinti A, Poletti G (1991) Il contributo delle specie marine del macrobenthos e della radiocontaminazione alla compressione del funzionamento dlla Sacca di Goro. In: Bencivelli S, Castaldi N (eds) Studio integrato sull'ecologia della Sacca di Goro. FrancoAngeli, Milano, Italy, p 205
104. Parisi V, Mezzadri MG, Poletti G, Cattani S (1994) Ricerche sul popolamento macrobentonico e sulla radiocontaminazione della Sacca di Goro. In: Bencivelli S, Castaldi N, Finessi D (eds) Sacca di Goro: Studio integrato sull'ecologia. FrancoAngeli, Milano, Italy, p 245

105. Bondavalli C (2003) Environ Pollut 125:433

106. Sivelli F (1999) Stusio della ripartizione di radionuclidi e variazioni pluriennali della radioattività in un ecosistema lagunare del Delta del Po: Sacca di Goro. PhD Thesis, University of Parma, Italy

107. De Wit R, Stal LJ, Lomstein BA, Herbert RA, Van Gemerden H, Viaroli P, Ceccherelli VU, Rodriguez-Valera F, Bartoli M, Giordani G, Azzoni R, Schaub B, Welsh DT, Donnelly A, Cifuentes A, Anton J, Finster K, Nielsen LB, Underlien Pedersen A-E, Turi Neubauer A, Colangelo M, Heijs SK (2001) Cont Shelf Res 21:2021

108. De Leo GA, Bartoli M, Naldi M, Viaroli P (2002) Mar Ecol PSZN 23:92

109. Cellina F, De Leo GA, Bartoli M, Viaroli P (2003) Oceanolog Acta 26:139

Hdb Env Chem Vol. 5, Part H (2006): 233–264
DOI 10.1007/698_5_021
© Springer-Verlag Berlin Heidelberg 2005
Published online: 30 November 2005

Estuary of the Danube

Nikolai Berlinsky (✉) · Yulia Bogatova · Galina Garkavaya

Odessa Branch,
Institute of Biology of Southern Seas National Academy of Sciences of Ukraine,
Pushkinskaya St. 37, Odessa, 65011, Ukraine
ibss@paco.net

Abstract The Danube is the second largest major European river, with a huge estuary located in two countries: Ukraine and Romania. The Danube watershed embraces 15 highly industrialized European countries that produce a high level of anthropogenic pressure. During the last 30 years they have influenced the river, estuary, and Black Sea. At present the Danube runoff is totally regulated by dams. This factor changed the hydrological regime. Another factor that has changed the hydrochemical regime is the oversupply of nutrients, primarily nitrogen and phosphorus. This affected the environment of the river, estuary, and northwestern shelf of the Black Sea. Eutrophication, "water blooming", and near-bottom hypoxia as a result of this process are developing in the northwestern part of the Black Sea. In the estuary, both water quality and bottom sediments have deteriorated, and the fish catch and biodiversity have decreased. At present a new source of eutrophication is bottom sediment in the shore zone of the sea.

Keywords Black Sea · Danube estuary · Eutrophication · River runoff · Nutrient

Abbreviations
MCL Maximum allowable contaminant levels
CE   Coefficient of export
A   Anthropogenic constituent of river's nutrient runoff

The estuary parts of the river, occupying the intermediate position between the river basins and receiving reservoirs, are so-called specific border zones: geoecosystems rich in water, land, and biological resources. The common area of the estuary in the world ocean consists of not more than 0.4%, but it has the highest biological productivity amongst all kinds of ecosystems and is responsible for more than 4% of oceanic primary production. Estuaries are very important in ecology as natural biological filters where nutrients accumulate [1]. The term "estuary" originates from the Latin word *aestuarium* and means low water river mouth, lagoon, liman, or gulf where the process of mixing of fresh and marine waters takes place. As for the Danube estuary, its ecosystem includes the river mouth and the shallow water coastal zone.

1
Geographical Location of the Danube Estuary

1.1
Location, Morphometry, Climate

The Danube is a delta-type river. In the lower Danube plain the river forms many branches. The width reaches 1.2 km, with an average depth of 5–7 m, and a velocity of 0.5–1.0 m s^{-1}. The Danube estuary includes the delta area of 5640 km^2 and the near-mouth coastal zone of 1360 km^2 [2]. The estuary is located in the territory of two countries, Ukraine in the north and Romania in the south. The Ukrainian part of the delta has 1240 km^2 (approximately 22% of the common area); the rest is Romanian. The total area of the Danube estuary is about 7000 km^2. The length of the main Kiliya branch is 115 km. The length of the marine border is 180 km, with an average width of marine coastal zone of 6–10 km. The top of the delta is the place where the river divides into two main branches: the left is the Kiliya branch and the right the Tulcha branch (Fig. 1). The elevation of the delta plain above sea level near Ismail (Ukraine) is 3.7 m and near the Sulina mouth (Romania) is 0.5 m [3]. The delta is divided into two parts: ancient riverine and young marine. The main Kiliya branch forms two inner and one outer marine deltas (Fig. 1).

The inner deltas have lateral branches; most of which are not functioning at present (Fig. 2). The top of the marine delta is near Vilkovo (Ukraine), where the Kiliya branch divides into two big branches: Ochakov and Starostambulsky.

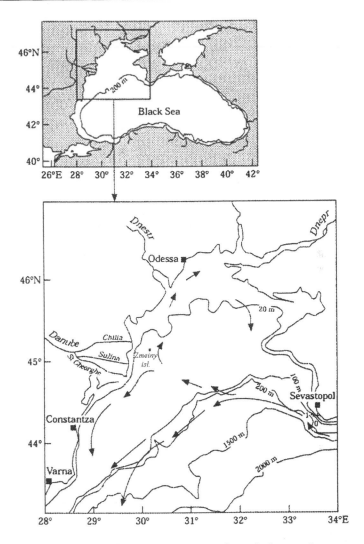

Fig. 1 Northernwestern part of the Black Sea. General trend of sea surface currents [34]

The system of the Ochakov branch includes six branches and that of Starostambulsky includes five branches. In the Romanian part, the Tulcha branch is 17 km long and is divided near Tulcha city into two branches: Sulina of 69 km long and St. George of 109 km long. Under the Decision of the Danube Commission the Sulina branch was transformed in 1867–1895 into an international shipping way with parallel moles into the open sea. At present, the total length of this channel is 87 km.

The delta composition includes many lakes, lagoons, and limans with a total area of about 1400 km^2. In the Ukrainian part they are Yalpukh, Kugurlui, and Kitai, with a total area of about 360 km^2. In the Romanian part

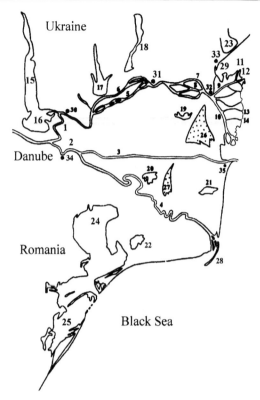

Fig. 2 The Danube estuary. Branches: 1 Kilia, 2 Tulcha, 3 Sulina, 4 St. George, 5 Sredny, 6 Kislitsky, 7 Solomonov, 8 Pryamoy, 9 Ochakovsky, 10 Starostambulsky, 11 Prorva, 12 Potapovsky, 13 Bistry, 14 Vostochny. Lakes: 15 Yalpuh, 16 Kugurluy, 17 Katlabuh, 18 Kitay, 19 Merhey, 20 Uzlina, 21 Rosu, 22 Dranov, 23 Sasik, 24 Raselm, 25 Sinoe. Sandy ridges: 26 Letya, 27 Karaorman, 28 Sakhalin island, 29 Zhebriayny bay. Settlements: 30 Ismail, 31 Kilia, 32 Vilkovo, 33 Primorskoe, 34 Tulcha, 35 Sulina

are the complexes Raselmn– Sinoe (740 km²) and the lakes complex of the inner part of the delta (270 km²). The State border between Ukraine and Romania passes through the channel of the Danube, Kiliya branch, and branches Pryamoi, Sredny, Starostambulsky, and Limba.

The modern delta is a huge wetland crossed by numerous small branches. The depth of these branches is not more than 1–2 or 3–4 m deep. The bottom consists of the river sand, and the upper layer is silt with rotten vegetation. Small lakes are formed on the islands between the Kiliya, Sulina, and St. George branches.

The Danube delta is a unique natural object with specific landscape, climate conditions, and very rich natural resources. The climate is moderate continental with a short winter and a long summer. Average annual temperature is about 10 °C. Average duration of the period with negative daily temperature is 50 days. The Black Sea and lakes influence and enrich the humidity and lessen the temperature contrast. The amount of precipitation is less than the evaporation in the delta. Usually the winter starts in the second half of December and continues up to the second half of February. The precipitation could be rain as well as snow. The spring is dry and cold. The summer (May–September) is hot. Summer precipitation consists of half of the annual total. The wind is moderate, on average $5\,\mathrm{m\,s^{-1}}$. The frequency of stormy winds (more than $15\,\mathrm{m\,s^{-1}}$) is 1–3%. The prevailing wind direction is north. The average annual water temperature in the Danube (in the delta) is $12.9\,^{\circ}\mathrm{C}$. The maximum is up to $28.4\,^{\circ}\mathrm{C}$ in July–August. An ice regime is observed once per 5 years. During the last 20 years it has been observed once per 7 years. The water level regime depends on wind direction and up and down welling in the sea coastal zone. The limit of level fluctuation is 20–40 cm, maximum 100 cm. In wintertime, ice blocking the delta can raise the level of water in the Danube to an extreme and provoke flooding. Spring tides usually come in two peaks: from the snow melting and from rain in a period from the second part of March up to the second part of June. In autumn there is a short flood because of rain. Minimum water level has been marked in summer-autumn periods, but sometimes it can be observed in winter. However, in the modern period this rule is violated because of climate change – there is a tendency for increasing temperatures in Europe. The level variability decreases approaching the sea. At the same time, salty marine water can penetrate the delta via small branches. The distance of penetration is equal to the size of these branches [3].

1.2
Evolution of the Estuary Zone

At the beginning of the Euxinus epoch in the northwestern part of the Black Sea there was a huge united delta of the Danube, Dniestr and Dniepr. Their waters flowed into the sea approximately on isobath 75–100 m of the modern period. At the end of this epoch transgression of the sea flooded this area [4, 5]. The modern Danube delta started to form about 5000 years ago. At first two branches, Ismail and Tulcha, were formed. Later the Tulcha branch was divided into two branches, St. George and Sulina. The first one formed was the Tulcha branch and later (V–I centuries BCE) the Sulina formed. Next (XVI–XVII centuries CE), the northern branches combined to form the Kiliya branch. Gradually Kiliya became the main branch with the most water runoff and formed two consecutive inner deltas in the shallow sea gulf. From the XVIII century Kiliya branch started to form a so-called advancing marine delta, which in the modern period is called Kiliya [4]. The process of development has four phases: with one branch (1740–1800), with a few branches (not more than 20 branches) (1800–1856), with numerous branches (40–60 branches, 1856–1956) and again a few branches (14 from 1957–1993) [3, 4]. More active growth of the delta had been marked in 1871–1922 with high wa-

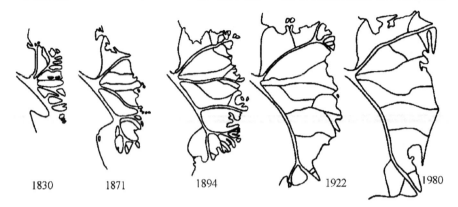

1830   1871   1894   1922   1980

Fig. 3 Kilia delta evolution in the Danube mouth [3]

ter runoff. The Kiliya area increased from 80 to 348 km^2 (Fig. 3). At present, the growth has become slower because of suspended matter reduction and an increase of the Black Sea level. In the period 1943–1980 the growth of the delta area was 1.1 km^2 year^{-1}.

In the process of land use the delta and hydrological regime have been exposed to powerful anthropogenic influence. As a result, many branches were dyked. The area of dyking lands is 430×10^3 ha in the Romanian part and more than 30×10^3 ha in the Ukraine.

In Ukrainian and Romanian parts of the delta shipping way channels have been constructed at different times. In the Romanian part there are the Sulina and St. George channels. The Sulina channel started functioning in 1858. In 1867–1895 twin moles 1400 m long straight into the sea were constructed in the mouth of Sulina. At present they are 8 km long because of permanent suspended matter inputs. Construction of the St. George channel was started in 1966 and is still unfinished at the present time; its approximate length is 104.6 km and it is 8.1–8.4 m deep.

In the Ukrainian part there is only one channel (4 m deep) between Zhebriyany bay and the Prorva branch. Using the channels in the Prorva branch was not successful because of the high level of sedimentation.

2
Hydrological Characteristic of the Estuary

2.1
Distribution of River Runoff in the Arms of the Danube.
Long-Term and Annual Variability

Specification of the Danube water balance is dependent upon the water runoff. The average annual water runoff in the period 1921–1993 was

203 km^3 year^{-1}, but for the 1921-1999 it was 204 km^3 year^{-1} [6]. During the last century the minimum river runoff was 134 km^3 year^{-1} in 1921 and maximum 313 km^3 year^{-1} in 1941. For the last decade the range of the water runoff has changed from 132.3 km^3 year^{-1} (1990) to 263 km^3 year^{-1} (1999) (Table 1, Fig. 4). Annually the impressments of the water in Ukraine are about 1.0 km^3.

Seasonal variability is characterized as 25% of the water in winter, 30% in spring, 25% in summer and 20% in autumn [2]. In winter the minimum is from the first decade of December to the second decade of February. The water runoff varies from 1.5 to 7.5 × 10^3 m^3 s^{-1}. The highest water level (10–13% of the annual water runoff) is in April–June. The least water runoff (5.5–6%) is in September–October. Maximum water runoff in the period of flood is 15–16 × 10^3 m^3 s^{-1}; in a period of low water it is 1.3–1.5 × 10^3 m^3 s^{-1}. The autumn rain often forms the autumn maximum in the second part of November or in the first part of December [7]. The windy water level variability is not more than 50 cm and maximum duration is 4 days.

The water runoff is distributed in three main branches: Kiliya, Sulina, and St. George. The largest share has gone to Kiliya and was 70% in 1895, 67% in 1905, and is 57.8% at present [3, 8]. Redistribution of the water runoff between the branches depends on natural processes of delta formation: delta mouth extension and silting. In the past the Kiliya branch was smaller than Sulina and St. George but after going out to the open sea this branch is

Table 1 Danube water runoff (km^3) (data from Hydrometeorological Observatory in Ismail)

Year	1990	1991	1992	1993	1994	1995	1996	1997	1998	1999	2000
Water runoff	132	198	172	154	181	230	236	224	221	263	215

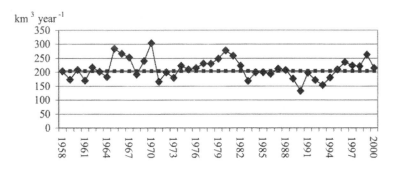

Fig. 4 Dynamics of the River Danube runoff

active and the water flow increased owing to the disappearance of small lateral branches [2]. Later, because of anthropogenic influences (hydrotechnical works, dredging for navigation) the direction of water runoff redistribution has changed (Table 2). In the Sulina branch, for the last 149 years the portion of water runoff has increased from 7–8 to 17% of the total water runoff.

At present, general river runoff (about 57%) is concentrated in the Kiliya delta in the several main branches (Table 3). In 1983–1984 in the Kiliya delta the arms Sredny and Zavodnensky were dead. Mouth lengthening is very important in the processes of delta development. The velocity of delta development and sandbar formation depends on water runoff or suspended matter. In the future, the process of mouth development limits sandbar formation. Examples are the Starostambulsky and Potapovo branches.

Reservoirs and wetlands in the delta are natural regulators of water runoff. In flood periods, part of the water accumulates in the wetlands and later in times of low water this water escapes. At present, as a result of dyking there is no more than one quarter of the delta under flooding. It is a great ecological and fishing problem because these areas were the natural habitats for development of juvenile fish.

Table 2 Water runoff distribution by the main branches of the Danube delta (% of average total Danube runoff)

Branches	1856	1871–1893	1895	1910	1928–1929	1958–1960	1980–1985	1986–1990
Kiliya	63	63	70	72	66	62.5	59	57
Tulcha	37	37	30	28	34	37.5	41	43

Table 3 Water runoff distribution by the main branches of the Kiliya delta (% of average total Danube runoff)

Branches	1984–1995	1942–1943	1958–1960	1966–1970	1976–1980	1986–1990
Ochakovsky	–	–	25.3	20.7	18.0	16.9
Belgorodsky	1.4	–	0.1	0.1	0.1	0.1
Polunochny	1.0	4.0	0	–	–	–
Prorva	10.0	4.6	6.1	7.7	7.6	7.6
Potapovo	5.7	20.0	15.0	8.2	4.3	3.1
Starostambul – beginning	–	–	37.2	40.5	40.7	–
Sredny	10.5	1.7	0.7	0.2	0.1	0
Bistry	–	6.6	10.2	12.4	14.3	16.5
Vostochny	–	1.0	1.5	1.7	2.3	–
Starostanbulsky – mouth	–	21.6	22.9	21.6	9.6	–

2.2
Hydrological Regime of the Sea Coastal Zone:
Climate, Temperature, Salinity, Fresh Water Influences

The Danube coastal zone is a part of the northwestern shelf of the Black Sea. The Danube coastal zone is bilateral: open sea area and semiclosed (Zhebryany bay). The main specific characteristic of this zone is the process of permanent mixing of fresh and salty water. The border of the area is changeable and depends on the season (flood, low water) and wind direction. Commonly, the distribution of fresh water in the sea takes the form of tongues in front of the main river branches. The inner border of the zone of mixing fresh and salty waters (2‰ salinity at the surface) is at a distance 0–4 km from the mouth. The outside border is 3–20 km from the mouth (depending on season and wind direction). Variability of salinity inside this zone is 4–17‰. In times of low water and under easterly winds salinity in this zone can achieve 15–17‰. The maximum horizontal gradient of salinity is at a distance of 2–8 km from the river mouth. Under easterly winds the mixing zone converges and under a west wind it widens [9].

In the sea the Danube water spreads on the surface. The layer of fresh water is 1–3 m to a maximum of 5 m deep. From 5 m depth and down marine water with salinity 17–18‰ is found. Between the upper riverine and marine water is a vertical gradient of salinity and temperature (in warm periods). These gradients limit vertical mixing between the layers. For example, the significant drop of vertical density can be 3–5 $kg\,m^{-3}$ per meter depth, which corresponds to the same change as for the open ocean over a depth of about 1000 m. The depth of the Danube coastal zone is 10–15 m and smoothly increases with distance from the seashore. A depth of 30 m is reached near Zmeiny island (a single island in the Black Sea, apart from the other three islands), some 40 km from the Danube (Fig. 1).

The range of seawater temperatures on the surface is 2 °C in January to 22 °C in August. The ice regime is one per 5 years. In Zhebryany bay it is more often, one per 3 years, especially close to the shore. The highest temperature in summer was 27 °C. The daily variability of the temperature under a moderate wind can achieve 6 °C. The vertical temperature structure changes seasonally. Before the spring warming the whole water column has the same temperature (2–7 °C, depending on winter air temperature). During the spring a surface warmed layer, about 5 m deep, is formed. In August, this layer deepens to 15–20 m with a maximum vertical gradient between upper and lower water masses of 3–5 $°C\,m^{-1}$. In the area with a depth less than 15 m the temperature is homogeneous. In December, winter vertical circulation (vertical convection) makes the water column homogeneous again (decreasing temperature from 10 to 2–4 °C) [10].

At the beginning of the 1970s, for the Danube coastal zone and for the whole northwestern part of the Black Sea, the development of near-bottom hypoxia (concentration of dissolved oxygen less than $2\,mg\,L^{-1}$) was noted. Near-bottom hypoxia starts to develop in shallow water (8–15 m deep) in July under the seasonal pycnocline (thermocline). Later, in June, following thermocline development, the near-bottom layer with the oxygen deficit slips to depths of 20–25 m, but seldom to the depths of 30–50 m at the border of the northwestern shelf. In this period the oxygen regime regenerates in the shallow water column because of turbulent diffusion. In the deep water, hypoxia and some times anoxia continues until November and is destroyed in the process of vertical winter circulation and storm activity [11]. In the winter period the bottom layer is saturated with oxygen. In summer in the shallow water disturbances of oxygen regime can occur under upwelling processes. Under the influence of west winds, surface water goes out to the open sea and water from the bottom layer with a deficit of oxygen rises to the surface near the shore. On average this upwelling phenomena happens five times per season. This effect had been noted in the coastal zone between the Danube and Dnepr in the Ukraine, near the shores of Romania and Bulgaria. As a result much fish and benthic marine organism mortality has been recorded during the last 30 years.

2.3
Physical and Chemical Processes in the Coastal Zone – Sedimentation of Suspended Matter

Suspended matter in the zone of mixing marine and riverine water is coagulated and sedimented. Coagulation or enlargement of small particles is more active and takes place in ranges of salinity between 2–6‰ [12]. Erasing of the temperature structure and mixing promotes this process. At the same time microbiological dissociation of detritus, decreasing the size and number of the suspended particles, takes place as well. The balance between coagulation and destruction depends on water pH and Eh. Thin particles are carried out into the sea with the fresh water surface layer, while rough and heavy particles (2–5 μkm) sink down to the bottom. The Danube coastal zone is a zone of avalanche sedimentation. A lot of suspended matter gets down to the bottom here. There is a geochemical barrier between the different conditions of sedimentation. It is a zone of rapidly changing environments. In the process of coagulation and organic matter flocculation more than 50% of the Fe, 80% of Mg, more than 50% of P, 10–70% of Al and also Zn, Cu, Ni, Co, Mb, and other trace elements get down to the sediment. During coagulation, ions of the common salinity composition of marine water: Cl, Na, Mg, S, Ca, and K partly come out of solution [12]. However, under specific conditions these ions can transfer from suspended matter to solution. In the mixing zone 10–60% of the suspended matter gets down to the bottom and forms the so-called liquid bottom. This tilt to the shore bottom stops suspended particles and forms sand bars and banks.

3
Hydrochemical Regime of the Estuary

3.1
Salt Composition, Oxygen, Dynamics of Suspended Matter, Nutrients, Oil and Heavy Metals in River Branches

Variability of the hydrochemical balance in the Danube estuary depends mainly on the river runoff, drainage and precipitation, seasonal temperature distribution, hydrobiont activity (especially phytoplankton, which use nutrients) and man-caused factors such as pollution from industry and agriculture. Natural cycling of river runoff and regulated flow in the Danube is also very important.

In the 1950–1960s the hydrochemical regime of the Danube delta was rather stable [2], as high water turbidity and current velocity hampered the process of photosynthesis and the hydrochemical regime was related to hydrological factors.

Starting from the 1970s the hydrological and hydrochemical balance has changed. Construction of water storages Gerdap 1, Gerdap 2 (Bulgaria), and Gapchikovo–Nadiamarosh (Hungary); the large dams Iron Gates I and Iron Gates II; and hydropower stations in the middle river all caused the decrease of current velocity and increased water transparency. Water storages also precipitate the suspended matter and contaminants. Increasing water transparency caused development of photosynthesis due to mass development of phytoplankton. Development of hydrobiological processes started to play an important role in the delta's hydrochemical regime.

At the same time, the anthropogenic pressure on river ecosystems has increased: the usage of mineral fertilizers in agriculture of the countries located in the catchment area of the Danube has increased and the input of the urbanized water discharge from cities and industrial plants has also increased.

The water quality in the delta area has changed. For instance, during the last 50 years mineralization – an important characteristic of fresh water – has increased by 1.5 times (Table 4) [13]. This increase was provoked by anthropogenic causes such as industrial pollution and city sewerage outputs. Mineralization changes ($230–520$ mg L^{-1}) in the delta region are correlated with river runoff. Thus, in the flood period the sum of main ions decreases. The principal ionic materials in the Danube water are bicarbonate and calcium. The level of bicarbonate is $140–250$ mg L^{-1}, and of calcium $40–75$ mg L^{-1}. The ratio of these ions in river water is practically constant all the year round. In recent years there has been a tendency for mineralization to decrease. In the Ukrainian part between Reni and Vilkovo, mineralization decreased from 409 to 362 mg L^{-1} (the results of monitoring by the Danube Hydrometeorological Observatory in Ismail during the period 1995–2000).

Table 4 Dynamics of mineralization in the Ukrainian part of the Danube

Year	Mineralization ($mg\,L^{-1}$) Variability	Average
1948–1950	226–397	287
1958–1959	253–344	289
1963–1965	235–334	296
1976–1978	275–475	374
1985–1989	295–506	372
1990–1994	297–521	409

During the last century the concentration of suspended matter has changed in the Danube delta mouth. In the period 1895–1922 suspended matter input into the Black Sea was estimated as 75.7×10^6 t year^{-1} [14] and in the 1921–1960 period it was 67.5×10^6 t year^{-1} [2]. Bulgarian authors estimated the input in this period as 66×10^6 t year^{-1} [15]. In the 1970–1980s, after putting into operation hydrotechnical constructions, the turbidity in the Danube decreased by one third and the annual Danube sediment runoff became about 44×10^6 t [13]. Romanian authors [16] estimated the current average sediment discharge of the Danube as $40–50 \times 10^6$ t year^{-1}, out of which $5–8 \times 10^6$ t year^{-1} is sandy material. For the period 1961–2000, according to data of Danube Hydrometeorological Observatory in Ismail, the suspended runoff was 39.2×10^6 t year^{-1}.

However, the level of turbidity in the Danube is very high and changes from $93–242$ g m^{-3}. Daily variability can change from a few grammes to $2–3$ kg m^{-3}. For the period 1995–1997 the average value of suspended matter in the Kiliya branch was 93 g m^{-3} with a variability of $15–215$ g m^{-3} [17, 18], but in periods of flood concentrations of suspended matter can be high. For example, in May, 2000 the maximal concentration was 528 g m^{-3}. In the delta branches the process of sedimentation of suspended matter has started because of a decrease in river velocity. The suspended matter composition has changed as well. If it was established earlier that 95% of the suspended matter in the Danube water was mineral particles, now 20–30% is organic matter because of phytoplankton development.

Oxygen concentrations and pH in the delta waters have also changed. Before flow regulation, photosynthesis was limited by the high level of turbidity and the concentration of oxygen was $7–12$ mg L^{-1}, saturation was not more than 80–85%, and pH 7.60–8.40; the maximum was in winter [2]. At present in summer during the phytoplankton vegetation period, oxygen concentration increases up to 15 mg L^{-1} (more than 120% saturation) and pH increases up to 8.80 [19].

Other important elements are the nutrients P, N, and Si ,which are vital in the life of organisms. Increase of nutrients in the water stimulates the eutrophication process, which causes the growth of phytoplankton. High concentrations of nutrients in the river can be the cause of the phytoplankton overgrowth, which can be noticed by eye. This process is called "water bloom".

In the period 1950-1960 the dynamics of nutrients were only linked with river runoff. Increasing P and Si concentrations depended on river runoff intensity, i.e., input from the drainage area [20, 21].

At the beginning of 1970-1980 nutrient concentrations in the delta area increased sharply. Increase of the N and P concentrations in the Danube water is linked with industrial intensity in all of Europe. During this period the FAO World Agricultural Development Plan "Green Revolution" was realized. If, in 1960 about 8×10^9 t of N and P were applied with fertilizers on fields all over the world, in 1970 the quantity of fertilizers increased to 42×10^9 t and in 1975 it was 65×10^9 t [22].

According to changes in nutrient values in the Danube delta runoff for the period 1958-2000 (Table 5) the following periods could be defined: I (1958-1960) – before regulated stream and eutrophication processes; II (1977-1985) – start of eutrophication processes; and III (1986-2000) – development of eutrophication processes.

The values of the period 1958-1960 can be used as a natural background and for analysis of the following variability.

The 1977-1985 period has the full water regime, more than average water runoff, of 227.7 km^3 per year. Nutrients (mineral compounds of nitrogen and phosphorus) increased because of washout from the catchment area (Table 5,

Table 5 Annual variability of nutrient runoff (mg L^{-1}) in the water of Kiliya delta in the period 1958-2000

Ingredient	1958– 1960[a]	1977– 1985[b]	1986– 1988[c]	1989– 1992[d]	1993– 1996[d]	1997– 2000[d]
NH_4^+	0.248	0.620	0.575	0.441	0.125	0.042
NO_2^-	0.012	0.044	0.160	0.118	0.074	0.015
NO_3^-	0.530	1.000	1.126	1.626	1.184	0.580
N_{min}	0.790	1.664	1.861	2.185	1.383	0.637
N_{org}	0.630	0.900	3.072	5.069	3.739	4.397
N_{total}	1.420	2.564	4.933	7.254	5.122	5.034
PO_4^{3-}	0.071	0.165	0.281	0.233	0.091	0.079
P_{org}	0.031	0.073	0.100	0.113	0.096	0.038
P_{total}	0.102	0.238	0.381	0.336	0.187	0.117
SiO_3^{2-}	4.375	3.980	2.571	2.979	2.356	2.120
River runoff (km^3 year^{-1})	179.4	227.7	204.7	169.7	195.1	230.9

[a] [20], [b] [23], [c] [18, 24], [d] [19, 25]

Fig. 5) This is the maximum of anthropogenic eutrophication development in the Danube. Ammonia nitrogen in the water of the delta increased by 2.5 times, nitrites by four times, nitrates by five times, and phosphate doubled compared with the period 1958–1960. Mineral forms of nitrogen during this period comprised 65% of the total nitrogen in the water (Fig. 5).

Slowing water velocity and increasing transparency with a surplus of nutrients caused the intensification of photosynthesis in water storages and development of water blooms in the delta (mainly diatomaceous phytoplankton) [26, 27]. This caused an increase of organic compounds and decrease of silicon concentrations. Thus, in the Danube delta concentrations of organic nitrogen increased by 1.5 times, phosphorus doubled, and silicon decreased by 1.1 times (Fig. 5). During this period, suspended matter and pollutants were accumulated in the bottom sediments of water storage areas.

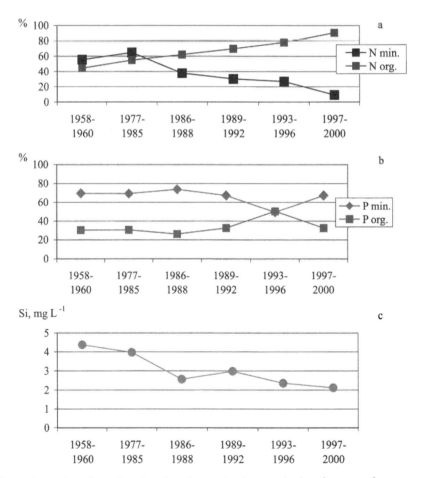

Fig. 5 Dynamics of **a** ratio mineral and organic nitrogen, **b** phosphorus, and **c** concentration of silicon (mg L^{-1}) in the Kilia delta

The long-term seasonal monitoring in the Danube delta (Kiliya branch) during the period of intensive eutrophication of river runoff (1986–2002) carried out by the Odessa branch of the Institute of Biology of Southern Seas, National Academy of Science of Ukraine, has helped to reveal some tendencies in quantitative characteristics of nutrient runoff. Irrespective of volumes of the Danube runoff (1986–1988 and 1993–1996 about average long-term runoff, 1989–1992 shallow period, and 1997–2000 full period) the general tendency of increase of the volumes of organic nitrogen in the nitrogen compound balance of the river runoff and decrease of silicon concentrations has been marked. Thus, at present, the concentration of organic nitrogen is much higher than the sum of mineral forms: ammonia nitrogen, nitrite, and nitrate. It comprises 87% of the total nitrogen in the Danube delta. For nearly a 50-year period silicon concentrations have decreased to a third of their former value, from $4.375 \, \text{mg} \, \text{L}^{-1}$ in 1958–1960 to $2.120 \, \text{mg} \, \text{L}^{-1}$ in 1997–2000. This correlates with the activity of the photosynthesis process and nutrient utilization by diatomaceous phytoplankton for cell construction. In the previous period water blooming occurred only in the water storage areas, while now it occurs permanently in the delta during the summer. As a result of phytoplankton decay labile organic matter has increased. Probably it is linked with partial sedimentation in the water storage areas and phytoplankton consumption.

This state of the environment has been predicted [28] according to the results of the Complexes International Expeditions "Blue Danube-88" and "Blue Danube-90" when the investigation was carried from the sea up to the source of the river (river head). Similar changes have been recorded earlier for other European rivers (Dnieper, Volga) after regulation of their runoff.

At present, the level of mineral compounds of nitrogen and phosphorus in the delta has decreased and returned to values of 1958–1960, i.e., to the period before the beginning of eutrophication. But this does not mean that the eutrophication level is less because, at the same time, the level of organic nitrogen has increased. This forms ammonia nitrogen, nitrites, and nitrates after mineralization and provokes eutrophication.

The Danube wetlands occupy large areas where the reeds *Phragmites australis* dominate. Wetlands are highly productive ecosystems whose formation, functioning, and characteristics are determined by the water regime. They are one of the factors forming the hydrochemical balance in the estuary. In wetlands, hydrobionts play an important role in nutrient utilization, and in mineralization of organic matter and pollutants, which are included into their trophic cycles. The major role belongs to bacteria. For instance, based on the average river runoff, the destruction velocity of organic matter caused by bacteria, unicellular plants, and zooplankton, is $1.5 \times 10^9 \, \text{t} \, \text{year}^{-1}$. The share of bacteria is 68%, zooplankton 20%, and phytoplankton 12%, with a minimum for the benthos [29]. Special investigations in Ukrainian wetlands stated that an area of 7100 ha could accumulate 75% of suspended matter coming with the Danube water, to utilize 50% of the phosphorus and 30% of the nitrogen in the process of photosynthesis. As the result of hydrobiological processes wetlands accumulate dissolved organic matter and silicon compounds [30]. High water plants (for example *Phragmites australis*) serve as natural biofilters, providing 93–97% water clearance, accumulating nutrients with their roots. It was calculated that the Danube wetland flora could extract 59.1 t of nitrate, 20.5 t of orthophosphate, 23.3 t of heavy metals, and 0.1 t of pesticides [17]. The cleaning effect of the Danube wetlands is much more effective than that of the Dnieper and Dniestr.

3.1.1
Contamination in the Danube Delta

At the beginning of 1970s the concentration of heavy metals and oil increased in the riverine water. The character of the distribution depends on the hydrological regime and anthropogenic power (industry, port activity, shipping ways, etc.). In the delta, the concentration of pollutants is much greater than in the river, because of their accumulation. In the period 1993–1997 it was determined that different pollutants chronically contaminate the Danube water in the delta. In general, they are oil, phenol and the heavy metals Cu, Ni, Cd, Pb, and Hg. In 72% of the samples, during the period of study, the oil concentration registered more than the maximum allowable contaminant levels (MCL). For riverine water the MCL is 0.050 mg L^{-1} [31]. Average concentration was also more than the MCL (Table 6) and the maximum was ten times more than the MCL. Chronic water pollution by oil is the reason for its accumulation in the bottom sediments. The concentration is at the level 0.1–4.5 mg g^{-1} dry weight (Table 6). This concentration could be the cause of bottom biocenosis degradation and sometimes of benthos organism mortality [32].

Table 6 Level of water and sediments pollution in the Kiliya delta during 1993–1997 (variability/average). The heavy metals are in dissolved form.

Level of measurement	Oil	Cu	Zn	Ni	Cd
Surface (mg L^{-1})	0.01–0.48 0.09	0–0.005 0.003	0–0.067 0.011	0–0.003 0.001	0–0.001 0.001
Near-bottom (mg L^{-1})	0.01–0.25 0.07	0–0.007 0.003	0–0.058 0.014	0–0.014 0.003	0–0.001 0
Bottom sediments (mg g^{-1} dry weight)	0.1–4.5 0.9	0.002–0.103 0.046	0.025–0.243 0.130	0.023–0.396 0.063	0–0.013 0.007

The cause of chronic water pollution by Cu and Zn is input from the soil used for gardening and viticulture. The average value of Cu and Zn, for the surface layer and also near-bottom, exceeds the MCL (MCL_{Cu} 0.001 mg L^{-1}, MCL_{Zn} = 0.010 mg L^{-1} for riverine water) (Table 6).

In the delta area high concentrations of phenol were registered: 0.013–0.120 mg L^{-1} at TCL_{phenol} =0.001 mg L^{-1}. These concentrations are mostly caused by water enrichment from products of natural destruction of the air-water vegetation (reeds, etc.).

It should be noted that concentrations of the heavy metals in particulate form exceeds concentrations in solution: Cd, Co, Ni, and Pb 1.1–1.5 times, Zn and Cu 4–5 times [31] (Table 7).

The unfavorable ecological situation in the region could be worsened due to accidents at industrial plants, which use chemical compounds. Thus, an accident at the gold mine in Romania (January–March, 2000) provided fluxes of pollution to the Danube (cyanide, heavy metals). In the Kiliya delta the following concentrations of contamination were noted: cyanide 0.044–0.116 mg L^{-1}, oil 0.075 mg L^{-1}, dissolved forms of Cu 0.006 mg L^{-1} and Zn 0.019 mg L^{-1}. The influence of the accident on marine ecosystems was not measured.

According to the National Report of the Ministry of Ecology and Natural Resources of Ukraine regarding Black Sea environmental conditions for the period 1996-2000, the input of contaminants into the Black Sea with the Danube runoff comprise: oil 53×10^{12} t, Cu 1.2×10^{12} t, and Zn 3.3×10^{12} t year^{-1} [33].

Investigation in the Danube delta of the EROS 2000 and 21 projects "The interaction between the Danube, Dniester, and Dnieper Rivers and northwestern Black Sea (1994-1998)" in July 1995 and April 1997 indicated that the dissolved concentrations of Cu, Cd, Co, Mn, Ni, Pb, and Zn were low and for some parameters (Pb, Co, Mn) lower than that in the Rhone, Huanghe, Mississippi, or Lena [34, 35]. It is possible that these differences in results are caused by fragmental studies in this region in the frame of the EROS program, while data [31] have been received during seasonal and long-term surveys.

Table 7 Heavy metal (particulate form) concentration (mg L^{-1})in the Kiliya delta during 1993–1997 (variability/average)

Level of measurement	Cu	Zn	Ni	Cd
Surface	0–0.016	0.004–0.060	0–0.013	0–0.002
	0.005	0.021	0.004	0.001
Near-bottom	0.001–0.026	0.008–0.070	0.001–0.016	0–0.002
	0.008	0.027	0.005	0.001

3.2
Anthropogenic Component of the Nutrients in River Runoff

Compounds of N and P are recognized as a main stimulator of the eutrophication process. It is a well-known fact that their content in the basin increases
as the result of adding industrial, communal and agricultural discharge, since
nitrogen and phosphorus are present in the composition of all types of discharges. It is impossible to divide these compounds into natural (entering water as the result of mechanical or chemical erosion, etc.) and anthropogenic by
the existing analytical methods. Nitrogen and phosphorus of anthropogenic
genesis are involved in the natural cycles and, later, it is impossible to define
the origin of the nitrogen and phosphorus.

For estimation of anthropogenic influence in limnology there are some
criteria and calculations. For example, there is the nutrient runoff module
(coefficient of export, CE) of nutrients. It is the ratio of mass of substances in-
putting into the sea by river runoff per year, to the drainage area. Thus, it is
possible to calculate the anthropogenic role using nitrogen and phosphorous
compounds and to estimate the man-caused pressure to the reservoir.

CE calculations for the Danube during the period 1958–2000 have been done
(Table 8). Man-caused pressure increased from 1977 to 1985. It was the peak of
fertilizer use in agriculture. Maximum CE of mineral nitrogen and phosphorus input was in 1986–1988: input of mineral nitrogen from the drainage area
consisted of 0.470 t and 0.070 t of phosphates from 1 km^2 $year^{-1}$.

Over 15 years (1977–1992) the input of mineral nitrogen from the catchment area to the Danube was 0.450–0.470 t km^2 $year^{-1}$. At the present time,
anthropogenic input of nitrogen and phosphorus has decreased. In the period
1997–2000 the input of mineral nitrogen was only 0.180 and mineral phosphorus 0.020 t km^2 $year^{-1}$.

The value of CE is closely related to the value of river runoff, while the
ratio Si : P and Si : N in the river runoff, which is not polluted by the anthro-

Table 8 Coefficient of export (t km^2 $year^{-1}$) of nitrogen and phosphorus compounds from
the Danube drainage area (817×10^3 km^2)

Period	N_{min}	N_{org}	P_{min}	P_{org}	River runoff (km^3 $year^{-1}$)
1958–1960	0.17	0.14	0.02	0.01	179.4
1977–1985	0.46	0.25	0.05	0.02	227.7
1986–1988	0.47	0.77	0.07	0.03	204.7
1989–1992	0.45	1.05	0.05	0.02	169.7
1993–1996	0.33	0.89	0.02	0.02	195.1
1997–2000	0.18	1.24	0.02	0.01	230.9

pogenic input of nitrogen and phosphorus, is constant. It does not depend upon the fluctuations of runoff because of the well-known fact that the content of silicon in the ecosystem is not connected with the anthropogenic factor. Si enters the basin only with the terrigenous discharge. It was proposed [36] that these ratios be used as empirical coefficients for calculation of the anthropogenic constituent of a river's nutrient runoff. For calculation of the anthropogenic constituent of the nitrogen and phosphorus runoff during the eutrophication period it is necessary to know the values of Si : P and Si : N ratio before development of eutrophication. For estimation of the anthropogenic constituent during the eutrophication period it is very important to have data on the river's nutrient discharge before the beginning of the eutrophication period, because then these are mainly connected with the natural fluctuations.

For calculation of the anthropogenic constituent the following formula should be used:

$$A = B - \text{Si dissolved}/K$$

where A refers to anthropogenic constituent, B amounts of this element during eutrophication period, Si dissolved to the amounts of dissolved silicon during this period, and K is the empirical coefficient before the beginning of eutrophication period (Si dissolved:analyzed element).

Calculations of the anthropogenic constituent (A) in the nutrient runoff of the Danube have been done for the period 1977–2000 using data from the delta of Kiliya branch. For calculation of the empirical coefficient, data on nutrient runoff of the Danube for the period 1958–1960, has been used, i.e., before the beginning of the eutrophication period (Table 9).

The maximal values of $A_{N \text{ min}}$. and $A_{P \text{ min}}$ were recorded in 1986–1988. During these years the maximal values of CE [25] were recorded. In the runoff of the organic compounds of nitrogen and phosphorus, $A_{N \text{ org}}$ and $A_{P \text{ org}}$ are continuing to increase and, at present, they comprise more than 90% for organic compounds of nitrogen and about 70% for phosphorus.

Table 9 Anthropogenic constituent (A) of N and P (%) in the Danube nutrient runoff (1977–2000)

Period	Nitrogen compounds		Phosphorus compounds	
	$A_{N \text{ min}}$	$A_{N \text{ org}}$	$A_{P \text{ min}}$	$A_{P \text{ org}}$
1977–1985	57	36	61	61
1986–1988	75	88	85	82
1989–1992	74	89	78	77
1993–1996	68	90	71	72
1997–2000	37	92	72	70

Estimation of the anthropogenic constituent of the river's nutrient runoff is important, i.e., it allows the possibility of eliminating natural fluctuations of the water runoff, permitting analysis and prognosis of the eutrophication development.

3.3
Hydrochemical Regime of the Coastal Zone

The hydrochemical balance of the Danube coastal zone was formed under different factors such as river input and specification of hydrological conditions – vertical stratification of water masses, the processes in the zone of mixing fresh and salty water masses. The result of these processes is the changing of water properties in both riverine (calcium– hydrocarbon) and marine (sodium–chlorine) water. Phyto- and zooplankton activities and their seasonal variations influence the concentration of dissolved oxygen, pH, forms of nitrogen, phosphorus and silicon [19, 37–41].

In the variability of hydrochemical parameters in the coastal zone the main factor is spatial distribution of fresh water in the sea. According to the salinity distribution, four main zones were established. The first is a zone of quasi-fresh water (0.2–2.0‰), the second is a zone of mixing fresh and salty water (2.0–6.0‰), the third is transforming riverine water zone (6.0–17.0‰) and the fourth is a zone of marine water (> 17.0‰). The borders of these zones and their areas depend on the river runoff intensity and wind direction. The river runoff forms an average location of transformed water border, and wind currents define the spatial distribution of fresh water in the sea. North winds provide the southern sea currents and fresh water transport along the shoreline, southern winds provide the northern sea current [2].

Transformation of the riverine water starts in the delta because the current velocity slows down in the branches. As a result of adsorption, particles together with sorbed dissolved matter sediment rapidly. This matter can contain the compounds of phosphorus (up to 50% of the total) as well as silicon (up to 30%) [12]. So, in the delta fresh water becomes cleaner and removal of some chemical compounds takes place.

There are more complicated processes in the sea coastal zone with low salinity. It was established that the area with a salinity of 2–6‰ is a result of decreasing current velocity where there was an intensive sedimentation of particles greater than 5–10 μkm from Stokes' law (Fig. 6). If the particles were less than 2–5 μkm they coagulated because of their neutralization in the high electrolyte content of marine water and the enlargement of clay particles. In this mixture of salinity, organic and mineral components of the river runoff transfer into the bottom sediments. The marine coastal zone with water salinity in the mixture of 2–6‰ is called the avalanche sedimentation or hydro-chemical border between riverine and marine water [12, 19, 39]. This interval of salinity on the surface is located around the 5th m isobath; in flooding periods it is around the 10th m isobath.

The behavior of dissolved nutrient components of the Danube runoff in the avalanche sedimentation (precipitation) zone can be conservative. In this case it depends on the distance from shore (source–river). In the case of non-conservative behavior of components the concentration can increase or decrease as a result of coagulation, flocculation, chelation, complex formation, sorption, and desorption processes.

Usually in the sedimentation zone the behavior of components such as ammonia nitrogen, nitrites, and nitrates is conservative, i.e., the concentration is linearly dependent and decreases with increase of salinity. The behavior of phosphorus compounds is non-conservative (Fig. 6). The phosphorus concentration increases with salinity in the mixing zone of 2–5‰. This happens

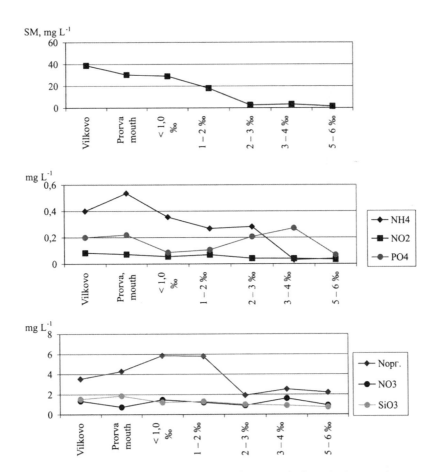

Fig. 6 Loss of suspended matter, nitrogen, phosphorus, and silicon in the Danube estuary (transect Vilkovo–Prorva mouth–nearshore area) in August 1998

because phosphates desorb from particles that have much more phosphate than the water [12].

The behavior of dissolved silicon in the mixing zone is estimated as both conservative and non-conservative. Some authors' point of view [42, 43] is that silicon concentration depends on silicon input in the estuary (the volume of river runoff), biological outputs (phytoplankton vegetation), and physical-chemical processes in front of the river mouth (sorption on and desorption from particles). The conservative behavior of silicon was established for low salinity in the Kiliya mouth coastal zone during the period 1990–1998, Fig. 6). The maximum output of dissolved matter in the sedimentation zone is in spring. Some 75% of ammonia nitrogen, about 50% of the nitrites, phosphates and silicon, and up to 40% of the nitrates transfer into the bottom sediments [39].

In the transformed riverine zone (6–17‰) there is a horizontal gradient of salinity (hydro-front). Just here sedimentation of dissolved mineral and organic compounds from water takes place because they can sorb onto the clay particles, which are the main components of the suspended sediments in the Danube. The location of the hydro-front is not constant and depends on the volume of river runoff and marine current intensity.

The most active hydro-biological processes are in the zone with a salinity of more than 8‰. Salinity in the range 5–8‰ is a limit for marine and fresh water organisms as well because it is linked to the impossibility of osmosis adaptation of hydrobionts [44, 45]. Active development of hydro-biological processes (photosynthesis) in warm periods provides the decrease of mineral compounds of nitrogen, phosphorus, silicon and increase of dissolved oxygen, pH, and dissolved and suspended organic matter. So, for example, the maximum value of dissolved oxygen during the period 1977–2000 (20 mg L^{-1}, 270% saturation), pH (9.60) and organic nitrogen (15.94 mg L^{-1}) had been fixed in the summer on the surface and were brought on by mass phytoplankton development (blooming) [39, 46].

In summer in conditions of temperature stratification of water masses without vertical diffusion there is a decrease of dissolved oxygen in the bottom layer, so-called near-bottom hypoxia and anoxia. It is linked with oxygen consumption of organic matter oxidation that can be allochthonous or autochthonous in nature. At the end of June the oxygen regime in the shallow water (not more than 15 m depth) is rehabilitated because of the descent of the thermocline down to the bottom and its final disappearance. At the same time in the open sea, in the area with a depth of more than 15 m, hypoxia development takes place in a condition of temperature stratification and with dissolved oxygen decreasing to 1.5–2.2 mg L^{-1} (15–20% saturation), pH < 8.0. In some regions of the river mouth coastal zone in a period of hypoxia development pH decreased to 7.3–7.4.

The variability of hydrochemical parameters in the river mouth coastal zone is wide and correlated with the volume of river runoff and photosynthesis development. Seasonal variability of nutrients in the river mouth coastal zone predominates. In spring, in the period of maximum river runoff, surface mineral and organic compounds of nitrogen, phosphorus, and silicon increase in the sea. In summer in low water the concentration of mineral compounds decreases because of photosynthesis in the process of utilization by phytoplankton. The reason for organic compounds decreasing is the mineralization process in conditions of high water temperature. In autumn there is some increase in the concentrations of phosphates, ammonium nitrogen, nitrites, nitrates, and silicon. This correlates with re-circulation of these compounds in the production–destruction cycle of nutrients in the river mouth coastal zone [47]. Near the bottom layer the same seasonal dynamic was established.

During the last 5 years in this zone the influence of Danube water has stopped and the nutrients have changed because of decreasing river water input (Table 10).

Construction of a large dam (Iron Gates I in 1972 and Iron Gates II in 1984) has decreased the transport of sediments. So, in 1990-2000 in the Kilia delta the range of suspended matter was 0–200 mg L^{-1} on the surface (annual average is 29.86 mg L^{-1}) and 0–280.0 mg L^{-1} near the bottom layer (annual average is 30.14 mg L^{-1}). The maximum was in summertime; that defined Danube runoff and bacteria and phytoplankton productivity. In the river mouth coastal zone a high level of pH on the surface 8.03–9.31 (annual average is 8.63) is typical; it is linked with development of photosynthesis processes. In the near-bottom layer the pH was 7.39–8.86 (annual average is 8.28). The minimum was in a period of hypoxia. Low pH on the surface had been noted in a flood period when a huge area of the sea was covered by riverine water. Low concentrations of phosphorus (mineral and organic) are defined by a decrease of river runoff in summer and by phytoplankton utilization. However, fast recycling of phosphorus in the warm period provides for development of biological processes, even at low phosphorus concentrations.

The main form of nitrogen in the Danube mouth coastal zone in a period of eutrophication development is organic nitrogen. The mineral

Table 10 Annual variability of the nutrient concentration (mg L^{-1}) in the zone of the Danube water influence

Period	NH_4^+	NO_2^-	NO_3^-	N_{org}	PO_4^{3-}	P_{org}	SIO_3^{2-}
50–60[a]	0.025	0.003	0.010	0.230	0.014	0.016	2.980
70–80[b]	0.445	0.005	0.042	0.441	0.029	0.025	2.367
80–90[c]	0.470	0.061	0.619	1.767	0.142	0.066	1.965
90–2000[c]	0.081	0.006	0.056	0.706	0.025	0.022	0.795

[a] [48] [b] [37] [c] [19, 39]

forms are 70–90% of the total except during the flooding period, when they are about 50%. On the surface the range of concentrations is very wide, 0.020–15.946 mg L^{-1}. The maximum was extremely high for marine water. It was marked in a period of water blooming, especially on the surface [39].

The decrease of anthropogenic pressure on the watershed explained the decreasing mineral forms of nitrogen in the river runoff and in the river mouth coastal zone. The predominant form of mineral nitrogen is nitrate. It is known that under near-bottom hypoxia reduction processes take place. At this time ammonia output from the bottom sediment occurs. At present, some improvement of the oxygen regime in the near-bottom layer has been noted. It provides the decrease of ammonia in the coastal zone.

The construction of large dams (Iron Gate I in 1972 and Iron Gates II in 1984) has decreased the transport of sediments and silica from the Danube River into the Black Sea. During a period of 50 years silica input from the Danube decreased by a half and the concentration in the sea coastal zone was decreased to a third (Table 10). This corresponds with diatom blooms [40, 46]. Silica is taken out of the water column and accumulates in the bottom sediments. Silica inputs from the bottom in the storm period (roiling) were marked, and also from bioturbation and under reducing conditions at the sea floor. However, the range of silicon variability is wide and covers 0.076–2.721 mg L^{-1} (average 0.865 mg L^{-1}) on the surface and 0.76–1.66 mg L^{-1} (average 0.850 mg L^{-1}) in the near-bottom layer. The maximum for silicon is in the flood period in spring because the main source of silicon in the sea nearshore zone is terrigenous runoff. The minimum for silicon in the sea is in winter.

Anthropogenic eutrophication development in the Danube, and later in the sea, provoked the increase of dissolved organic matter (DOM) in the coastal zone both allochthonous and native. DOM (by oxidizability) is one of the indices of organic matter that influences the dissolved oxygen balance, especially in the near-bottom layer. Production process activity provides an increase of autochthonous organic matter because of the decrease of suspended matter with river runoff. It was established that the oxidizability dynamic directly depends on the river runoff. The maximum oxygen concentration was 12.6 mg L^{-1} on the surface and 9.0 mg L^{-1} near the bottom layer in spring. The minimum concentration was 1.9 mg L^{-1} on the surface in the autumn–winter period in a condition of maximum salinity. DOM accumulation in the near-bottom layer, with mineralization in a stratified condition, promotes hypoxia development in the sea shore zone.

3.3.1
Contamination in the Coastal Zone of the Black Sea

Coastal zone waters are polluted by heavy metals and oil; their concentrations are congruent with their values in the Danube delta branches (Table 11) [31].

Table 11 Level of water and sediment pollution in the sea costal zone during 1993–1997 (variability/average)

Level of measurement	Oil	Cu	Zn	Ni	Cd
Surface ($mg L^{-1}$)	0.04–0.34 0.10	0–0.004 0.002	0–0.050 0.011	0–0.003 0.001	0–0.001 0
Near-bottom ($mg L^{-1}$)	0.03–0.39 0.10	0–0.006 0.003	0–0.065 0.013	0–0.011 0.003	0–0.012 0.001
Bottom sediment ($mg g^{-1}$ dry weight)	0.1–5.2 1.8	0–0.202 0.049	0.051–0.516 0.139	0.020–0.144 0.051	0–0.017 0.006

Average data for the period for all parameters is less than the maximum allowable contaminant levels (MCL) for marine water. The bottom sediments of the estuary are also polluted by heavy metals and concentrations are more than those in the water by up to a factor of ten. Accumulation of heavy metals in the sediments is linked to the sorption processes on suspended matter and sedimentation. In the sea coastal zone maximum pollution was noted in the surface layer and near-bottom, because of active processes of adsorption and sedimentation in the region of mixing of fresh and marine waters.

Investigations in the nearshore of the Danube delta during the EROS 2000 and 21 projects in July 1995 and April 1997 indicated that in the near-bottom layer the values of Zn and Ni were higher than in the surface layer [35].

4
Role of the Danube River Runoff in the Process of Eutrophication of the Black Sea

The Black Sea's geographical position and morphometric features make it a classic example of an "ecological target" that has been influenced by human activities. The Black Sea is distant from the ocean, and the territory of its catchment basin (2.3 E6 km^2) greatly exceeds its own area ($423 \times 10^3 km^2$). Within the Black Sea catchment area are located the territories of 26 developed European countries and about 300 big and small rivers, of which the Danube is the main river. The Danube is the second European river after the Volga. It provides up to 40% of the fresh water input into the Black Sea and up to 80% into its northwestern part. Intensive economic development and exhaustive nature management has led to considerable ecological pressure on the Black Sea ecosystem. The development of the eutrophication process in the Black Sea was mainly determined by the increasing amounts of nutrients in the river runoff. It is important to note that in a marine environment, the areas affected by eutrophication are extremely large and difficult to control; they are open spaces without precise limits.

The Danube fresh water influence over the northwestern part of the Black Sea is very strong. This influence is marked in the Romanian and Bulgarian shelf and sometimes spreads up to the Bosphorus. In a year of heavy rainfall the area of the Danube influence occupies 70% of the northwestern part of the Black Sea. In a poor water year this area decreases down to 20–30%. The total area of the Danube's direct influence on the Black Sea, defined according to the area of freshwater phytoplankton presence, is not less than 10^5 km [49] (Fig. 7).

Before eutrophication of the Danube (1950s and 1960s), runoff into the sea annually discharged up to 940×10^3 t per year of the dissolved mineral compounds of N, P, and Si and the discharge of organic compounds was insignificant. During the period of intensive eutrophication (1970s and 1980s) Danube waters discharged into the sea nearly three times more mineral compounds of N and P, and 1.6 times more organic compounds (Table 12).

Monitoring of the Romanian part of the delta at the Sulina mouth (one of three main branches of the Danube) during eutrophication (1980–1990) revealed that the Danube emits $600–800 \times 10^3$ t year^{-1} of total mineral nitrogen, $23–32 \times 10^3$ t year^{-1} phosphate and $150–300 \times 10^3$ t year^{-1} silicon [50, 51].

Loss of the substances – about 75% of ammonia nitrogen and about 40–50% of nitrites, phosphates, nitrates, and silicon [39] – occurs at the

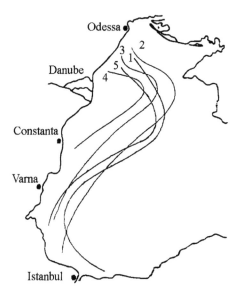

Fig. 7 Danube influence on the Black Sea: 1 isohaline 17‰, 2 phosphates 0.03 mg L^{-1}, 3 nitrates 0.04 mg L^{-1}, 4 pH 8.75, 5 outside border of blooming (during the period 1973–1987) [49]

Table 12 Entrance of nitrogen, phosphorus, and silica (10^3 t year^{-1}) with the Danube runoff into the northwestern part of the Black Sea

Period	N		P		Si	Total	Total	Total
	Mineral	Organic	Mineral	Organic		N_{min}, P_{min}, Si	N_{org}, P_{org}	
1950–1960[a]	141.8	113.0	12.7	5.6	784.9	939.4	118.6	1058
1970–1980[b]	376.9	564.6	44.3	18.8	716.3	1137.5	583.4	1721
1990–2000[b]	187.6	1054	18.0	13.8	572.5	778.1	1067.8	1846

[a] [21], [b] [19, 38, 39]

coastal zone of the Danube in spring (maximum runoff). This causes an increase of nutrients in the area influenced by the Danube waters. Maximum concentrations of the mineral compounds of nitrogen and phosphorus have been registered in the area of the Danube influence in the 1970s and 1980s during development of eutrophication in the northwestern part (Fig. 7).

Algal blooms were recorded during this period at the surface layer, caused by intensive development of phytoplankton. The dissolved oxygen concentrations in this area were 150–200% of saturation and pH was 8.6–9.3. The total phytoplankton biomass was more than 400×10^3 t in an area of about 40×10^3 km^2 during the summer period [49]. Under the conditions of density and temperature stratification of water masses in summer time, decay of dead phytoplankton leads to oxygen lack – near-bottom hypoxia.

The area of hypoxia directly depends on flooding intensity and on the time of the beginning of flood – an inverse negative relationship. If the peak of the flood is in April the main mass of Danube water is driven out of the northwestern part of the Black Sea by the north wind and hypoxia is absent. If the peak of flood is in May the main mass of Danube water remains in the northwestern part of the Black Sea because the general wind direction is changing to the south. So, in this case Danube water is fortified by nutrients located in the area between the Danube and Dniestr. This area is formed by quasi-stationary topography vortexes. Just here, hypoxia is forms later [52]. Hypoxia is the consequence of anthropogenic eutrophication of the sea and leads to mass mortality of the bottom organisms. On average, almost half of the area of the Romanian continental shelf bottoms, down to 30 m depth, were annually affected by mortalities of benthic organisms during 1972–1981 [53]. In 1973–1990 in the northwestern part of the Black Sea the zone of hypoxia occupied $3500–40 \times 10^3$ km^2 (Fig. 8). It has caused the mass mortality of 60×10^6 t of bottom animals and 5000 ton of fish, especially juveniles [54].

Fig. 8 Near-bottom hypoxia development in northwestern part of the Black Sea

Anthropogenic eutrophication of the Black Sea has resulted in transformation of some habitats and significant changes in the biota in terms of population, species, biocenoses, and ecosystems.

The structure of phytoplankton has been changed. The proportion of peridiniales phytoplankton increased compared to diatoms. The density and biomass of zooplankton increased due to non-predated zooplankton development – *Noctilluca milliaris* and allied species – jelly combfish *Mnemiopsis leydy* and *Beroe ovata*. The catch of mass planktofagus fishes such as khamsa and sprat increased. The increase of density and biomass of phytoplankton defined the transparency decrease and as a result the decrease of solar radiation penetrating the water column. It caused degradation of algae bottom biocenoses, silting of mussel biocenoses, and a decrease in the diversity of species inhabiting the shelf biotopes. In estuary regions new benthic biocenoses have formed – the mollusc *Mya arenaria* and polychaete *Nereis succinea*. These species are capable of inhabiting silted sediments and have pelagic larvae with life periods exceeding the duration of hypoxia [55]. The decrease of biological diversity on all systematic levels and in all biotopes of the Black sea shelf became a characteristic feature for the 1980s and 1990s. It could be an indicator of the anthropogenic impact on ecosystems [56, 57].

Chronic algal blooms at the zone of the Danube water influence led to accumulation of the mineral and organic compounds of nitrogen and phosphorus in the bottom sediments of the Danube coastal zone (Table 13). Here their concentrations are a dozen times higher than in the water column [19, 39].

During the period of reduction conditions at the border of water–bottom sediments, fluxes of ammonia nitrogen, phosphates, and silicon were ob-

Table 13 Mean concentrations ($mg\,L^{-1}$) of hydrochemical parameters in the porewaters of bottom sediments of the Danube coastal zone for the period 1979–2000

Period	DOM	NH_4^+	NO_2^-	NO_3^-	N_{org}	PO_4^{3-}	P_{org}	SIO_3^{2-}
1979–1992	52.41	3.19	0.03	0.47	2.83	0.76	0.78	9.18
1994–1997	45.03	3.12	0.03	0.17	7.16	0.14	0.10	4.97
1998–2000	15.63	2.95	0.01	0.06	9.74	0.16	0.17	7.71

Table 14 Danube input of nutrients and nutrient supply (t day^{-1}) of the shelf area into the northwestern Black Sea

Parameter	Danube input (t day^{-1})	Flux area Nearshore zone 2400 km^2	Offshore 45 600 km^2
Summer 1995			
P	37	17	145
NO$_3^-$	1381	–	
NH$_4^+$	64	66	64
Si	770	270	770
Spring 1997			
P	44	6	0
NO$_3^-$	760	–5	60
NH$_4^+$	52	43	100
Si	850	160	0

served from the bottom sediments. This is an additional source of eutrophication for marine waters. These fluxes are compared with the nutrients coming from the Danube.

Benthic fluxes of nutrients and metals measured in the coastal zone of the northwestern part of the Black Sea in two cruises in summer 1995 and spring 1997 proved that there is a cycling and nutrient output from the bottom sediments to the water column [58] (Table 14).

Two zones were discovered: areas of high benthic fluxes nearshore and of low benthic fluxes offshore. The high flux area of about 2400 km^2 within 20 km of the coast is strongly influenced by river inputs of organic matter and nutrients, which cause high productivity and sedimentation rates nearshore and, hence, trigger the high benthic fluxes. The anticyclonic circulation nearshore, driven by river influx and wind, keeps water with nutrient and particle loads nearshore. Benthic fluxes are about one order of magnitude higher in this zone than offshore on the shelf. However, due to the large extension of the low flux area (45 600 km^2), both zones provide, on average, similar total benthic fluxes of P, N, and Si [58].

Monthly average hypoxia development from the area of 40×10^3 km^2 of bottom sediments results in 80×10^3 t ammonium, 20×10^3 t phosphates, and 90×10^3 t silica coming into the near-bottom layer of the sea [59, 60]. Eutrophication in the northwestern part of the Black Sea was transformed into a new stage.

Benthic nutrient recycling and benthic fluxes are an important factor in sustaining high productivity of the ecosystem as, even with reduction of nutrients in the river input, the sea will have an additional amount of nutrients that support a significant level of eutrophication in the ecosystem.

References

1. Safjyanov GA (1987) Estuary Mysl Moscow (in Russian)
2. Hydrologia ustjevoi oblasti Dunay (1963) Hydrologia estuary of the Danube. Hydrometeoizda, Moscow (in Russian)
3. Mikhailov VN (1997) Ustja rek Rossii I spdelnih stran. Russia, proshlor, nastoyacshee I buducshee. Russian and adjacent countries river mouths. The past, present and future. GEOS, Moscow (in Russian)
4. Samoilov IV (1952) Ustja rek estuary of the river. Geogrpggiz, Moscow (in Russian)
5. Strakhov NM (1937) Istoricheskaya geologia. Historical geology. Uchpedgiz, Moscow (in Russian)
6. Mikhailova M, Levashova E (2001) Vodnie resursi. Water Resources 2:203 (in Russian)
7. Timchenko VM, Novikov BI (1993) In: Ecologo-gidrologicheskaya haracteristica Dunaya I pridunaiskih vodoemov v predelah Ukraini. Ecologo-hydrological characteristic of the Danube and near Danube basins in Ukraine. Naykova dumka, Kiev, p 7 (in Russian)
8. Mikhilov VN, Povalishnikova ES, Morozov VN (2001) Vodnie resursi. Water Resources 2:189 (in Russian)
9. Mikhailov VN et al. (1988) Vodnie resursi. Water Resources 2:189 (in Russian)
10. Bolshakov VS (1970) Transformatsiya rechnih vod Chernom more. Transformation of river waters in the Black Sea. Naykova dumka, Kiev (in Russian)
11. Berlinsky NA (1989) Water Resources 4:112 (in Russian)
12. Gordeev VV (1983) Rechnoi stok v okean I cherti ego geohimii. River runoff to the ocean and the feature of its geochemistry. Nayka, Moscow (in Russian)
13. Kharchenko TA et al. (1999) Gidrobiologichesky zhurnal. Hydrobiol J 6:3 (in Russian)
14. Constantineanu M (1958) Hidrobiologia (in Russian)
15. Pechinov D (1968) 11 Anniversary conf Bulgarian Academy of Science
16. Panin NS et al. (1996) Report of EROS-2000 (Romania)
17. Alexandrov BG et al. (2001) Proceeding of the international symposium on the problems of regional seas, 25–28 Sept 2001, Istanbul, Turkey
18. Garkavay GP, Bogatova JI, Bulanaya ZT (1997) 32 Conference International Association for Danube Research (IAD), Vienna, Austria
19. Garkavay GP, Bogatova JI, Berlinsky NA (1998) In: Ekosistema vzmorja ukrainskoy delti Dunaya. Ecosystem of the Danube mouth coastal zone in Ukrainian part. Astroprint, Odessa, p 21 (in Russian)
20. Almazov AM, Maistrenko JG (1961) In: Dunai I Pridunaiskie vodoemi v predelh SSSR. The Danube and Danube waterpools in the USSR territory. Isd Acad Sci Ukraine, Kiev, p 13 (in Russian)
21. Almazov AM (1962) Gidrohimiya ustevih oblastei rek. Hydrochemistry of river mouths. Isd Acad Sci Ukraine, Kiev (in Russian)
22. Topping G, Mee L, Sarikaya H (1998) Land-based sources of contaminants to the Black Sea. In: Black Sea pollution assessment. United Nations, New York
23. Enaki GI (1987) Gidrohimichesky rejim sovetskogo uchastka Dunaya. Hydrochemical regime of Soviet Part of the Danube. In: Gidrobiologicheskie issledovaniya Dunaya n pridunaiskih vodoemov. Hydrobiological investigations in the Danune and Danube waterpools. Naykova dumka, Kiev, p 14 (in Russian)
24. Enaki GI, Zhuravleva LA (1993) Gidrohimichesky rejim. Hydrochemical regime. In: Gidroecologiya ukrainskogo uchastka Dunaya I soprdredelnih vodoemov. Hydroecology of Ukrainan part of the Danube and adjacent waterpools. Naykova dumka, Kiev, p 23 (in Russian)

25. Bogatova JI (2002) Meteorologia, Klimatologia ta gidrologia. Meteorol Climatol Hydrol 46:250 (in Russian)
26. Ivanov AI (1987) Fitoplankton sovetskogo uchastka Dunaya i zalivov perednego kraya Kiliskoi delti. Phytoplankton of the Soviet part of the Danube and gulfs of the front of the Kilia delta. Hydrobiological investigations in the Danune and Danube waterpools. Naykova dumka, Kiev, p 44 (in Russian)
27. Ivanov AI (1993) Vodorosli planktona I bentosa. In: Gidroecologiya ukrainskogo uchastka Dunaya I soprdredelnih vodoemov. Hydroecology of Ukrainan part of the Danube and adjacent waterpools. Naykova dumka, Kiev, p 7 (in Russian)
28. Kharchenko TA, Lachenko AV (1993) Vodnie resursi. Water Resources 4:514 (in Russian)
29. Alexandrov BG et al. (1997) 32 Conference International Association for Danube Research (IAD), Vienna, Austria
30. Bogatova JI, Garkavaja GP (2000) 33 Conference International Association for Danube Research (IAD), Osijek, Croatia
31. Ryasintseva NI et al. (1998) In: Ekosistema vzmorja ukrainskoy delti Dunaya. Ecosystem of the Danube mouth coastal zone in Ukrainian part. Astroprint, Odessa, p 63 (in Russian)
32. Mironov OG, Milovidova NJ, Kiruchina LN (1986) Gidrobiologichesky journal. Hydrobiol J 6:76 (in Russian)
33. Ministry of Environment of Ukraine (2002) National Report of Ministry of Environment of Ukraine about Black Sea condition (1996–2000). Astroprint, Odessa (in Ukranian)
34. Guieu C, Martin JM et al. (1998) Estuar Coast Shelf Sci 47:471
35. Guieu C, Martin JM (2002) Estuar Coast Shelf Sci 54:501
36. Maximova MP (1979) Vodnie resursi. Water resources 1:35 (in Russian)
37. Garkavaya GP, Bulanaya ZT, Bogatova JI (1985) 25 Conference International Association for Danube Research (IAD), Bratislava, CSSR
38. Garkavaya GP, Bogatova JI, Bulanaya ZT (1991) Sovremennie tendentsii izmenei gidrohimicheskih uslovii cevero-zapadnoi chasti Chernogo moria. Modern tendencies in hydrochemistry variability in northwestern part of the Black Sea. In: Izmenchivostj ecosisitemi Chernogo moria. Variability of the Black Sea ecosystem. Nayka, Moscow, p 299 (in Russian)
39. Garkavaya GP, Bogatova JI, Berlinsky NA (2000) In: Ecologicheskaya bezopastnostj pribrezhnoi I shelfovoi zon I komplexnoe ispolzovanie resursov shelfa. Ecological safety and complex using in shelf resources. Sevastopol, p 133 (in Russian)
40. Nesterova DA (1998) In: Ekosistema vzmorja ukrainskoy delti Dunaya. Ecosystem of the Danube mouth coastal zone in Ukrainian part. Ecosystem of the Danube mouth coastal zone in Ukrainian part. Astroprint, Odessa, p 159 (in Russian)
41. Polyshyk LN, Nastenko EV (1998) In: Ecosystem of the Danube mouth coastal zone in Ukrainian part. Astroprint, Odessa, p 203 (in Russian)
42. Millero FJ (1981) In: River inputs to ocean systems. UNEP and UNESCO, Switzerland, p 116
43. Morris AW, Bale AJ, Howland RJM (1981) Estuar Coast Shelf Sci 2:205
44. Chlebovic VV (1974) Kriyicheskaya solenostj biologichaskih protsessov. Critical salinity of biological process. Nauka, Leningrad (in Russian)
45. Alexandrov BG (1998) In: Ecosystem of the Danube mouth coastal zone in Ukrainian part. Astroprint, Odessa, p 245 (in Russian)
46. Nesterova DA (2001) Algologia 4:502 (in Russian)
47. Gershanovich DE, Elizarov ÅÅ, Sapognikov VV (1990) Bioproduktivnostj okeana Ocean bioproductivity. Agropromizdat, Moscow (in Russian)

48. Rogdestvensky AV (1979) Himicheskie osnovi bioproduktivnosti Chernogo morya. Chemical basics Black Sea bioproductivity. Naykova dumka, Kiev, p 34 (in Russian)
49. Zaitsev Yu P et al. (1989) Gidrobiologicheski journal. Hydrobioll J 25:21 (in Russian)
50. Cociasu A, Dorogan L, Humborg C, Popa L (1996) Mar Pollut Bull 32:32
51. Cociasu A, Petranu A, Michnea PE (eds) (1998) In: Black Sea pollution assessment. United Nations, New York, p 131
52. Berlinsky NA, Dichanov Yu M (1991) Ecologia morya. Mar Ecol 38:11 (in Russian)
53. Gomoiu MT (1983) Rapp Comm In Mer Medit 3:203
54. Zaitsev Yu P (1992) Fish Oceanogr 1:180
55. Zaitsev Yu P, Alexandrov BG. In: Ekosistema vzmorja ukrainskoy delti Dunaya. Ecosystem of the Danube mouth coastal zone in Ukrainian part. Ecosystem of the Danube mouth coastal zone in Ukrainian part. Astroprint, Odessa, p 304 (in Russian)
56. Zaitsev Yu P, Mamaev V (1997) Marine biological diversity in the Black Sea. A study of change and decline. Black Sea environmental series. United Nations, New York
57. Zaitsev Yu P (1998) Gidrobiologicheski journal. Hydrobiol J 6:3 (in Russian)
58. Friedrich J et al. (2002) Estuar Coast Shelf Sci 54:369
59. Garkavaya GP, Bogatova JI (2001) Nauchnie zapiski Ternopolskogo universiteta. Scientific notes of Ternopol Teachers University 3:188 (in Russian)
60. Berlinsky NA, Garkavaya GP, Bogatova JI (2003) Mar Ecol 63:17 (in Russian)

Hdb Env Chem Vol. 5, Part H (2006): 265–301
DOI 10.1007/698_5_024
© Springer-Verlag Berlin Heidelberg 2005
Published online: 8 November 2005

Role of Particle Sorption Properties in the Behavior and Speciation of Trace Metals in Macrotidal Estuaries: The Cadmium Example

J.-L. Gonzalez[1] (✉) · B. Thouvenin[1] · C. Dange[1] · J.-F. Chiffoleau[2] ·
B. Boutier[2]

[1]Biogéochimie and Ecotoxicologie Department, IFREMER, Z.P. de Brégaillon, B.P. 330,
83507 La Seyne sur Mer, France
gonzalez@ifremer.fr

[2]Biogéochimie and Ecotoxicologie Department, IFREMER, Rue de l'Ile d'Yeu, B.P. 21105,
B.P. 21105, 44311 Nantes, France

Abstract The role of particles in the fate and speciation of trace metals in macrotidal estuaries was studied using a surface complexation model (MOCO). Cadmium was selected as the target metal contaminant due to its reactivity in estuaries: cadmium behavior is mainly controlled by heterogeneous processes (sorption/desorption) related to salinity and suspended matter gradients.

Various scenarios of suspended matter distribution according to salinity were simulated. The impact of surface properties (specific surface area, density of surface sites, acido-basic properties, and complexation constant) was evaluated using data collected on particles from the Gironde, Loire, and Seine estuaries.

Our results show that particle surface properties, evaluated on the basis of various parameters, are instrumental in "non-conservative" contaminant speciation in the estuarine environment. Their evaluation enables us to understand and simulate, to a large extent, the fate of "Cd-type" contaminants (whose behavior is controlled by competition between sorption and desorption processes). The natural variations of these properties can be responsible for significant modifications of the Cd speciation in the macrotidal estuaries where salinity and SM gradients are very strong.

Keywords Cadmium · Macrotidal estuaries · Modeling · Particles · Sorption · Speciation · Turbidity maximum

Abbreviations

ABT	Potentiometric acid-base titration
BET	Brunauer, Emmett, Teller
%CdD	Percentage of dissolved Cd
CEC	Cation exchange capacity in $mol\,g^{-1}$ of SM
GSC	Global sorption capacity of particles
K	Complexation constant
K_{a1} and K_{a2}	Intrinsic surface acid-base constants of active surface sites
K_d	Partition coefficient
K_m	Global intrinsic complexation constant
POC	Particulate organic carbon
RSD	Relative standard deviation
SA	Specific particle surface area in $m^2\,g^{-1}$
SM	Suspended matter
SOH_{tot}	Total density of active surface sites in $mol\,g^{-1}$ of SM

1
Introduction

Estuaries are transition zones between continental hydrographic networks and the marine environment. These essential interfaces come in a vast number of shapes and sizes, and can be classified according to their degree of fresh–salt water mixing (well mixed: fjords; moderately stratified: macrotidal estuaries; highly stratified: deltas or microtidal estuaries). The estuary type is primarily defined by the balance of power between the upstream river (river flow) and the downstream ocean (spring tidal range) [1–3].

Climatic and geological factors are also responsible for wide estuary variations. Each estuary has its own hydrologic and morphologic characteristics, meaning that, biologically speaking, specific communities can exist within each environment.

Due to their particular location and characteristics, these environments are often highly sought after as residential areas, fishing zones, or for the development of port installations and industrial activities. The Escaut and Seine estuaries constitute good examples of this on a European level: they both feature dense populations and rank among Europe's biggest industrial centers. They are also Europe's most highly contaminated estuaries. Anthropic activities all have a potentially significant and short-term impact on ecosystem structure and functioning, leading to the weakening, and possible eventual death, of the traditional economic interests associated with these environments.

Just 5% of the estuaries located along the French coast spanning the English Channel and Atlantic are actually the mouths of major rivers, i.e., with mean water flows exceeding $50 \, m^3 \, s^{-1} \, year^{-1}$ [4]. The Seine, Loire, and Gironde are ranked as France's three largest estuaries (Fig. 1), with mean annual flows of over $400 \, m^3 \, s^{-1}$.

The peculiar character of these macrotidal estuaries is forged by a large input of particles from their upstream rivers, coupled with a tidal amplitude > 4 m along the adjacent coasts, resulting in a high-impact dynamic tide (up to 150 km upstream of the mouth in the case of the Seine), the presence of a turbidity maximum, and extensive mud deposit zones.

The contaminant content of estuarine particles is not only the result of pollutant inputs; it may also be significantly modified and regulated by reactions occurring on the surface of particles (sorption/desorption), which also play a significant role in speciation and contaminant bioavailability.

Contaminants may be present in the aquatic environment in dissolved, colloidal, or particulate form. The distinction between these three forms is fixed arbitrarily by filtration (and ultrafiltration) according to size [5]. Within these operationally defined "physical" categories, contaminants may belong to various "chemical" species, according to their nature and type of association.

Fig. 1 Location of the Seine, Loire, and Gironde estuaries

Contaminant transition from one of these chemical species to another is controlled by various physicochemical and biological factors, including pH, redox potential, salinity, water concentrations of various complexation agents, sedimentologic characteristics, and particle geochemistry. The role of particles is particularly important in macrotidal estuaries, which are often characterized by the presence of a turbidity maximum [6]. Intra- and interestuary concentrations of suspended matter (SM) also vary widely (from a few milligrams to several grams per liter), according to river flow and tidal conditions. The repercussions of SM on biogeochemical cycles in the estuarine environment has already been underlined by Turner and Millward [7], along with the impact of salinity and particles on trace metal particulate-dissolved phase distribution [8].

Contaminant behavior during estuarine transit can be globally described as conservative (meaning the element stays in the same phase throughout estuary crossing) or non-conservative (meaning the element undergoes a process which modifies its distribution).

Cadmium is an excellent example of non-conservative behavior, and one of the most reactive metal contaminants found in the estuarine environment.

If the longitudinal evolution of dissolved Cd concentrations is plotted according to salinity, a bell-shaped curve is obtained [9–17]. Most field and laboratory studies [8, 18] agree this is due to the desorption of particle-associated Cd triggered by the formation of chlorocomplexes.

The dissolved Cd peak is even more marked in macrotidal estuaries, where particle residence time is long and SM concentrations are very high.

Conversely, excess dissolved Cd (versus simple dilution) is minor or non-existent in microtidal estuaries such as the Rhone, Danube, or Lena rivers [19–22]. Similar situations can also be encountered in macrotidal estuaries during periods of low turbidity. Chiffoleau et al. [16] demonstrated that in the Seine estuary, virtually no dissolved Cd maximum exists when SM concentrations are low.

The solubilization of Cd in the estuarine environment is also governed by other more-or-less influential processes, i.e., early diagenesis in sediments, in particular in fine-matter deposit zones, and the dissolution of Cd associated with oxides (Fe, Mn) and/or organic matter in the turbidity maximum.

Due to its particular characteristics, Cd was chosen to "explore" the role of particle properties in the macrotidal estuary contaminant cycle.

This paper aims to explore the impact of particle surface properties (specific surface area, density of surface sites, acido-basic properties and complexation constant) – which we determined for three macrotidal estuaries (Seine, Loire, and Gironde) – on the behavior and speciation of "reactive" trace metals during estuarine transit.

We will be assessing this impact using a surface complexation model (MOCO) to reproduce Cd speciation during estuarine transit, taking into account the surface properties of particle samples taken from our three study estuaries in various scenarios.

2
Seine, Loire, and Gironde Estuaries: Main Characteristics

The functioning, morphology, and dynamics of our three study estuaries have already been extensively described in various works [4, 23-30]. Their main characteristics are recapped in Table 1.

The Seine catchment area alone represents approximately 40% of French economic activity and 30% of the French population. This estuary has undergone major successive engineering works and most of the downstream sections are dammed. The Seine estuary is the most highly anthropized of the three study estuaries, particularly compared with the Gironde estuary. The Gironde's catchment area features a very moderate population and degree of urbanization, meaning that it is close to a natural equilibrium.

The distribution of suspended matter measured during various cruises in the three estuaries [12, 13, 16, 31, 32] is shown on Fig. 2 as an example. This data highlights the space-time variability of SM distribution accord-

Table 1 Main morphologic and hydrodynamic characteristics [4, 23-30]

	Seine	Loire	Gironde
Catchment surface area	79×10^3 km^2	115×10^3 km^2	74×10^3 km^2
Tidal length	160 km	90 km	170 km
River flow	440 m^3 s^{-1} (average) 40 m^3 s^{-1} (min) 2500 m^3 s^{-1} (max)	850 m^3 s^{-1} (average) 80 m^3 s^{-1} (min) 6000 m^3 s^{-1} (max)	990 m^3 s^{-1} (average) 200 m^3 s^{-1} (min) 5000 m^3 s^{-1} (max)
Flushing time	2-30 d	1-20 d	Few days to several weeks
Particle residence time	Few days to several months	One week to several months	One to several months
Mean annual particulate discharge	5×10^5 t	1×10^6 t	20×10^6 t
Average variation of SM concentrations	1 mg L^{-1}-2 g L^{-1}	1 mg L^{-1}-5 g L^{-1}	1 mg L^{-1}-10 g L^{-1}

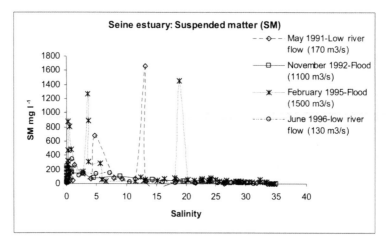

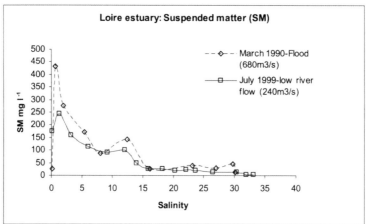

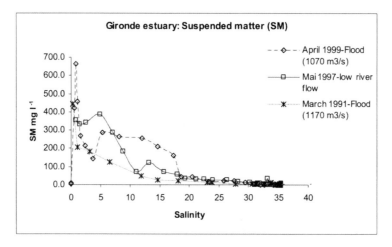

Fig. 2 Distribution of suspended matter along the salinity gradient [12, 13, 16, 31, 32]

ing to salinity. Although these variations are due to multiple factors, the most influential are estuary shape, river flow, particulate input, and tidal conditions [6].

SM concentrations in the turbidity maximum may be as high as a few tens of milligrams per liter in surface waters and several grams per liter in bottom waters. According to the tidal coefficient, fluid mud lenses containing up to several hundred grams of SM per liter may be formed close to the bottom. This fluid mud layer, which can grow to a thickness of several meters, is subsequently subjected to accumulation and erosion cycles: it accumulates when the tidal coefficient drops and during neap tide, fractions into several lenses when the tidal coefficient rises, and erodes completely during spring tides [33].

3
Particle Geochemical Characteristics and Surface Properties

The geochemical and mineralogical characteristics of particles taken from the three study estuaries were determined in conjunction with an evaluation of parameters representing their surface properties. Most of this data, published by Boutier et al. [11, 12], Chiffoleau et al. [13, 16, 17], and Dange [30], is presented in Table 2.

Certain surface properties of natural particles – strongly related to their mineralogic characteristics – allow us to assess their reactivity in the aquatic environment. As most of these properties cannot be measured directly, they must be evaluated using various experimental approaches. The relevant parameters (Table 2), some of which will be used for simulations, are as follows: specific particle surface area (SA in $m^2 g^{-1}$); cation exchange capacity (CEC in $mol g^{-1}$ of SM); total density of active surface sites ([-SOH$_{tot}$] in $mol g^{-1}$ of SM); mean intrinsic surface acid-base constants of these sites (K_{a1} and K_{a2}); and global intrinsic complexation constant (K_m) of these sites with regards to Cd.

The techniques implemented to evaluate these parameters and their limits are described in Gonzalez et al. [34] and Dange [30].

Briefly, the specific surface area of the particles is measured via nitrogen sorption using the BET method (Coulter SA 3100). Cation exchange capacity is estimated using ammonium as an exchangeable cation [35, 36].

The density of active surface sites and intrinsic surface acid-base constants of these sites (K_{a1} and K_{a2}) are evaluated using potentiometric acid-base titration (ABT). Surface acid-base constants are determined by adjusting the experimental data obtained by ABT using FITEQL 3.2 [30, 37–39].

The global intrinsic complexation constant of these sites with regards to Cd is obtained by experiments based on [109]Cd (particle samples taken throughout the estuary).

Table 2 Particle geochemical characteristics and surface properties: mean values and relative standard deviation (RSD)

	Seine			Loire			Gironde		
	Mean	RSD (%)	n	Mean	RSD (%)	n	Mean	RSD (%)	n
CaCO3 (%)	26.8	13	24	7.9	11	5	8	3	11
Al (%)	3.4	32	45	7.7	28	25	8.9	20	30
Fe (%)	2.1	26	45	4.0	23	25	4.4	12	30
Mn (μg g^{-1})	479	27	45	1056	29	25	865	19	30
COP (%)	5.1	70	45	4.3	44	25	1.6	49	30
SA (m^2 g^{-1})	6.0	32	21	22.8	20	6	37.2	17	11
CEC (mol g^{-1})	1.69×10^{-4}	30	24	2.37×10^{-4}	39	6	3.47×10^{-4}	26	11
[− SOH_tot] (mol m^{-2})	2.98×10^{-5}	70	9	7.94×10^{-6}	13	6	3.47×10^{-6}	26	8
K_{a1}				3.16×10^{-6}	10	6	1.10×10^{-5}	9	8
K_{a2}	2.57×10^{-5}	13	10	1.91×10^{-7}	8	6	3.80×10^{-7}	7	8
Log K_m	0.48	52	7	−1.51	45	4	−1.21	22	5

Data from Boutier et al. [11, 12], Chiffoleau et al. [13, 16, 17] and Dange [30]
n number of measurements, K_{a1} and K_{a2} acidity intrinsic constants of surface sites, K_m "global" intrinsic stability constant of surface sites. These parameters are associated with equilibrium.
Gironde and Loire estuaries (single "average" type of amphoteric surface site):

$$- SOH_2^+ = - SOH + H^+ \quad K_{a1} \tag{1}$$

$$- SOH = - SO^- + H^+ \quad K_{a2} \tag{2}$$

Seine estuary (single "average" type of non-amphoteric surface site):

$$- SOH = - SO^- + H^+ \quad K_{a2} \tag{3}$$

Surface complexation reaction in "average" sites:

$$- SOH + Cd^{2+} = - SOCd^+ + H^+ \quad K_m \tag{4}$$

The associated methodology is described in Dange [30].

Particles from the Seine estuary are characterized by their high carbonate and particulate organic carbon (POC) content. Loire and Gironde particles showed fairly similar characteristics, except for POC.

Relative standard deviation showed POC to be the most variable particle fraction in all three estuaries. Particles from the Gironde were the most homogeneous; this may be due to the fact that samples were taken exclusively from fluid mud, where the finest estuary particles homogenize.

Using this dataset, we then evaluated average particle mineralogic composition (calcite, quartz, organic matter, clays) [30]. Loire and Gironde particles were shown to have a similar composition, but Seine particles had a very high carbonated fraction versus quartzose and argillaceous components due to the estuary's calcareous-type catchment area.

Seine particles were shown to have the lowest average specific surface area of the three particle types, due mainly to their large size [30]; they also had the lowest Al, Fe, and Mn concentrations, and a high carbonate content.

Cation exchange capacity is a function of the geochemical composition of particles, and in particular the characteristics of the clay minerals they contain. Gironde samples showed the highest values (Table 2), due to their high argillaceous content.

The density of active surface sites, and the proton exchange capacity (acidity constants) of these sites, can be evaluated using data from particle acidobasic titration. Seine particles had the highest density of surface sites, while Loire particles had approximately twice as many surface sites per square meter as Gironde particles. Despite the similar character of particles from the latter two estuaries, this could reflect the higher Mn and POC concentrations.

Experimental results have demonstrated that virtually no protons bind to Seine particles. Their global acido-basic properties are mainly controlled by the organic fraction of the particles and the carboxylic and phenolic functions (i.e., characteristic of the humic substances). Conversely, a significant quantity of protons bind to suspended matter in the Loire and Gironde estuaries.

The acido-basic properties of the particles can be modeled for two types of sites (Table 2): (i) amphoteric surface sites (hydroxyls), typical of sites associated with Fe and Mn oxy-hydroxides and with the argillaceous phase (fractions representing a significant proportion of Loire and Gironde particles), or (ii) non-amphoteric sites associated with the organic fraction, which largely control the acido-basic behavior of Seine particles.

Seine particle POC content is fairly similar to that of the Loire particles; the considerable differences observed in acido-basic behavior are actually due to the high argillaceous fraction of Loire particles.

These assumptions were tested by modeling experimental data (acidobasic titrations) using the FITEQL model [37] which enabled us to evaluate the acidity constants of the reactions studied (Table 2).

The data obtained showed Loire and Gironde particles to have very similar acido-basic properties, as suggested by their acidity constants (Table 2).

The K_{a2} value estimated for Seine samples is representative of carboxylic and/or phenolic sites, and confirms the predominant role of organic matter in the reactivity of these particles [30].

Depending on pH and total surface site density (-SOH$_{tot}$), the acidity constants of surface sites largely determine -SOH site density (Eqs. 1, 2, 3 in Table 2), responsible for Cd surface complexation (Eq. 4 in Table 2). We obtain:

$$[- SOH] = [- SO^-][H^+]/K_{a_2}$$ (5)

and

$$[- \text{SOH}] = [- \text{SOH}_2^+]K_{a_1}/[H^+] \tag{6}$$

By calculating $- \text{SOH}_{\text{tot}} = [- \text{SOH}] + [- \text{SO}^-] + [- \text{SOH}_2^+]$, according to Eq. 5 and Eq. 6 we obtain:

$$- \text{SOH}_{\text{tot}} = [- \text{SOH}](1 + (K_{a_2}/[H^+]) + ([H^+]/K_{a_1})) \tag{7}$$

therefore:

$$[- \text{SOH}] = - \text{SOH}_{\text{tot}}/(1 + (K_{a_2}/[H^+]) + ([H^+]/K_{a_1})), \tag{8}$$

for amphoteric surface sites (i.e., Loire and Gironde particles).
   For non-amphoteric sites (i.e. Seine particles), we obtain:

$$[- \text{SOH}] = - \text{SOH}_{\text{tot}}/(1 + (K_{a_2}/[H^+])) \tag{9}$$

Adjustment of [109]Cd sorption data [30], whereby Cd sorption is considered as a formation of inner-sphere complexes (Table 2), enabled us to estimate the intrinsic complexation constant of surface sites in terms of Cd (K_m). Like most of the properties mentioned above, these constants were similar for Loire and Gironde particles. Seine particles showed the highest mean values.
   In terms of sorption parameters, comparison with data collected in other works was problematic, as parameter values vary widely according to chosen experimental techniques and models (this is particularly true for acidity intrinsic constants and surface complexation constants). The comparison of values obtained in all three estuaries versus values estimated on particles from different environments gave comparable results [30]; any differences observed are largely due to particle mineralogic composition. In terms of Cd, the complexation constants of particles of various origins obtained using a similar modeling technique were compiled by Muller and Duffek [40]. In our study, Loire and Gironde complexation constants were comparable to compiled data.

4
Behavior of Cd in Macrotidal Estuaries

Contamination by trace metals, and their behavior in the three estuaries, has been widely studied [10–17, 32, 34, 41, 42].
   The high physicochemical gradients encountered in the estuarine environment (suspended matter, salinity, pH, and major elements) are responsible for the phase changes undergone by Cd during its oceanward transit.
   The majority of macrotidal estuaries experience a rise in dissolved Cd concentrations as soon as salinity rises, followed by a drop in highly saline waters [9–17].

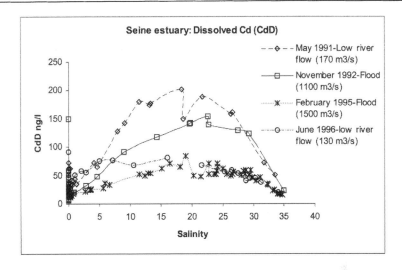

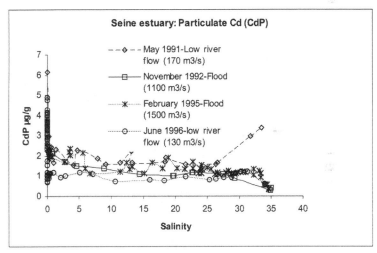

Fig. 3 Dissolved and particulate Cd in the Seine estuary [13, 16]

Maximum dissolved Cd can be studied with various flow regimes and in different seasons within the same estuary. To illustrate this point, the distribution of dissolved and particulate Cd corresponding to the previously illustrated SM profiles is presented in Figs. 3, 4, and 5.

We observed this typical behavior in our study estuaries, crystallized by a more-or-less visible "dissolved Cd bump". Variations in the scope of this phenomenon within one or several estuaries may be conditioned by several factors: flow rates, Cd concentrations at the upstream and downstream limits, contamination levels, the size of the turbidity maximum and its position versus the salinity gradient, or particle sorption properties. This paper aims to examine the impact of the latter two factors.

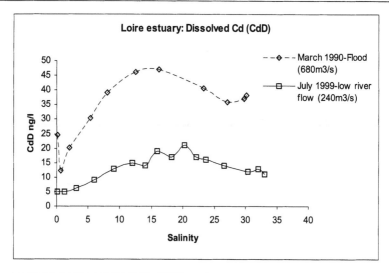

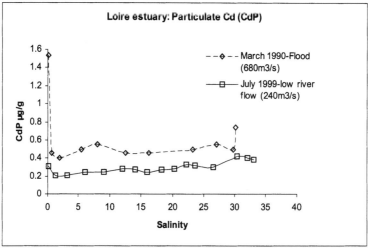

Fig. 4 Dissolved and particulate Cd in the Loire estuary [12]

Temporal variations in Cd fluxes entering the estuary, or the presence of a sporadic Cd source, could explain the dissolved Cd bump. But the majority of studies assign this characteristic evolution (bell-shaped curve) to Cd desorption from particles entering the estuary, caused by the formation of highly stable dissolved chlorocomplexes when chlorinity (salinity) rises. This hypothesis was confirmed by laboratory experiments in which [109]Cd was added to samples of raw river water (dissolved phase + particles) mixed with raw water taken from the open sea [8, 18]. Turner [8] demonstrated that the Cd partition coefficient (K_d) in various estuaries evolves as a function of salinity. K_d differs widely with low salinity in the three study estuaries due to variations in the characteristics of particles enter-

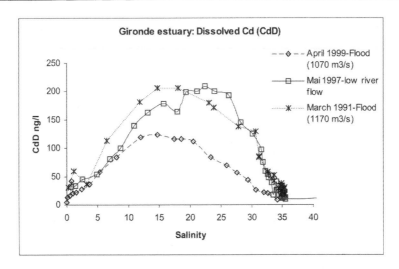

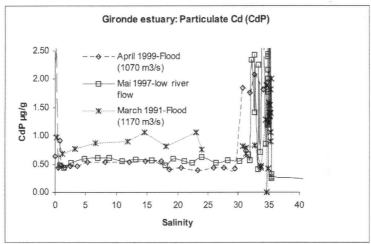

Fig. 5 Dissolved and particulate Cd in the Gironde estuary [31, 32]

ing the estuary and river water composition. But K_d evolution as a function of salinity remains similar, and merges to comparable values when salinity is high.

Interestingly, some campaigns (Figs. 3, 4, and 5) have observed high particulate Cd concentrations associated with high salinity (up to 3 $\mu g\,g^{-1}$ in the Seine estuary). This seemingly contradictory behavior is generally thought to be associated with high chlorophyll a concentrations [12, 16, 32], and may suggest that particles of phytoplanktonic origin could have powerful Cd sorption properties.

The global behavior of Cd at the fresh/salt water interface means that macrotidal estuaries can flush significant amounts of dissolved Cd into adja-

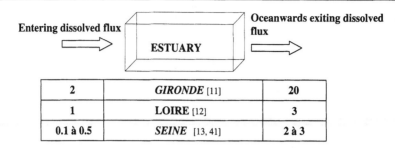

2	*GIRONDE* [11]	20
1	**LOIRE** [12]	3
0.1 à 0.5	*SEINE* [13, 41]	2 à 3

Fig. 6 Impact of estuarine processes on oceanwards Cd fluxes (in t year^{-1}) [11–13, 41]

cent coastal waters. Figure 6 illustrates the extent of this phenomenon, which actually reflects the differences in quantities of dissolved Cd entering and exiting the estuary.

In Table 3, mean dissolved Cd concentrations in estuarine waters are compared with likely concentrations in various other aquatic environments.

On the whole, concentrations near the coast are higher than those in the ocean; this underlines the fact that Cd inputs are primarily of continental origin. The differences in surface water and bottom water concentrations pinpointed in the Atlantic and Pacific Oceans illustrate the "nutrient-like" be-

Table 3 Dissolved Cd concentrations (ng L^{-1}) in various aquatic environments

Oceanic waters		Atlantic	10–45	[43]
		Pacific	80–110	[43]
	Surface waters			[44]
		Arctic	7–33	[44]
		Atlantic	0.2–16	[44]
		Pacific	3–67	[44]
Coastal and		Marennes-Oléron Bay	13–23	[45]
estuarine waters		Gironde estuary	10–393	[10, 11]
		Loire estuary	11–61	[12]
		Rhône (microtidal) estuary	48 (average)	[10]
	Seine estuary			
		Upstream limit (Poses weir)	5–73	[41]
		Estuary	5–202	[16, 17]
		Scheldt estuary	450 (max)	[46]
		Hudson estuary	618 (max)	[47]
Porewaters		St-Laurent estuary	6–582	[48]
		Fjord	< 10	[49]
		Equatorial Pacific	112–267	[50]
		Californian Shelf	12–280	[51]
		Marennes-Oléron Bay	4–170	[52]
		West-Gironde mud patch	60–400	[52]
		Seine (intertidal mud patch)	19–1328	[53]

havior of Cd. Differences in average concentrations in Atlantic and Pacific waters are owed to the fact that Pacific waters are older.

Concentrations within each estuary are extremely variable and characterized by wide space–time variations related mainly to hydrodynamics. Catchment area characteristics and variations in inputs of anthropic origin can lead to widely differing mean concentrations between estuaries.

In porewaters, the concentrations are very variable, even within the same environment. These concentrations are controlled by the speciation of the particulate Cd which integrates the sedimentary column and of the diagenetic conditions which in the sediments vary temporally and according to the depth.

With the aim of correlating the degree of contamination of various estuaries, Table 4 offers a comparison of unindustrialized estuaries (Amazon, Lena), featuring minimal observable concentrations, versus a selection of European and American estuaries of various sizes, with differing degrees of contamination.

These comparisons are questionable, however, as the various cruises took place at different and sometimes far-distant periods. In addition, the majority of cruises were one-offs (one cruise lasting a few days), whereas others involved pluriannual monitoring. Finally, even if identical techniques and strategies are used, it is still difficult to draw firm conclusions by comparing concentrations in various sites. For example, comparing maximum dissolved Cd in the Gironde and Seine estuaries could lead us to conclude that the Gironde is far more contaminated by Cd than the Seine, whereas measurements of particulate Cd in the turbidity maximum suggest the contrary. The explanation for this lies in the different characters of estuary particles, e.g., Seine particles are much richer in carbonates and organic matter than the highly argillaceous Gironde particles, and hence have different sorption capacities. This hypothesis is corroborated by the particle characteristics presented above.

The data presented in Table 4 with the aim of characterizing estuary contamination shows that:

- Dissolved and particulate Seine concentrations at Poses weir (upstream limit) are actually higher than the minimum levels previously described in the literature. They largely exceed levels in the Loire estuary, which can be considered as an example of a little-contaminated European estuary.
- Measurements at Poses weir are in the same range as those observed in the Gironde or Scheldt estuaries, known to be contaminated by Cd.
- The particulate Cd content of the Seine mixing zone is very high versus other estuaries. Only the Scheldt has higher levels.
- Maximum dissolved Cd measured in the Seine fresh–salt water mixing zone is very high versus the Lena, Amazon, or Loire. For the past 5 years, it has been similar to Gironde levels.

Table 4 Cd concentrations in various estuaries

	Upstream zone		Downstream zone (mixing zone)		Refs.
	Cd D (ng L^{-1})	Cd P (mg kg^{-1})	Max Cd D (ng L^{-1})	Cd P (mg kg^{-1})	
Amazon	50				[9]
Lena	3–12		34		[20]
Mississipi	18–20		33		[54]
St Laurent	11–16	0.3–5	25	0.2–0.5	[55]
Scheldt	20	8.8	90	3.7	[56]
Ebre	40	5.8			[57]
Rhône					
1995	5–38	0.2–1.5			[58]
1996		55	0.8		[19]
Loire	24	1.5	60	0.47	[12]
Gironde					
1984–1985	20–100	6–14	400	1	[11]
1994	40–70	1.8–7	130	0.53	[14]
Seine					
1990–1992	5–73	2.8–7.2	202	1.7	[41]
1994–1998	5–87	1.3–12	82–127	0.8–1.5	[16, 17, 42]

(Cd D dissolved Cd, Cd P particulate Cd)

The Seine estuary can hence be classified as one of the most highly Cd-polluted estuaries.

In our comparison (Table 4), average Cd concentrations in suspended matter in both the upstream and mixing zones are significantly higher in the Seine estuary than in the other study estuaries. There may be several reasons for this: the fraction of Cd associated with the crystalline phase may be larger because of geological reasons, higher sorption capacity (as per estimated K_m values), or higher Seine contamination, contrary to the data shown in Table 4 (especially when compared with dissolved Cd concentrations measured in the Gironde estuary).

These various "possibilities" must, in some way, be responsible for measured particulate Cd concentrations. The purpose of this paper is to assess the impact of particle characteristics and behavior, through the study of sorption properties.

5
Presentation of the Model and Simulation Conditions

We used a surface complexation model (MOCO) to highlight the role of particle surface properties (specific surface area, density of surface sites, acido-basic properties, complexation constant) in Cd speciation in the macrotidal estuarine environment. This model had already been used to simulate Cd behavior in the three study estuaries. In our previous studies [34, 59], model results were compared with the numerous dissolved and particulate Cd measurements performed in situ. The model has also previously been applied to other metal cations such as Co, Cs, and Hg [30, 60].

This type of model considers Cd sorption to particles as a formation of complexes with functional surface groups [34, 61-64]. Dissolved and particulate Cd species are computed by solving equilibrium equations incorporating various dissolved ligands and particles simultaneously. Included dissolved ligands are chlorides, hydroxides, and sulfates. Complexation constants are derived from Comans and Van Dijk [18].

Table 5 shows the reactions processed by the model.

For the purposes of this study, particles were considered "globally" (use of mean sorption properties evaluated from particles from the various estuaries shown in Table 2). This type of approach assumes that the various

Table 5 Reactions processed by the MOCO model to simulate Cd speciation

	Log K	
$Cd^{2+} + Cl^- = CdCl^+$	2.0	$[Cl^-]$ calculated from salinity
$Cd^{2+} + 2Cl^- = CdCl_2$	2.6	
$Cd^{2+} + 3Cl^- = CdCl_3^-$	2.4	
$Cd^{2+} + 4Cl^- = CdCl_4^{2-}$	1.7	
$Cd^{2+} + OH^- = Cd(OH)^+$	3.9	$[OH^-]$ calculated from pH
$Cd^{2+} + 2OH^- = Cd(OH)_2$	7.6	
$Cd^{2+} + 3OH^- = Cd(OH)_3^-$	8.7	
$Cd^{2+} + 4OH^- = Cd(OH)_4^{2-}$	8.6	
$Cd^{2+} + SO_4^{2-} = CdSO_4$	2.4	$[SO_4^{2-}]$ calculated from salinity
$Cd^{2+} + 2SO_4^{2-} = Cd(SO_4)_2^{2-}$	3.4	with presumed conservativity
$Cd^{2+} + 3SO_4^{2-} = Cd(SO_4)_3^{4-}$	3.1	
$Cd^{2+} + 4SO_4^{2-} = Cd(SO_4)_4^{6-}$	-0.7	

Surface reactions: Cd sorption to the surface of a "global" particle. Apparent stability constants were calculated from intrinsic constants($_{int}$), taking into account the electrostatic effects of the surface charge as estimated by the Gouy-Chapman model.

$S - OH_2^+ = S - OH + H^+$	$K_{a1(int)}$	$K_{a1(int)}$ and $K_{a2(int)}$ are the acidity
$S - OH = S - O^- + H^+$	$K_a2(int)$	constants determined experimentally
"Global" particle		
$S - OH + Cd^{2+} = S - OCd^+ + H^+$	$K_{m(int)}$	$K_{m(int)}$ is determined experimentally

Complexation of cadmium with various dissolved ligands: Each equilibrium is defined by a complexation constant (stability constants) for each dissolved complex derived from Comans and Van Dijk [18]. Values at 25 °C are used (no temperature correction). The activity coefficients of the various species are calculated using the Davies equation reactions reach equilibrium quasi-instantaneously, and that they are completely reversible.

The choice of modeled processes was based on studies on Cd biogeochemistry in the estuarine environment, which suggest that most Cd behavior can be explained by its stability in the form of chlorocomplexes and rapid desorption during estuarine transit [11, 14, 18, 65].

A fuller description of the model, including parameter selection methods and validation for the various study estuaries, was published previously [30, 34, 42, 59].

We simulated various particle distribution scenarios within a given estuary, using the surface properties of particles taken from the three study estuaries. These "classic" scenarios (Fig. 7) were defined on the basis of suspended matter distribution measurements according to salinity, presented in Fig. 2.

The distribution of SM concentrations in the three study estuaries (Fig. 2) can be roughly grouped into three scenarios: migration of the turbidity maximum towards the highest salinities, the presence of a second, smaller and more "thinly spread" turbidity maximum, and very high SM peaks within the estuary (Fig. 7). These scenarios were used as the basis for our simulations.

The required input data was as follows (Table 5):

- Concentrations of the various included dissolved ligands: chlorides (calculated on the basis of salinity), hydroxides (calculated on the basis of pH, which will be considered as equal to 7.8 throughout the estuary), sulfates (calculated on the basis of salinity, considered as conservative and based on concentrations of 3×10^{-4} M and 0.02 M at the river and ocean boundaries, respectively)
- SM concentrations in our classic situations (Fig. 7)
- Parameters representing the surface properties of natural particles (Table 2).

All simulations used mean sorption parameter values (SA, [-SOH$_{tot}$], K_{a1}, K_{a2} and K_m), evaluated for particles from the three study estuaries. To test the natural variability of these parameters within the same estuary, we also did simulations taking into account standard deviation from mean values (Table 2).

The model computes concentrations of various species (dissolved and particulate) at equilibrium (Table 5) for a given total Cd concentration.

We chose a total Cd concentration of 10^{-9} M, which is representative of data collected in the three study estuaries.

For greater clarity, the various dissolved Cd species (free Cd and Cd complexed with chlorides, sulfates, or hydroxides) are not shown.

The results are expressed as a percentage of total dissolved Cd (sum of all dissolved species calculated by the model, see Table 5).

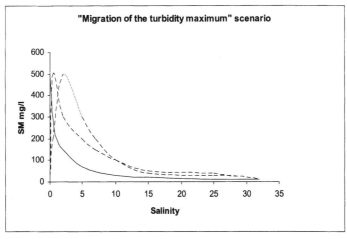

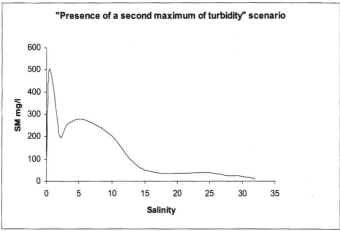

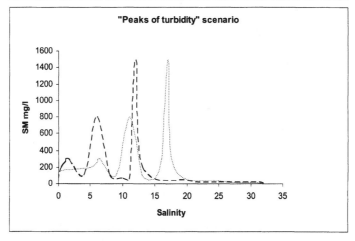

Fig. 7 Typical scenarios defined according to the distribution of suspended matter versus salinity (measurements presented in Fig. 2)

6
Results and Discussion

6.1
Global Sorption Capacity of Particles (GSC)

Particle sorption capacity depends on: the exchange capacity of protons in surface sites, ascertained using acidity constants K_{a1} and K_{a2}; total surface site concentrations and specific particle surface areas; and the complexation constant of surface sites in terms of Cd.

In order to compare the sorption characteristics of particles from the three study estuaries, and interpret the results obtained, we evaluated their global sorption capacity (GSC). For this purpose, -SOH site concentration (in $mol\,L^{-1}$) was computed with various salinities, at constant pH of 7.8 (corresponding to the simulation pH) and a constant SM concentration ($100\,mg\,L^{-1}$). We based our calculation on the sorption parameters (mean, minimum, and maximum) of each particle type (Table 2), i.e., specific surface area, density of surface sites, and acid-base constants of these sites (K_{a1} and K_{a2}).

The concentration of "complexating" surface sites (Eq. 4, Table 2) thus obtained was multiplied by the global intrinsic complexation constant of the study sites in terms of Cd.

This parameter is a good indicator of the importance of the particles sorption capacities and it makes it possible to compare particles of different nature.

According to our results (Fig. 8), Loire particles possess the highest concentrations of complexating surface sites due to their acidity intrinsic properties. Seine particles have the lowest concentrations, as their surface sites are non-amphoteric (Eq. 3, Table 2).

Computing GSC with respect to Cd underlined the impact of the surface site complexation constant. The high value of this constant in Seine particles means that despite a very weak concentration of complexating sites, computed GSC reactivity was similar to that obtained for the other two particle types (Fig. 8).

Differences exceeding one order of magnitude were observed in GSC computed on the basis of average, minimum, or maximum sorption parameters. Using average and maximum parameters, Loire particles were shown to have the highest GSC, whereas Gironde particles had the GSC highest global reactivity using minimum parameters. The reactivity of Seine particles was very similar to that of Gironde particles using average parameters, and similar to Loire particles using maximum parameters. Interestingly, using average and maximum parameters, Seine particles had the highest GSC with zero salinity; reactivity decreased very strongly when salinity increased, and was similar to that of the other particles at around salinity 1.

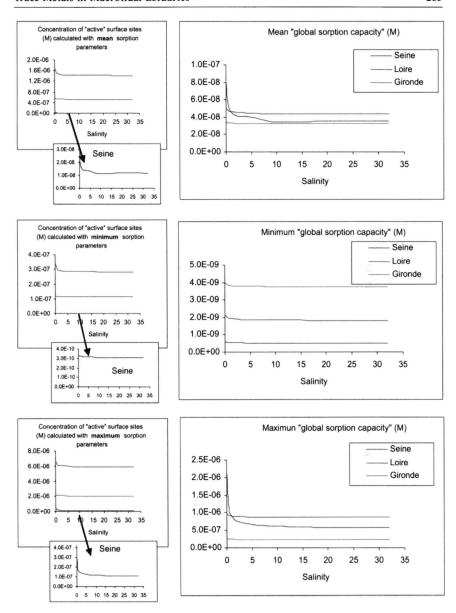

Fig. 8 Concentration of "reactive" surface sites and "global sorption capacity" of particles (see text)

This phenomenon stems from the fact that Seine particle surface sites are non-amphoteric (Eq. 3, Table 2). Salinity increases cause conditional equilibrium constants (K_{a1}, K_{a2}) to rise, due to a drop in activity coefficients (this drop is most significant at an ionic force of zero to a few hundredths). According to Eq. 9, when pH is constant, the increase in K_{a2} leads to a reduction in sites capable of complexing Cd. This reduction increases as K_{a2} value is low. As a result, using average and maximum parameters, the GSC of Seine particles was far greater when salinity neared zero, whereas practically no difference was observed with other particle types. In the case of the Loire and the Gironde (amphoteric sites), the increase in K_{a2} when salinity rises is countered by that of K_{a1} (Eq. 8); the phenomenon is therefore virtually non-existent.

The comparison of GSC values shows that within the same estuary the differences (between the average, minimum, and maximum values) can be very significant. These differences are the result of the space and temporal variations of the nature of the particles.

6.2
Simulation of Migration of the Turbidity Maximum

Migration of the turbidity maximum, according to flow rates and tidal conditions, and within a relatively weak salinity range, is frequently encountered in macrotidal estuaries.

Using average properties, the results obtained were virtually identical notwithstanding particle type (Fig. 9), in particular for the Seine and Loire estuaries. Differences were significant with zero salinity and a low SM content (25 mg L^{-1}). As observed in the evaluation of GSC, Seine particles were shown to have far greater sorption properties with zero salinity.

With minimum parameters (Fig. 10), all Cd is considered to be in dissolved form, notwithstanding particle type, except in zones where SM concentrations are high (characterized by minimal %CdD values). Despite the major differences in GSC (Fig. 8) resulting from reduced sorption capacities, differences in Cd partition were negligible except in the turbidity maximum. Differences in particle origin were nil when salinity reached 5.

Using maximum sorption parameters (Fig. 11), the evolution of dissolved Cd percentages according to salinity was virtually the same for Seine and Loire particles, whose global reactivity is comparable. Dissolved Cd percentages were highest with Gironde particles, which are characterized by the lowest GSC. The turbidity maximum had a very marked impact in comparison to previous simulations, with extremely low dissolved Cd percentages at salinities of 0–5, in particular for Seine and Loire particles. Unlike in the simulations using average or minimum properties, we observed significant differences in dissolved Cd percentages even when salinity was high. This is due to the fact that chloride impact is weakened when particle sorption capacity rises. As for average properties, a shift in maximum dissolved Cd (around 80% for Gironde particles and 60% for Loire and the Seine particles) was observed with higher salinities and migration of the turbidity maximum.

To illustrate this scenario, our simulation results are also shown in terms of concentrations (Figs. 12–14).

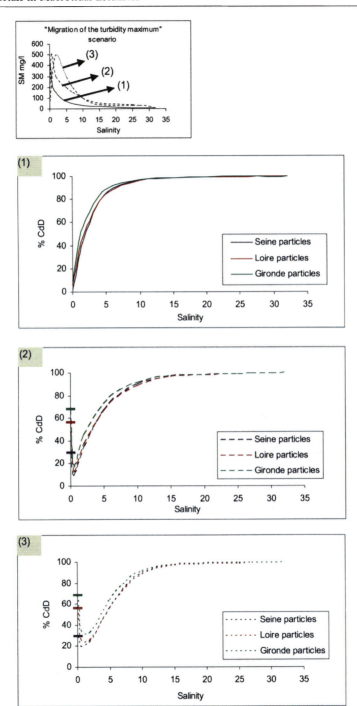

Fig. 9 Percentage of dissolved Cd (%CdD): "Migration of the turbidity maximum" scenario. Mean sorption properties simulation

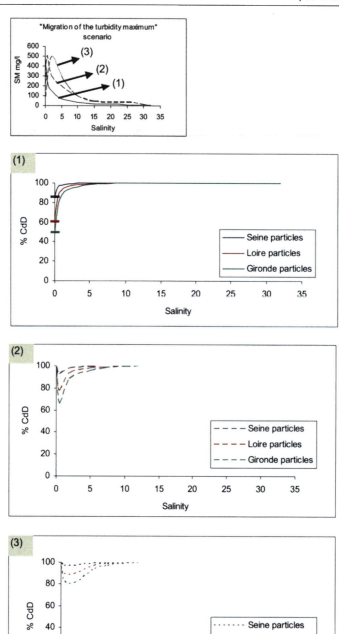

Fig. 10 Percentage of dissolved Cd (%CdD): "Migration of the turbidity maximum" scenario. Minimum sorption properties simulation

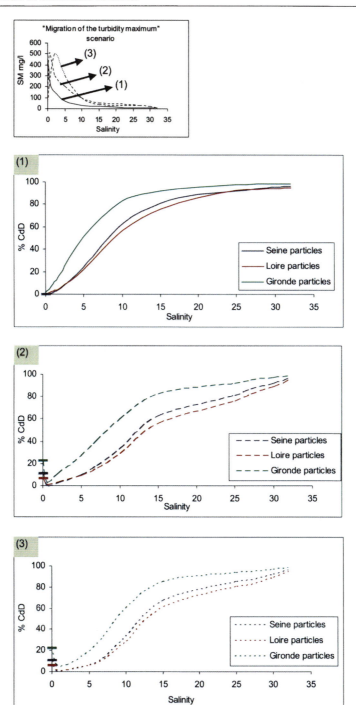

Fig. 11 Percentage of dissolved Cd (%CdD): "Migration of the turbidity maximum" scenario. Maximum sorption properties simulation

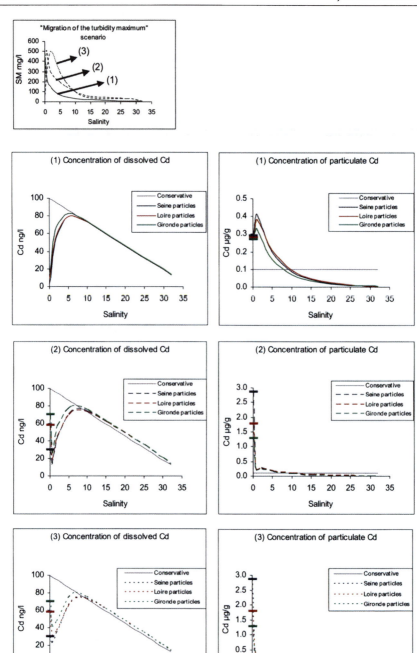

Fig. 12 Concentration of dissolved and particulate Cd: "Migration of the turbidity maximum" scenario. Mean sorption properties simulation

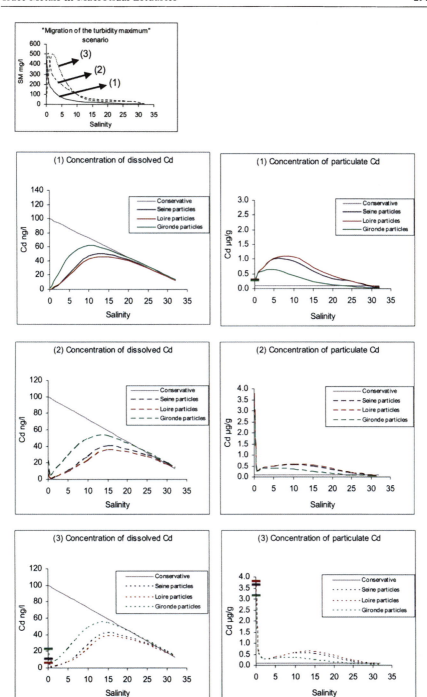

Fig. 13 Concentration of dissolved and particulate Cd: "Migration of the turbidity maximum" scenario. Maximum sorption properties simulation

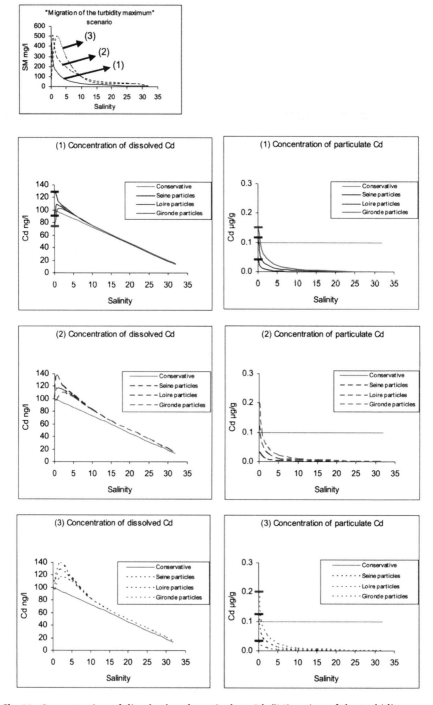

Fig. 14 Concentration of dissolved and particulate Cd: "Migration of the turbidity maximum" scenario. Minimum sorption properties simulation

For the purpose of these simulations, we used the following dissolved Cd concentrations: $100\ ng\ L^{-1}$ at the upstream limit (0 salinity) and $5\ ng\ L^{-1}$ at the marine boundary (salinity 35). In terms of particulate Cd, particles entering the estuary were considered as having an adsorbed Cd concentration of $0.1\ \mu g\ g^{-1}$. The model divides "total" Cd into various species (Table 5) for a given salinity. Total Cd is calculated on the basis of SM, sorbed Cd, and dissolved Cd concentrations.

Particulate cadmium computed by the model only represents the fraction adsorbed to the surface of particles. This fraction cannot be compared directly with total particulate cadmium concentrations measured in the study estuaries, which comprises a "sorbed" fraction (potentially "desorbable" in the physicochemical conditions encountered in the study estuaries) and a "non-exchangeable" fraction (Cd incorporated in the crystalline matrix, co-precipitated in various mineral phases, etc.).

Typical Cd behavior in macrotidal estuaries was well reproduced using average and maximum sorption properties (Figs. 12 and 13), and featured the classic bell-shaped CD D curve. Maximum dissolved Cd concentrations remained similar notwithstanding particle type or the position of the turbidity maximum: approximately $80\ ng\ L^{-1}$ using average properties, with greater variations (30–$50\ ng\ L^{-1}$) according to particle type using maximum properties. Using minimum properties (Fig. 14), maximum dissolved Cd concentrations were restricted to very low salinity zones.

In all cases, maximum dissolved Cd migrated towards higher salinities at the turbidity peak.

6.3
Simulation of the Presence of a Second Turbidity Maximum

This scenario aimed to evaluate the impact of a second turbidity maximum in more saline waters. This situation, which stems from hydrodynamic factors, can be encountered in macrotidal estuaries such as the Loire and Gironde.

Using average properties (Fig. 15), particle types only differed significantly with zero salinity. The highest turbidity peak was clearly correlated with a minimum percentage of dissolved Cd. Adding a second turbidity maximum resulted in a slight sag in the Cd % curve.

Using minimum properties (Fig. 16), particle influence was only visible at the first peak; this was particularly true for Loire and Gironde particles, which have the highest GSC.

The impact of two turbidity maxima is major in the maximum sorption property scenario (Fig. 17). Dissolved Cd percentages remain low until salinity reaches about 10, in particular for Loire and Seine particles which have high GSC.

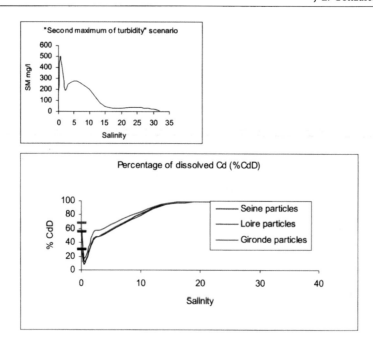

Fig. 15 "Second maximum of turbidity" scenario. Mean sorption properties simulation

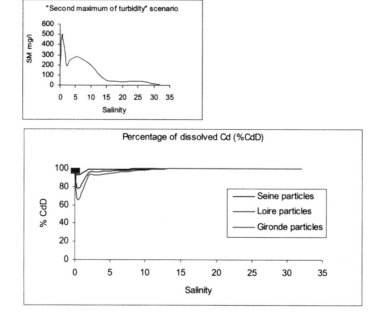

Fig. 16 "Second maximum of turbidity" scenario. Minimum sorption properties simulation

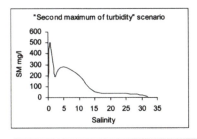

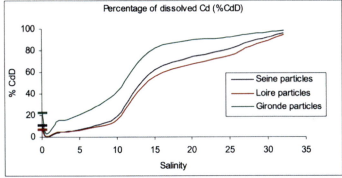

Fig. 17 "Second maximum of turbidity" scenario. Maximum sorption properties simulation

6.4
Simulation of Turbidity Peaks

The presence of major turbidity peaks, in any salinity range, is often encountered when mud deposit zones are eroded.

Using average and maximum sorption properties, turbidity peaks were shown to have a high impact, even in highly saline waters (Figs. 18 and 20). Differences in particle types were far more marked when maximum properties were used. The impact of turbidity peaks on particles with low GSC was minimal (Fig. 19), especially with high salinity. In this case, Cd behavior proved to be quasi-conservative.

7
Summary and Conclusions

We selected Cd to explore the impact of particle surface properties on the fate and speciation of contaminants in macrotidal estuaries due to its high reactivity. The behavior of Cd – which is highly sensitive to dissolved/particulate exchanges – was simulated using sorption properties representative of particles from different estuaries to demonstrate that:

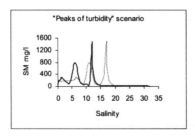

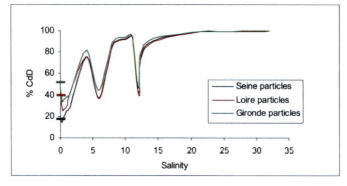

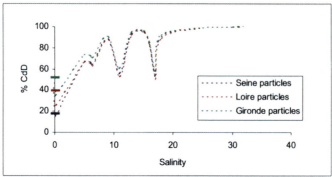

Fig. 18 Percentage of dissolved Cd (%CdD): "Peaks of turbidity" scenario. Mean sorption properties simulation

- Even in the case of weak sorption properties and virtually conservative Cd behavior within the estuary, major differences according to particle origin can be observed when SM concentrations are high and salinity is close to zero (Fig. 10), especially if GSC are significantly different (around factor 7) (Fig. 8).
- Using average properties (representing the overall measurements conducted on various samples), simulation of the dissolved/particulate partition gives comparable results notwithstanding particle origin, due to the fact that GSC is relatively similar (Fig. 8). The scenarios simulated as a whole highlighted the instrumental role of particles in all salinity ranges.

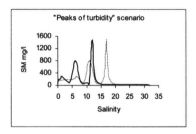

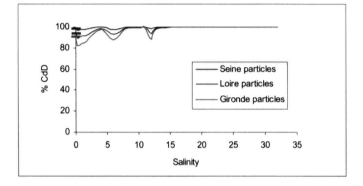

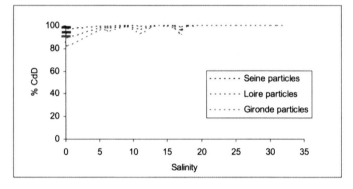

Fig. 19 Percentage of dissolved Cd (%CdD): "Peaks of turbidity" scenario. Minimum sorption properties simulation

- When sorption capacity is high, variations in particles have a far greater impact on Cd behavior, even in high salinity ranges.

The evaluation of GSC highlighted the relative uselessness of standard characteristics (POC, Al, Fe, Mn, SA, etc.) employed to ascertain particle sorption capacity to contaminants. More explicit parameters, i.e., nature and density of surface sites, acidity intrinsic properties and complexation constants, and their natural variabilities, are necessary to efficiently evaluate particle reactivity with regards to a given element in various physicochemical conditions.

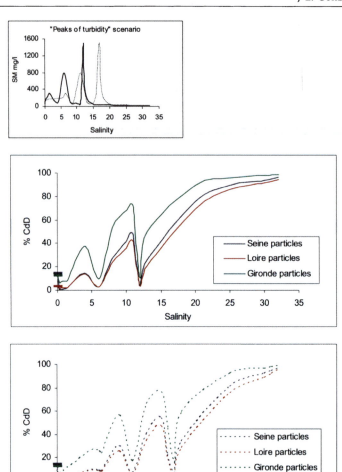

Fig. 20 Percentage of dissolved Cd (%CdD): "Peaks of turbidity" scenario. Maximum sorption properties simulation

The comparison of Cd measurements (Table 4) showed average Cd concentrations in suspended matter in both the upstream and mixing zones to be significantly higher in the Seine estuary than in the other study estuaries (Loire and Gironde). One of the reasons put forward for this could be the higher sorption capacity of Seine particles (as per the estimated K_m value). If we consider the average GSC of particles taken from the various estuaries (Fig. 8), it is interesting to note that Seine particles only have significantly higher sorption capacities than particles from the other two study estuaries when salinity is close to zero. When salinity exceeds 5, the average reactivity of Seine particles decreases strongly, to around that of Gironde particles. The variations we observed in Seine particle sorption capacity versus water salinity could partially explain the differences in particulate Cd concentrations measured in the upstream and downstream zones.

The impact of particle surface properties is maximized when salinity is around zero and SM concentrations are high. The importance of the particulate phase is relative when salinity increases, due to the key role of chlorides in Cd speciation. This suggests that particle sorption properties have a major impact on the speciation of Cd-type elements, whose behavior is controlled by competition between the particulate phase and a major ligand associated with the dissolved phase (e.g., the ion chloride in the case of Cd). Hence, in macrotidal estuaries, space–time variations in terms of salinity gradient and SM quantity and type will contribute to dissolved/particulate partition and to contaminant fluxes to the ocean/continent interface.

From a modeling viewpoint, the results obtained suggest that it is possible to simulate (and forecast) the behavior of contaminants such as Cd in complex environments by evaluating various parameters experimentally. These parameters are used to characterize processes (surface complexation constant and acidity constants of surface sites) and particles (density of surface sites). Their value must be determined with sufficient accuracy, and their variability within the environment must be known.

Our study exclusively related to the role of particles in processes close to equilibrium, and did not take into account the dynamic aspect of processes, in particular the temporal evolution of turbidity and salinity gradients, which is of major importance in macrotidal estuaries. "Slow" processes, driven by particles, may also regulate the dynamics and speciation of trace metals in the estuarine environment, e.g., desorption and processes related to the mineralization of certain mineral phases (Fe and Mn oxides and organic matter), in particular in fine matter deposit zones and maximum turbidity zones.

Our results nevertheless showed that particle surface properties, evaluated on the basis of various parameters, are instrumental in non-conservative contaminant speciation in the estuarine environment. Their analysis enables us to understand and simulate, to a large extent, the fate of non-conservative contaminants whose behavior is controlled by competition between sorption and desorption processes.

References

1. Pritchard DW (1955) Proc Am Soc Civ Engin 8:1
2. Pritchard DW (1967) In: Lauff GH (ed) Estuaries. Washington American Association for Advancement of Science 83:158
3. Officer CB (1976) Physical oceanography of estuaries (and associated coastal waters). Wiley, New York
4. Guézennec L, Romana LA, Goujon R, Meyer R (1999) Seine-Aval: un estuaire et ses problèmes, Programme Scientifique "Seine Aval", Fascicule no 1. IFREMER, Plouzané (France)

5. Stumm W, Morgan JJ (1981) Aquatic chemistry, 2nd edn. Wiley, New York
6. Uncles RJ, Stephens JA, Smith RE (2002) Cont Shelf Res 22:1835
7. Turner A, Millward GE (2002) Coast Shelf Sci 55:857
8. Turner A (1996) Mar Chem 54:27
9. Boyle EA, Edmond JM, Sholkovitz JR (1982) Geochim Cosmochim Acta 41:1313
10. Elbaz-Poulichet F, Huang WW, Martin JM, Zhu JX (1987) Mar Chem 22:125
11. Boutier B, Jouanneau JM, Chiffoleau JF, Latouche C, Phillips I (1989) La contamination de la Gironde par le cadmium, Rapports Scientifiques et Techniques de l'IFREMER. IFREMER, Plouzané (France)
12. Boutier B, Chiffoleau JF, Auger D, Truquet I (1993) Coast Mar Sci 36:133
13. Chiffoleau JF, Cossa D, Auger D, Truquet I (1994) Mar Chem 47:145
14. Krapiel AML, Chiffoleau JF, Martin JM, Morel FMM (1997) Geochim Cosmochim Acta 61:1421
15. Thouvenin B, Gonzalez JL, Boutier B (1997) Mar Chem 58:147
16. Chiffoleau JF, Auger D, Chartier E, Michel P, Truquet I, Ficht A, Gonzalez JL, Romana LA (2001a) Estuaries 24:1029
17. Chiffoleau JF, Claisse D, Cossa D, Ficht A, Gonzalez JL, Guyot T, Michel P, Miramand P, Oger C, Petit F (2001b) La contamination métallique, Programme scientifique "Seine Aval", Fascicule no 8. IFREMER Plouzané (France)
18. Comans RNJ, Van Dijk CPJ (1988) Nature 336:151
19. Elbaz-Poulichet F, Garnier JM, Guang DM, Martin JM, Thomas AJ (1996) Coast Shelf Sci 42:289
20. Guieu C, Huang WW, Martin JM, Yong Y (1996) Mar Chem 53:255
21. Guieu C, Martin JM (2002) Estuar Coast Shelf Sci 54:501
22. Garnier JM, Guieu C (2003) Mar Environ Res 55:5
23. Glangeaud L (1938) Bulletin de la Société de Géologie de France 8:599
24. Gallenne B (1974) Thèse de doctorat de 3ème cycle, Université de Nantes
25. Phillips I (1980) Thèse de doctorat de 3ème cycle, Université de Bordeaux I
26. Avoine J (1981) Thèse de doctorat de 3ème cycle, Université de Caen
27. Jouanneau JM (1982) Thèse de doctorat d'état, Université de Bordeaux I
28. El Sayed M (1988) Thèse de doctorat d'état, Université de Bretagne Occidentale
29. Le Hir P, Ficht A, Silva Jacinto R, Lesueur P, Dupont JP, Lafite R, Brenon I, Thouvenin B, Cugier P (2001) Estuaries 24:950
30. Dange C (2002) Thèse de doctorat, Université de Reims Champagne-Ardenne
31. Boutier B, Chiffoleau JF, Gonzalez JL, Lazure P (1996) 5ème Colloque International d'Océanographie du Golfe de Gascogne, La Rochelle, France
32. Boutier B, Chiffoleau JF, Gonzalez JL, Lazure P, Auger D, Truquet I (2000) Oceanologica Acta 23:745
33. Allen GP (1971) CR Acad Sc 273:2429
34. Gonzalez JL, Thouvenin B, Dange C, Fiandrino A, Chiffoleau JF (2001b) Estuaries 24:1041
35. Sposito G (1984) The surface chemistry of soils. Oxford University Press, New York
36. Garnier JM, Martin JM, Mouchel JM, Thomas AJ (1993) Estuar Coast Shelf Sci 36:315
37. Westall JC (1982) FITEQL. A computer program for determination of chemical equilibrium constants from experimental data. Technical Report 82–01. Department of Chemistry, Oregon State University, Corvallis
38. Gulmini M, Zelano V, Daniele PG, Prenesti E, Ostacoli G (1996) Anal Chim Acta 329:33
39. Wang F, Chen J, Forsling W (1997) Water Res 7:796
40. Muller B, Duffek A (2001) Aquat Geochem 7:107

41. Cossa D, Meybeck M, Idlafkih Z, Bombled B (1994) Etude pilote des apports en contaminants par la Seine. IFREMER Report DEL 94-13
42. Gonzalez JL, Thouvenin B, Chiffoleau JF, Miramand P (1999) Le cadmium: comportement d'un contaminant métallique en estuaire. Programme scientifique "Seine Aval", Fascicule no 10. IFREMER, Plouzané (France)
43. Bruland KW, Franks RP (1983) In: Wong CS, Boyle EA, Bruland KW, Burton JD, Goldberg ED (eds) Trace metals in sea water. NATO Conference Series IV, Mar Sci 9:395
44. Cossa D, Lassus P (1989) Le cadmium en milieu marin Biogéochimie et ecotoxicologie. IFREMER Report 16
45. Gonzalez JL, Boutier B, Chiffoleau JF, Auger D, Noel J, Truquet I (1991b) Oceanol Acta 6:559
46. Valenta P, Duursma EK, Merks AG, Ruetzel H, Nuernberg HW (1986) Sci Total Environ 53:41
47. Klinkhammer G, Bender ML (1981) Estuar Coast Shelf Sci 12:629
48. Gobeil C, Silverberg N, Sundby B, Cossa D (1987) Geochim Cosmochim Acta 51:589
49. Westerlund SFG, Anderson LG, Hall POJ, Iverfeldt A, Van Der Loeff RMM, Sundby B (1986) Geochim Cosmochim Acta 50:1289
50. Klinkhammer G, Heggie DT, Graham DW (1982) Earth Planet Sci Lett 61:211
51. McCorkle DC, Klinkhammer GP (1991) Geochim Cosmochim Acta 55:161
52. Gonzalez JL (1994) Evaluation des flux de contaminants à l'interface eau-sédiments en zone littorale Oceanis 3:23
53. Gonzalez JL, Boutier B, Auger D, Chartier E (2004) Les vasières intertidales et subtidales sont elles des sources de contaminants métalliques? In: Contribution à l'étude de la dynamique et de la spéciation des contaminants. Programme Scientifique Seine-Aval Report
54. Shiller AM, Boyle EA (1991) Geochim Cosmochim Acta 11:3241
55. Cossa D (1990) Chemical contaminants in the St. Lawrence estuary and Saguenay Fjord. In: El-Sabh MI, Silverberg N (eds) Oceanography of a large-scale estuarine system: The St. Lawrence. Springer, Berlin Heidelberg New York, p 239
56. Baeyens W, Decadt G, Dedeurwaerder H, Dehairs F, Gillain G (1984) Symposium on contaminant fluxes through the coastal zone, Nantes, France
57. Guieu C, Martin JM, Thomas AJ, Elbaz-Poulichet F (1991) Mar Poll Bull 22:176
58. Thomas AJ, Huang WW (1996) Les métaux traces particulaires et dissous (Cd, Cu, Ni, Pb et Zn) dans le Rhône à Arles durant le cycle annuel juin 1994–mai 1995: origines, concentrations et flux. Agence de l'Eau Rhône-Méditerranée-Corse Report
59. Gonzalez JL, Dange C, Thouvenin B (2001a) Hydroécologie Appliquée 1:37
60. Laurier FJG, Cossa D, Gonzalez JL, Breviere E, Sarazin G (2003) Geochim Cosmochim Acta 18:3329
61. Stumm W, Kummer R, Sigg L (1980) Croat Chem Acta 53:291
62. Davis JA, Kent DB (1990) Surface complexation modelling in aqueous geochemistry. In: Hochella MF, White AF (eds) Mineral-water interface geochemistry, reviews in mineralogy. Mineralogical Society of America, Washington, p 177
63. Dzombak DA, Morel FMM (1987) J Hydraul Eng 113:430
64. Dzombak DA, Morel FMM (1990) Surface complexation modeling: Hydrous ferric oxide. Wiley, New York
65. Van der Weidjen CH, Arnoldus MJHL, Meurs CJ (1977) Neth J Sea Res 11:130

Subject Index